CW00794372

GRAVITATIONAL COLLAPSE AND SPACETIME SINGULARITIES

Physical phenomena in astrophysics and cosmology involve gravitational collapse in a fundamental way. The final fate of a massive star when it collapses under its own gravity at the end of its life cycle is one of the most important questions in gravitation theory and relativistic astrophysics, and is the foundation of blackhole physics.

General relativity predicts that continual gravitational collapse gives rise to a spacetime singularity, which may be hidden inside an event horizon or visible to external observers. This book investigates these issues, and shows how such visible ultra-dense regions arise naturally and generically as an outcome of dynamical gravitational collapse. Quantum gravity may take over in these regimes to resolve the classical spacetime singularity. The quantum effects from a visible extreme gravity region could then propagate to external observers, providing a useful laboratory for quantum gravity, and implying interesting consequences for ultra-high energy astrophysical phenomena in the universe.

This volume will be of interest to graduate students and academic researchers in gravitation physics and fundamental physics, as well as in astrophysics and cosmology. It includes a review of recent research into gravitational collapse, and several examples of collapse models are worked out in detail.

PANKAJ S. JOSHI conducts research at the Tata Institute of Fundamental Research, Mumbai. His research interests include gravitation physics, spacetime structure and quantum gravity, and cosmology and relativistic astrophysics. He has published many research papers and books in these areas, and has held visiting faculty positions in several countries, lecturing and doing research on these topics.

Professor Joshi has an excellent international reputation for his work in the field of gravitation theory. His extensive analysis of general relativistic gravitational collapse has been widely recognized as providing significant insights into the final end states of a continual collapse, formation of visible singularities, and nature of cosmic censorship and blackholes.

CAMBRIDGE MONOGRAPHS ON MATHEMATICAL PHYSICS

General editors: P. V. Landshoff, D. R. Nelson, S. Weinberg

S. J. Aarseth *Gravitational N-Body Simulations*
J. Ambjørn, B. Durhuus and T. Jonsson *Quantum Geometry: A Statistical Field Theory Approach*
A. M. Anile *Relativistic Fluids and Magneto-Fluids*
J. A. de Azcárrage and J. M. Izquierdo *Lie Groups, Lie Algebras, Cohomology and Some Applications in Physics*†
O. Babelon, D. Bernard and M. Talon *Introduction to Classical Integrable Systems*
F. Bastianelli and P. van Nieuwenhuizen *Path Integrals and Anomalies in Curved Space*
V. Belinkski and E. Verdaguer *Gravitational Solitons*
J. Bernstein *Kinetic Theory in the Expanding Universe*
G. F. Bertsch and R. A. Broglia *Oscillations in Finite Quantum Systems*
N. D. Birrell and P. C. W. Davies *Quantum Fields in Curved Space*†
M. Burgess *Classical Covariant Fields*
S. Carlip *Quantum Gravity in* $2+1$ *Dimensions*
J. C. Collins *Renormalization*†
M. Creutz *Quarks, Gluons and Lattices*†
P. D. D'Eath *Supersymmetric Quantum Cosmology*
F. de Felice and C. J. S. Clarke *Relativity on Curved Manifolds*†
B. S. DeWitt *Supermanifolds*, 2nd edition†
P. G. O. Freund *Introduction to Supersymmetry*†
J. Fuchs *Affine Lie Algebras and Quantum Groups*†
J. Fuchs and C. Schweigert *Symmetries, Lie Algebras and Representations: A Graduate Course for Physicists*†
Y. Fujii and K. Maeda *The Scalar Tensor Theory of Gravitation*
A. S. Galperin, E. A. Ivanov, V. I. Orievetsky and E. S. Sokatchev *Harmonic Superspace*
R. Gambini and J. Pullin *Loops, Knots, Gauge Theories and Quantum Gravity*†
M. Göckeler and T. Schücker *Differential Geometry. Gauge Theories and Gravity*†
C. Gómez, M. Ruiz Altaba and G. Sierra *Quantum Groups in Two-dimensional Physics*
M. B. Green, J. H. Schwarz and E. Witten *Superstring Theory, volume 1: Introduction*†
M. B. Green, J. H. Schwarz and E. Witten *Superstring Theory, volume 2: Loop Amplitudes, Anomalies and Phenomenology*†
V. N. Gribov *The Theory of Complex Angular Momenta*
S. W. Hawking and G. F. R. Ellis *The Large-Scale Structure of Space-Time*†
F. Iachello and A. Arima *The Interacting Boson Model*
F. Iachello and P. van Isacker *The Interacting Boson–Fermion Model*
C. Itzykson and J.-M. Drouffe *Statistical Field Theory, volume 1: From Brownian Motion to Renormalization and Lattice Gauge Theory*†
C. Itzykson and J.-M. Drouffe *Statistical Field Theory, volume 2: Strong Coupling, Monte Carlo Methods, Conformal Field Theory, and Random Systems*†
C. Johnson *D-Branes*
P. S. Joshi *Gravitational Collapse and Spacetime Singularities*
J. I. Kapusta *Finite-Temperature Field Theory*†
V. E. Korepin, A. G. Izergin and N. M. Boguliubov *The Quantum Inverse Scattering Method and Correlation Functions*†
M. Le Bellac *Thermal Field Theory*†
Y. Makeenko *Methods of Contemporary Gauge Theory*
N. Manton and P. Sutcliffe *Topological Solitons*
N. H. March *Liquid Metals: Concepts and Theory*
I. M. Montvay and G. Münster *Quantum Fields on a Lattice*†
L. O' Raifeartaigh *Group Structure of Gauge Theories*†
T. Ortín *Gravity and Strings*
A. Ozorio de Almeida *Hamiltonian Systems: Chaos and Quantization*†
R. Penrose and W. Rindler *Spinors and Space-Time, volume 1: Two-Spinor Calculus and Relativistic Fields*†
R. Penrose and W. Rindler *Spinors and Space-Time, volume 2: Spinor and Twistor Methods in Space-Time Geometry*†
S. Pokorski *Gauge Field Theories*, 2nd edition
J. Polchinski *String Theory, volume 1: An Introduction to the Bosonic String*†
J. Polchinski *String Theory, volume 2: Superstring Theory and Beyond*†
V. N. Popov *Functional Integrals and Collective Excitations*†
R. J. Rivers *Path Integral Methods in Quantum Field Theory*†
R. G. Roberts *The Structure of the Proton*†
C. Rovelli *Quantum Gravity*
W. C. Saslaw *Gravitational Physics of Stellar and Galactic Systems*†
H. Stephani, D. Kramer, M. A. H. MacCallum, C. Hoenselaers and E. Herlt *Exact Solutions of Einstein's Field Equations*, 2nd edition
J. M. Stewart *Advanced General Relativity*†
A. Vilenkin and E. P. S. Shellard *Cosmic Strings and Other Topological Defects*†
R. S. Ward and R. O. Wells Jr *Twistor Geometry and Field Theories*†
J. R. Wilson and G. J. Mathews *Relativistic Numerical Hydrodynamics*

† Issued as a paperback

Gravitational Collapse and Spacetime Singularities

PANKAJ S. JOSHI
Tata Institute of Fundamental Research,
Mumbai, India

CAMBRIDGE UNIVERSITY PRESS
Cambridge, New York, Melbourne, Madrid, Cape Town, Singapore, São Paulo

Cambridge University Press
The Edinburgh Building, Cambridge CB2 8RU, UK

Published in the United States of America by Cambridge University Press, New York

www.cambridge.org
Information on this title: www.cambridge.org/9780521871044

First published 2007

Printed in the United Kingdom at the University Press, Cambridge

A catalogue record for this publication is available from the British Library

ISBN 978-0-521-87104-4 hardback

To my parents,
Arunadevi Shantilal Joshi
and
Shantilal Ramshankar Joshi

Contents

Preface *page* ix

1 Introduction **1**

2 The spacetime manifold **10**
- 2.1 The manifold model 10
- 2.2 The metric tensor 21
- 2.3 Connection 24
- 2.4 Non-spacelike geodesics 29
- 2.5 Spacetime curvature 32
- 2.6 The Einstein equations 38
- 2.7 Exact solutions 43

3 Spherical collapse **60**
- 3.1 Basic framework 62
- 3.2 Regularity conditions 69
- 3.3 Collapsing matter clouds 71
- 3.4 Nature of singularities 79
- 3.5 Exterior geometry 87
- 3.6 Dust collapse 90
- 3.7 Equation of state 129

4 Cosmic censorship **135**
- 4.1 Causal structure 136
- 4.2 Spacetime singularities 149
- 4.3 Blackholes 161
- 4.4 Higher spacetime dimensions 169
- 4.5 Formulating the censorship 175
- 4.6 Genericity and stability 190

5 Final fate of a massive star **210**
- 5.1 Life cycle of massive stars 213
- 5.2 Evolution of a physically realistic collapse 215

5.3 Non-spherical models 225
5.4 Blackhole paradoxes 235
5.5 Resolution of a naked singularity 238

References 255
Index 269

Preface

The physical phenomena in astrophysics and cosmology involve gravitational collapse in a fundamental way. The final fate of a massive star, when it collapses under its own gravity at the end of its life cycle, is one of the most important questions in gravitation theory and relativistic astrophysics today. The applications and basic theory of blackholes vigorously developed over the past decades crucially depend on this outcome.

A sufficiently massive star many times the size of the Sun would undergo a continual gravitational collapse on exhausting its nuclear fuel, without achieving an equilibrium state such as a neutron star or white dwarf. The singularity theorems in general relativity then predict that the collapse gives rise to a spacetime singularity, either hidden within an event horizon of gravity or visible to the external universe. The densities and spacetime curvatures get arbitrarily high and diverge at these ultra-strong gravity regions. Their visibility to outside observers is determined by the causal structure within the dynamically developing collapsing cloud, as governed by the Einstein field equations. When the internal dynamics of the collapse delays the horizon formation, these become visible, and may communicate physical effects to the external universe. These issues are investigated here, and the treatment is aimed at showing how such visible ultra-dense regions arise naturally and generically as the outcome of a dynamical gravitational collapse in Einstein gravity. While it predicts the existence of visible singularities; classical general relativity may no longer hold in these very late stages of the collapse, and quantum gravity may take over to resolve the classical spacetime singularity. The quantum effects from a visible, the extreme gravity region could then propagate to outside observers to provide a useful laboratory for quantum gravity. Blackholes need not form in such a scenario and there may be interesting consequences for ultra-high energy astrophysical phenomena in the universe.

The general theory of relativity, which has strong experimental support, is used here, and its basics and useful features of spacetimes are reviewed. The necessary tools are developed as needed, but a prior familiarity with general relativity would help. It is a pleasure to thank many friends and colleagues for numerous discussions and work as cited, on the themes described here. Special thanks are due to R. Goswami and I. H. Dwivedi for their ideas and help and for our studies together. A. Mahajan and S. Khedekar helped with the manuscript.

1
Introduction

Gravitation theory and relativistic astrophysics have gone through extensive developments in recent decades, following the discovery of quasars in the 1960s, and other very high energy phenomena in the universe such as gamma ray bursts. Compact objects such as neutron stars and pulsars also display intriguing physical properties, where the effects of strong gravity fields are seen to play a fundamental role. When the masses and energy densities involved in the physical phenomena are sufficiently high, as is the case in the situations above, it has become increasingly clear that the strong gravitational fields, as governed by the general theory of relativity, play an important and much more dominant role. This gravitational dynamics must be taken into account for any meaningful description of these observed ultra-high energy objects.

A similar situation involving very strong gravitational fields, and which may be connected to some of the above phenomena, is that of a massive star undergoing a continual gravitational collapse at the end of its life cycle. This happens when the star has exhausted its nuclear fuel that provided a balance against the internal pull of gravity. This phenomenon, dominated essentially by the force of gravity, is fundamental to basic theory and astrophysical applications in blackhole physics that have received increasing attention in past decades, and also in cosmology. In the past two decades, there have been extensive investigations of gravitational collapse models within the framework of Einstein's theory of gravity, and these have provided useful insights into the final fate of a massive star.

This book is about the phenomena of gravitational collapse. Such a collapse of massive matter clouds is at the heart of the physics and astrophysics of happenings, some of which are mentioned above, where extremely high mass and energy densities are involved. For example, several models to explain gamma ray bursts are in terms of a collapsar, where the gravitational collapse of a single massive star is invoked to understand such a burst of ultra-high energy. Apart from blackhole physics, gravitational collapse is the key physical process that is fundamental to the formation of a star itself from interstellar clouds or nebulae, in the formation of galaxies and clusters of galaxies, and in structure formation in the universe as a whole. In general, gravitational collapse of a massive matter cloud would play an important role in the physical processes and a variety of happenings on a cosmic scale that involve the force of gravity in an important manner.

A continual gravitational collapse for a massive star would be the situation when the entire matter cloud collapses and shrinks under the force of its own gravity. Therefore, gravity overtakes and dominates the other three fundamental forces of nature, in particular the weak and strong nuclear forces, which generically provide the outward pressure in a star to balance it against the inward pull of gravity of the cloud, in addition to the usual thermal pressures. For massive stars, typically such a collapse takes place when the star has exhausted its nuclear fuel, and when there is no supporting force left against the force of its own gravity, which is ever present.

The final outcome of such a collapse depends on the initial mass of the star. A star with a mass lower than about two to three solar masses will stabilize as a white dwarf or neutron star after losing some of its original mass. In these cases, after an initial collapse of the cloud when the star has exhausted its nuclear fuel, the star again stabilizes at a much smaller radius due to internal balancing forces provided by either electron or neutron degeneracy pressures. For heavier stars that are several solar masses, they may again settle to a neutron star final state if the star could throw away the excess mass in the process of its evolution. However, for more massive stars, none of the above internal pressures can achieve the required balance, and a continual gravitational collapse becomes inevitable. The collapse then must proceed towards creating a spacetime singularity, as predicted by the singularity theorems of general relativity theory, which may be hidden within a blackhole or which may be visible to external observers. A spacetime singularity is a region where the physical parameters such as mass, energy densities, and the spacetime curvatures go to their extreme values and blow up, so that the usual laws of physics break down at such a singularity.

In such extreme regions, however, where the length and time scales are comparable to the Planck length and time, quantum effects become important. These must necessarily be taken into account and combined with the effects of gravity. At present, we have no mechanism or complete theory to deal with such quantum effects and the intense force of gravity together in a unified manner, namely a quantum gravity theory. However, it is widely believed that a quantum gravity theory, dealing with all forces of nature in a unified way, would take over from purely classical general relativity when the collapse reaches extreme matter densities and spacetime curvatures in its very advanced later stages. In these stages of collapse, it is very likely that when the quantum effects are incorporated together with the gravitational force, the classical spacetime singularity may be resolved, and may no longer exist in the full theory.

Gravitational collapse is thus a key phenomena for many astrophysical processes for stars or other larger systems in the universe. In particular, the very advanced stages of collapse of a massive star are occurrences in nature where the effects of both gravity and the quantum would be combined. Even if the final spacetime singularity, as predicted by classical general relativity, may be resolved, possibly through quantum gravity effects, such a collapse will necessarily give rise to spacetime regions of ultra-high mass densities and curvatures, where the physical effects will be extreme.

The important physical issue would then be whether such extreme gravity regions formed in the gravitational collapse of a massive star are visible to external observers in the universe. An affirmative answer here would mean that the physical phenomena of the gravitational collapse of a massive star could provide a very good laboratory to study quantum gravity effects in the cosmos, and this may help towards generating clues for an, as yet, unknown theory of quantum gravity. A laboratory similar to that provided by the early universe is then created in the later stages of the continual collapse of a massive star. An additional feature would be that, whereas the early universe was a unique event that happened only once, the collapse phenomena would continue to occur whenever a sufficiently massive star in the universe died on exhausting its nuclear fuel. If such ultra-strong gravity regions become visible to external observers in the spacetime, an opportunity to observe the quantum gravity effects in the universe is provided.

The answer to this is determined by the causal structure of spacetime in the vicinity of a spacetime singularity. This is actually decided by the dynamics of the gravitational collapse of the matter cloud, as it evolves from a regular initial data, defined on an initial surface, from which the collapse develops. This dynamical evolution is governed by the Einstein equations. In other words, it is only the study of the collapse dynamics of the matter clouds that would decide the visibility or otherwise of the ultra-strong gravity regions. If, as the collapse evolves, the event horizons of gravity develop much before the spacetime singularity forms, then these extreme gravity regions are hidden away from the external universe, and a blackhole forms as the collapse outcome. On the other hand, if such horizons are delayed or fail to develop during collapse, as governed by the internal dynamics of the collapsing cloud, then the scenario where the extreme gravity regions are visible to external observers occurs, and a visible naked singularity forms.

The importance of gravitational collapse processes in relativistic astrophysics was realized when Datt (1938) and Oppenheimer and Snyder (1939) used general relativity to study the dynamical collapse of a homogeneous spherical dust cloud under its own gravity. This model gave rise to the concept of a blackhole. The term *blackhole* itself was popularized only later in the 1960s. The above work established that, under idealized conditions, a collapsing cloud of matter with zero pressure will necessarily give rise to a blackhole. Such a blackhole is a region of spacetime from which no light or matter can escape away to faraway external observers, and which necessarily covers the spacetime singularity or the regions of extreme physical conditions from the external universe. Specifically, in order to create a blackhole as the final state of gravitational collapse of the star, an event horizon must develop in the spacetime earlier than the time when the final spacetime singularity forms. Such an event horizon is a one-way membrane such that light or matter can fall into the region covered by it, but cannot escape away. If the event horizon developed prior to the formation of the singularity, neither the singularity nor the collapsing matter that has fallen within it would be observable to an external observer, and a blackhole is said to have formed as the final endstate of the collapsing star. All the matter of the star is then supposed to be crushed into the infinite density singularity at the center of the blackhole.

How early and when the horizon will actually develop in a realistic collapse is determined by the dynamics of the collapsing matter, the physical conditions within the star, and the dynamical evolution of the cloud as governed by the Einstein equations of gravity. Investigations in high energy astrophysics have already used the concept of a blackhole quite extensively. However, the actual understanding of the phenomena of gravitational collapse, and the conditions under which it can lead to the blackhole formation, or otherwise, within the framework of general relativity has progressed only relatively recently.

Further to the early studies mentioned above, it was generally assumed that the final endstate of collapse of a massive star will be a blackhole only. However, several important questions remained unanswered. For example, what would be the effects of non-zero pressures, which would be certainly important in the later stages of collapse, towards determining the collapse endstate, or, how will an inhomogeneous cloud collapse, say with a physically realistic density profile that is higher at the center and decreases slowly as one moves away from the center of the star? Early work on gravitational collapse focused only on simple models with idealized conditions, assuming a totally homogeneous density within the star, zero pressures, and so on, which would not be physically realistic. For example, a realistic star must have non-zero internal pressures, and its density would be typically higher at the center, as compared with its outer layers.

These physical issues and important questions have been crucial to the foundations of blackhole physics. But, not much attention could be paid to them, mainly due to the complexity of the equations of general relativity. This is because, in general, the Einstein equations are non-linear, second order partial differential equations that are quite difficult to solve. Therefore, the only model available until the late 1960s for the dynamical gravitational collapse of a massive matter cloud was that of a homogeneous, pressureless spherical cloud. In addition, not much attention was paid to these issues by the general relativists of the 1940s and 50s, who, by and large, did not consider such ultra-high energy phenomena to be physically realistic or of much astrophysical significance.

As indicated above, it was only the discovery in the 1960s of very high energy astrophysical phenomena that generated a keen theoretical interest in the continual gravitational collapse processes. However, mathematical difficulties and the complexity of gravity theory did not allow much progress. Then, the cosmic censorship hypothesis was introduced by Penrose (1969), which conjectured that the outcome of any generic gravitational collapse of a massive star must lead necessarily only to a blackhole formation as the collapse final state. This hypothesis thus suggested that the extreme and ultra-strong gravity regions, or the spacetime singularity, must always necessarily be covered within an event horizon of gravity, and that the external observers should never be able to see the singularity. This assumption means that whatever the physical conditions and forces within the massive stars may be (for example, they may be inhomogeneous in their density distribution, the pressures may be non-zero, or they may not be totally spherical and so on), the outcome of their continual collapse must give rise to a blackhole only. In other words, this amounts to an assumption of the nature of the allowed dynamical evolutions of the collapsing clouds, namely that the Einstein equations must permit only those evolutions that create the event horizon necessarily much prior to the formation of the final singularity or the ultra-strong gravity regions. Then, the singularity would be necessarily hidden within the horizon, which is a one-way surface, not allowing it to be seen by any external observers.

The cosmic censorship conjecture thus implies that no ultra-strong gravity regions forming in continual collapse will be visible to outside observers. That is, no naked singularity will develop in the collapse, and the event horizon developing in the dynamical collapse will always manage to cover these. Hence, the outcome of any gravitational collapse is necessarily a blackhole, and external observers can never see any ultra-strong gravity regions forming in the collapse, as indicated in Fig. 1.1.

As yet, a specific mathematical formulation for cosmic censorship that has been properly defined does not exist. Then, a proof of the same would have to be obtained within the framework of Einstein's gravity theory. The cosmic censorship assumption nevertheless provided a major impetus to developments in blackhole physics, and two parallel streams of developments took place. On one hand, the theoretical properties of blackholes were developed extensively, using cosmic censorship as the basic assumption, thus

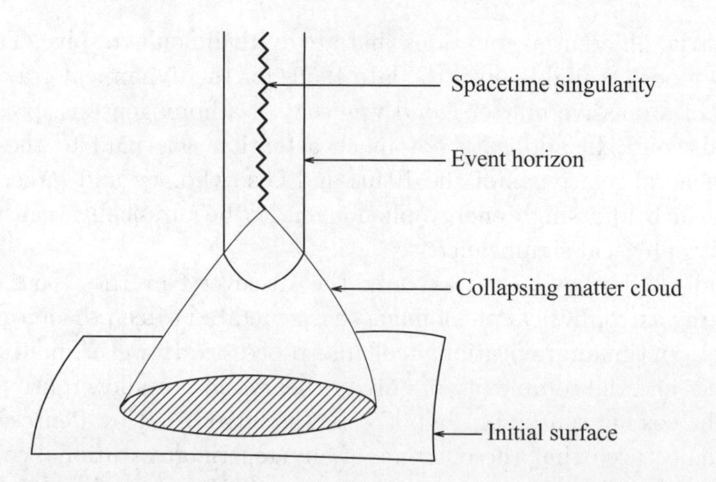

Fig. 1.1 The final outcome of a generic gravitational collapse must be a blackhole according to the cosmic censorship conjecture. Then, the eventual spacetime singularity of the collapse has to be preceded by the event horizon of gravity.

creating the laws of blackhole thermodynamics and related aspects (see for example, Hawking and Ellis, 1973). On the other hand, efforts to establish the censorship hypothesis continued, as it was clear all along that this assumption was absolutely fundamental to the theory and applications in blackhole physics, and so it needed a rigorous formulation and proof within the framework of general relativity. It is widely recognized that a proof of the censorship conjecture would place blackhole physics and its applications on a sound footing, whereas its failure would actually throw blackhole dynamics and related applications into serious doubt. Hence, the validity, or otherwise, of the cosmic censorship conjecture has remained an issue of crucial importance for all these years. The efforts to prove it have not succeeded for the past three decades, and there are even serious difficulties in formulating any rigorous mathematical version or a statement for this conjecture.

The theme that the only way out of this impasse is to study rigorously the dynamical gravitational collapse phenomena within the framework of Einstein's theory of gravity is proposed and developed here. This has been investigated extensively in the last couple of decades, and some of the issues that have been addressed include: what is the outcome of a continual gravitational collapse under physically realistic conditions, as governed by the Einstein equations? Will it be necessarily a blackhole as hypothesized by the censorship conjecture, or would it give rise to a naked singularity, where ultra-strong gravity regions forming in collapse are visible to external observers? In the latter case, would it be possible to observe the quantum gravity effects taking place in these visible ultra-strong gravity regions? Some of these issues are discussed here.

A detailed study of the collapse phenomena may be the only way towards any possible physically realistic formulation of censorship, if one exists. Such a study and investigation of collapse could also lead to novel physical insights and possibilities emerging out of the intricacies of the gravitational force. It would appear that beyond the studies so far, mainly of static and stationary solutions modeling blackholes, investigating dynamical evolutions as permitted by the Einstein equations would offer new insights into the nature of gravity. This is an arena that has been explored less, and which needs to be investigated carefully in detail.

To this end, gravitational collapse scenarios with non-zero pressures and more realistic equations of state for classes of general matter fields are considered here. A general formalism is developed to treat the spherical collapse from regular initial data. These considerations also point to why it has not been possible so far to make any definite progress on the censorship conjecture. It is seen that it is first necessary to acquire a deeper and more extensive understanding of the dynamical evolutions and gravitational collapse processes in general relativity. Recent work on studying and understanding the final fate of dynamical gravitational collapse in gravitation theory is discussed. General matter fields are considered so as to include important physical features in the collapse, such as inhomogeneities in matter distribution, non-vanishing pressures, different forms for the equations of state of the collapsing matter, and other such aspects. It is seen that in spherical gravitational collapse, given the matter initial data on an initial surface from which the collapse develops, there are the rest of the free initial data such as the velocities of the collapsing shells, and the classes of the dynamical evolutions as permitted by the Einstein equations, which lead to the final state that is either a blackhole, or a naked singularity that is a visible ultra-strong density and curvature region forming in the collapse not covered by an event horizon. The nature of the outcome depends on the regular initial data from which the collapse evolves, and the allowed dynamical evolutions in the spacetime, as permitted by the Einstein equations.

After the basics of the structure and properties of spacetimes and the essentials of relativity theory are summarized in Chapter 2, the above issues are discussed in Chapter 3. Collapsing dust clouds, which generalize and include as a special case the Oppenheimer–Snyder dust collapse models, and which give an idea of the possible outcomes of gravitational collapse in terms of a blackhole or a naked singularity, are also discussed in Chapter 3. The Oppenheimer–Snyder dust collapse scenario is included here as a special case when the cloud is homogeneous. It is seen, however, that a more realistic density profile with a density higher at the center and decreasing as one moves away from center, gives rise to a naked singularity as the collapse endstate

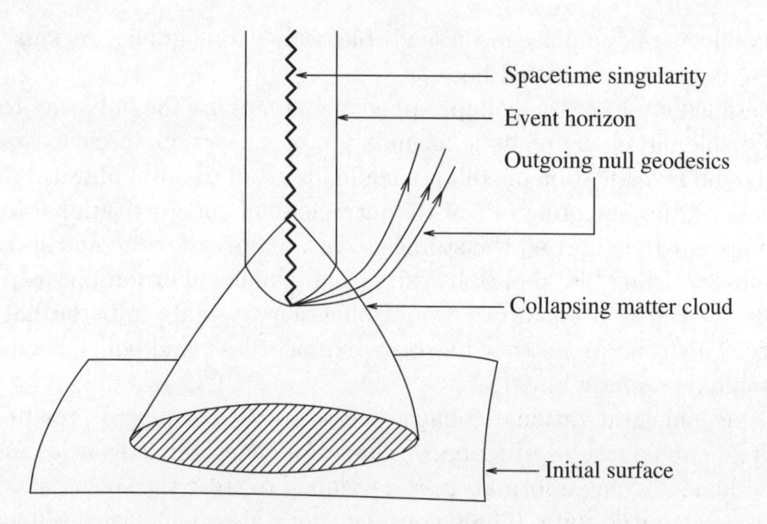

Fig. 1.2 If the collapsing cloud is inhomogeneous, with a density higher at the center, the trapped surface formation and event horizon in the collapse are delayed to give rise to a naked singularity, where the ultra-strong gravity regions are visible to outside observers.

(see Fig. 1.2). In general, it is seen that the collapse outcome depends on the nature of the initial matter profiles and the evolutions allowed by the Einstein equations. The structure of this spacetime in the homogeneous density case gives rise to the basic notion and concept of a blackhole. The dust collapse picture provides a concrete background to the possible final states of a continual gravitational collapse.

Chapter 4 then studies several useful aspects of spacetime structure, singularities and collapse, as related to the cosmic censorship hypothesis possibilities and the structure of naked singularities developing in gravitational collapse. It is pointed out that while the cosmic censorship does not hold in general relativity in the obvious sense of ruling out naked singularities from all physically realistic gravitational collapse models, any definite formulation of this hypothesis will depend on a detailed analysis of stability and genericity aspects related to collapse scenarios, and the naked singularities and blackhole phases developing as final outcomes of the gravitational collapse. Several possibilities towards any plausible formulation are discussed.

In light of the results available so far and the emerging scenario, the key physical issue is the possible final state of a massive star. The basic problem to be addressed is: what will the final outcome of the gravitational collapse of a massive star be when it collapses freely at the end of its life cycle on exhausting its nuclear fuel under the force of its own gravity? Under realistic astrophysical conditions, will it turn into a blackhole, or does it terminate as a naked singularity? Are there any observable consequences in the latter case?

These physical questions underlie many considerations here on gravitational collapse.

While theoretical properties of blackholes have been studied rather extensively, the naked singularity solutions in general relativity, arising out of dynamical collapse studies are relatively less understood as yet. It is sometimes asked how a naked singularity could arise in the collapse, allowing the light to escape from the extreme gravity regions even when the gravity fields are so strong. Some of these aspects are discussed in Chapter 5. Also explored and pointed out here are the physical features, such as the role of inhomogeneities and spacetime shear, that lead to a naked singularity rather than a blackhole as the collapse endstate. The physics that possibly causes a naked singularity in the collapse, rather than a blackhole, is examined. While it may be stated that a good understanding now exists on spherical collapse in general for a generic matter field, non-spherical collapse remains major uncharted territory. This is also closely related to the stability and genericity aspects of collapse outcomes, and these issues are discussed here.

The information loss paradox and related issues have highlighted some of the important problems with the blackhole paradigm, which also include the existence of an infinite density spacetime singularity at the center of a blackhole, leading to an instability even at the classical level, and uncertainties of the correctness, or otherwise, of the cosmic censorship conjecture. Under such a situation, a possibility worth considering could be the avoidance or delay of trapped surfaces formation as the star evolves, collapsing under gravity. This is the case when a collapse evolution to a naked singularity takes place, where the trapped surfaces do not form early enough or are avoided in the spacetime. In that case, in the late stages of the collapse, the star could radiate away most of its mass. This could then offer a way out of the blackhole conundrums, whilst also resolving the singularity problem.

As such, the outcomes of a continual collapse, namely the blackhole and naked singularity, are very different from each other in nature. The naked singularity, which is more like an event than an object in many cases, could have quite different physical properties compared with a blackhole. Therefore, the implications of the visibility of the ultra-high density and curvature regions to a faraway observer in the spacetime need to be investigated. Such a scenario offers an intriguing possibility that the quantum gravity effects may become observable during the later final stages of the collapse. This is because the ultra-strong gravity regions where quantum gravity effects take place are now no longer hidden under the event horizon, but are visible and can, in principle, communicate with external observers. This may offer interesting connections and pointers towards observational effects of quantum gravity arising from gravitational collapse. These possibilities are discussed in Chapter 5, where some implications of loop quantum gravity formalism from such a perspective are indicated.

2
The spacetime manifold

Here, the essential fundamentals of general relativity and related mathematical aspects are described. For further details, see texts such as Weinberg (1972), Misner, Thorne, and Wheeler (1973), and Wald (1984). Other necessary techniques are developed in later chapters as necessary. While defining vectors, tensors, and other quantities, we use both a local and a coordinate free global approach, and indicate how to make a transition from one to the other representation, which is useful in several situations.

In Section 2.1 the manifold model for spacetime is introduced. Basic definitions of a differentiable manifold, and various topological and orientability properties are discussed. The metric tensor and related aspects are considered in Section 2.2, and the connection on a spacetime is considered in Section 2.3. Timelike and null geodesics play a basic role in the considerations here on gravitational collapse. These are a special set of non-spacelike trajectories that represent the motion of freely falling material particles and light rays, and they clarify many properties of a spacetime. These are discussed in Section 2.4. The spacetime curvature is considered in Section 2.5, and the Einstein equations governing the dynamics of matter in the spacetime are discussed in Section 2.6. Many exact solutions have been found to the Einstein equations so far; however, the Schwarzschild and Vaidya geometries are particularly relevant to gravitational collapse scenarios, and Section 2.7 discusses these.

2.1 The manifold model

The universe is modeled as a four-dimensional spacetime M in general relativity, together with an indefinite Lorentzian metric tensor g, which has the signature $(-,+,+,+)$. Conditions ensuring physical reasonability to the spacetime model are generally assumed. These include the space and time orientability, and necessary topological regularity conditions such as the Hausdorffness and connectedness. Here, this basic model of the space-time universe that underlies Einstein's theory of gravitation is specified. The manifold model for the universe naturally incorporates the observed continuity of space and time at the classical level, and the basic principle of general relativity where the locally flat regions combine to produce a globally curved continuum. This implies that a smooth change of coordinates is possible when a transition is made from one coordinate patch to another.

2.1.1 Differentiable manifolds

The *n-dimensional Euclidian space* R^n is a collection of all n-tuples $(x^1, \ldots, x^n)$ such that $-\infty < x^i < \infty$, $i = 1, \ldots, n$, and which has the natural Euclidian metric. An *open ball* of radius r around any point x in R^n is the set of all points y such that $\mid x - y \mid < r$, where the modulus denotes the positive definite distance as defined by the Euclidian metric on R^n. The *open sets* in R^n are sets which can be expressed as a union of such open balls.

Basically, an n-dimensional differentiable manifold is a set that is locally similar to an open set of R^n. Therefore, locally Euclidian patches are glued together smoothly to obtain a space which need not be Euclidian globally.

An *n-dimensional, C^∞, real differentiable manifold* is a set M, together with a collection $\{u_\alpha, \phi_\alpha\}$, called an *atlas* for M. Here, the u_α values are subsets of M and the ϕ_α values are one–one maps of a given u_α onto an open subset in R^n, which satisfy the following.

(1) The sets u_α form a cover for M, that is, any given p in M must be in a u_α for some value of α, and

$$M = \bigcup_\alpha u_\alpha. \tag{2.1}$$

(2) Whenever two neighborhoods u_α and u_β intersect, that is, $u_\alpha \cap u_\beta \neq \phi$, then the map $\phi_\alpha \circ \phi_\beta^{-1}$ from R^n to R^n, which takes points of $\phi_\beta(u_\alpha \cap u_\beta)$ to points of $\phi_\alpha(u_\alpha \cap u_\beta)$, is infinitely differentiable in a continuous manner (a smooth C^∞-function) as a mapping between two open subsets of R^n (see Fig. 2.1).

Alternatively, it is possible to consider the map $\phi_\beta \circ \phi_\alpha^{-1}$, and the same condition again holds. Each u_α is called a local coordinate neighborhood or a *chart* where $p \in u_\alpha$ has coordinates of $\phi_\alpha(p)$ in R^n. The condition (2) above ensures that whenever an event $p \in M$ undergoes a coordinate

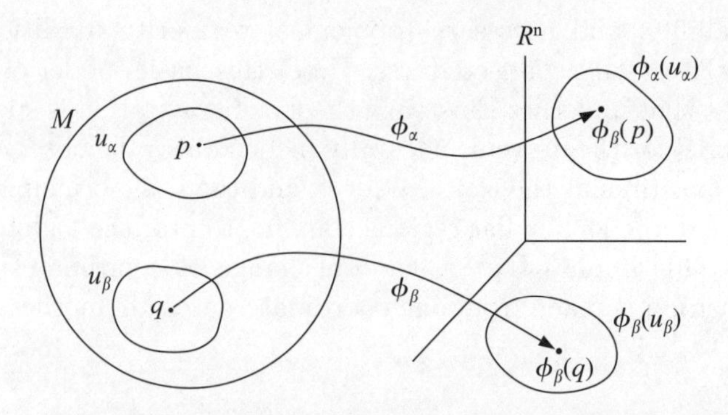

Fig. 2.1 All events p and q in the manifold have neighborhoods which are homeomorphic to subsets in R^n. The points $p, q \in M$ have coordinates of $\phi_\alpha(p)$ and $\phi_\beta(q)$. Whenever the neighborhoods in M intersect, there should be a smooth change of coordinates.

change, the change is necessarily smooth. That is, if $\{x^i\}$ and $\{y^i\}$ are local coordinates of $p \in M$ in u_α and u_β respectively, then the functions $x^i = x^i(y^1, \ldots, y^n)$ are C^∞-functions from R^n to R^n. A *maximal* or *complete atlas* is chosen for the spacetime manifold M, that is, if $\{u_\alpha, \phi_\alpha\}$ is an atlas for M, one selects for M the atlas that consists of all other atlases that are compatible with $\{u_\alpha, \phi_\alpha\}$. This implies that their union with $\{u_\alpha, \phi_\alpha\}$ is also a C^∞-atlas.

This implies that one has included all possible, mutually compatible coordinate systems for the manifold M. A C^r-manifold is defined in a similar way, where it is required that the transition functions $\phi_\alpha \circ \phi_\beta^{-1}$ are r-times continuously differentiable, where a continuous function is denoted by C^0.

The Euclidian plane R^2, or Euclidian space R^n, is, in itself, a manifold as it is covered by a single chart R^n, where ϕ would be the identity map with the coordinate range $-\infty < x^i < \infty$ for $i = 1, \ldots, n$. Another example of such a manifold is the two-sphere S^2 defined by

$$S^2 = \{(x^1,\ x^2,\ x^3) \in R^3 \mid (x^1)^2 + (x^2)^2 + (x^3)^2 = 1\}. \tag{2.2}$$

The six hemispherical open sets $O_i^\pm$ for $i = 1, 2, 3$ are given by $O_i^\pm = \{(x^1, x^2, x^3) \in S^2 \mid \pm x^i > 0\}$, which cover S^2. Each $O_i^\pm$ is mapped onto the open disk $\{(x, y) \in R^2 \mid x^2 + y^2 < 1\}$ by the projection maps such as $f_1^+(x^1, x^2, x^3) = (x^2, x^3)$. The overlap functions $f_i^\pm \circ (f_j^\pm)^{-1}$ are C^∞-functions in their domain of definition. Thus, S^2 is a two-dimensional, C^∞-manifold that cannot be covered by a single coordinate system. Similarly, the sphere S^n in n-dimensions is also a differentiable manifold.

2.1.2 Vectors and one-forms

A function $f : M \to R^n$ is called *differentiable* if the map $f \circ \phi_\alpha^{-1}$ is a C^∞-map for all charts ϕ_α as a map from R^n to R^n; C^r-functions can be defined similarly (Spivak, 1965).

Suppose now M and M' are two differentiable manifolds with ϕ_α and ψ_α denoting charts of M and M' respectively. A map $h : M \to M'$ is called *C^r-differentiable* if $\psi_\alpha \circ h \circ \phi_\alpha^{-1}$ is always C^r-differentiable as a map from R^n to R^n for all α. If the dimension of M is n and that of M' is n' with $n > n'$, then the map h cannot be one–one. However, if h is one–one, onto, and continuous from M to M' such that h^{-1} is also a continuous map, then h is called a *homeomorphism.* If a homeomorphism and its inverse are both C^r-maps, then it is called a *C^r-diffeomorphism.*

A *C^k-curve* in M is a C^k-map from an interval of R into M. A *vector* (or a contravariant vector) $(\partial/\partial t)_{\lambda(t_0)}$, tangent to a C^k-curve $\lambda(t)$ at a point $\lambda(t_0)$, is an operator from the space of all smooth functions on M into R:

$$\left(\frac{\partial}{\partial t}\right)_{\lambda(t_0)}(f) = \left(\frac{\partial f}{\partial t}\right)_{\lambda(t_0)} = \lim_{s\to 0}\frac{f[\lambda(t+s)] - f[\lambda(t)]}{s}, \tag{2.3}$$

where s denotes a small increment of the parameter t. This is $d(f \circ \lambda)/dt$, which is the derivative of f in the direction of $\lambda(t)$ with respect to parameter t. If $f = t$, where t is the parameter along the curve,

$$\left(\frac{\partial}{\partial t}\right)_{\lambda}(t) = 1. \tag{2.4}$$

If the x^i values are local coordinates in a neighborhood of $p = \lambda(t_0)$, then

$$\left(\frac{\partial f}{\partial t}\right)_{\lambda(t_0)} = \frac{dx^i}{dt}\frac{\partial f}{\partial x^i}\,|_{\lambda(t_0)}, \tag{2.5}$$

where a repeated index means summation over the values $1, \ldots, n$. (This *summation convention* is used throughout.) Therefore, every tangent vector at $p \in M$ is expressed as a linear combination of the coordinate derivatives, which are $(\partial/\partial x^1)_p, \ldots, (\partial/\partial x^n)_p$. Conversely, any linear combination of these operators that are partial derivatives with respect to coordinates can be chosen, namely, $V^i(\partial/\partial x^i)_p$, with the values of V^i being any numbers. It is then possible to find a curve which admits this linear combination as a tangent (see for example, Wald, 1984). The vectors $(\partial/\partial x^j)_p$ are linearly independent (if not, then there are numbers V^i such that

$$V^i\left(\frac{\partial}{\partial x^i}\right)_p = 0, \tag{2.6}$$

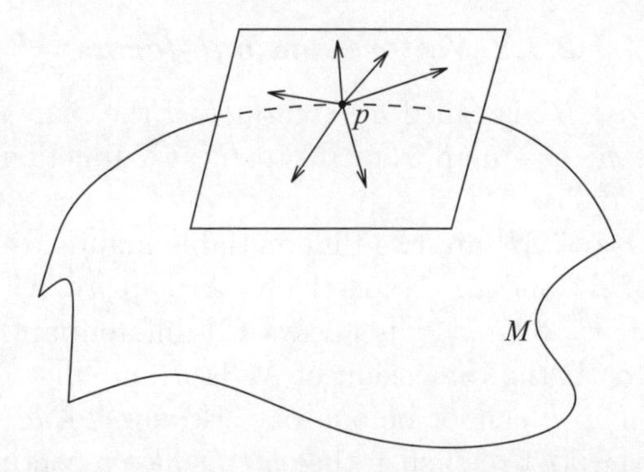

Fig. 2.2 The tangent space T_p at a point $p \in M$, which gives the set of all directions at that point.

with at least one V^i being non-zero, and applying this to the coordinate functions $x^1, \ldots, x^n$ gives $V^i = 0$ for all i, a contradiction). Therefore, the vectors $(\partial/\partial x^j)$ span the vector space T_p, the space of all tangent vectors at p (see Fig. 2.2). The vector space structure here is defined by

$$(\alpha \boldsymbol{X} + \beta \boldsymbol{Y})f = \alpha(\boldsymbol{X}f) + \beta(\boldsymbol{Y}f), \tag{2.7}$$

for $\alpha, \beta \in R$ and $\boldsymbol{X}, \boldsymbol{Y} \in T_p$, where $\boldsymbol{X}$ and $\boldsymbol{Y}$ are vectors at p; T_p is also called the *tangent space* at p. The basis $\{(\partial/\partial x^i)_p\}$ is called a *coordinate basis* of T_p. A general basis is denoted by $\{\boldsymbol{e}_i\}$, where $i = 1, \ldots, n$, are linearly independent vectors. Then, for any vector $\boldsymbol{V} \in T_p$,

$$\boldsymbol{V} = V^i \boldsymbol{e}_i, \tag{2.8}$$

where the numbers V^i are called the components of $\boldsymbol{V}$ with respect to the basis $\boldsymbol{e}_i$. In a coordinate basis, $V^i = dx^i/dt$. Again, $\{\partial/\partial x^i\}$ forms a basis of T_p which means the dimension of T_p is n.

For the tangent space T_p at $p \in M$, the vector space of all dual vectors at p, also called *covariant vectors* or one-forms at p, can be naturally defined. A one-form $\boldsymbol{\omega}$ at p is a real-valued linear functional on T_p, denoted by $\boldsymbol{\omega}(\boldsymbol{X}) \equiv \langle \boldsymbol{\omega}, \boldsymbol{X} \rangle$, and the linearity condition implies

$$\langle \boldsymbol{\omega}, \alpha \boldsymbol{X} + \beta \boldsymbol{Y} \rangle = \alpha \langle \boldsymbol{\omega},\ \boldsymbol{X} \rangle + \beta \langle \boldsymbol{\omega}, \boldsymbol{Y} \rangle. \tag{2.9}$$

Given a tangent space basis $\{\boldsymbol{e}_a\}$, a unique set of one-forms $\{\boldsymbol{e}^a\}$ is given by the condition that the one-form $\boldsymbol{e}^b$ maps a vector $\boldsymbol{V}$ into V^b, that is, the bth component of $\boldsymbol{V}$ in the basis $\boldsymbol{e}_a$. Therefore,

$$\langle \boldsymbol{e}^b, \boldsymbol{V} \rangle = V^b, \tag{2.10}$$

where $a, b, \ldots$ and $i,\ j, \ldots$ denote indices for vectors and tensors.

From the above,

$$\langle \boldsymbol{e}^a, \boldsymbol{e}_b \rangle = \delta^a_b, \tag{2.11}$$

where the right-hand side is the Kronecker delta function. The linear combinations of one-forms $\boldsymbol{\omega}$ and $\boldsymbol{\eta}$ are defined by

$$\langle \alpha\boldsymbol{\omega} + \beta\boldsymbol{\eta}, \boldsymbol{V} \rangle = \alpha\langle \boldsymbol{\omega}, \boldsymbol{V} \rangle + \beta\langle \boldsymbol{\eta}, \boldsymbol{V} \rangle, \tag{2.12}$$

with $\alpha, \beta \in R$. Then, $\{\boldsymbol{e}^a\}$ is a basis for the space of all one-forms at p because any one-form $\boldsymbol{\omega}$ can be written as $\boldsymbol{\omega} = \boldsymbol{\omega}_a \boldsymbol{e}^a$ with

$$\omega_a = \langle \boldsymbol{\omega}, \boldsymbol{e}_a \rangle. \tag{2.13}$$

Therefore, the set of all one-forms at the event p forms a vector space at p, the *dual* of T_p, and is denoted by T^*_p. The basis $\boldsymbol{e}^a$ is a *dual basis* to $\boldsymbol{e}_a$. If $\boldsymbol{\omega} \in T^*_p$ and $\boldsymbol{V} \in T_p$, then

$$\langle \boldsymbol{\omega}, \boldsymbol{V} \rangle = \langle \omega_a \boldsymbol{e}^a, V^b \boldsymbol{e}_b \rangle = \omega_a V^b \delta^a_b = \omega_a V^a. \tag{2.14}$$

A *vector field* $\boldsymbol{V}$ on a manifold M is an assignment of a tangent vector $\boldsymbol{V}_p$ at each $p \in M$. The vector field is said to assign vectors smoothly if, for each smooth function f on M, the function $\boldsymbol{V}(f)$, the directional derivative of f along the vector $\boldsymbol{V}_p$, is also smooth on M at each point p. The coordinate basis vector fields $\partial/\partial x^i$ are smooth, and so a vector field will be smooth provided that its coordinate components V^i are smooth functions. Given two vector fields $\boldsymbol{V}$ and $\boldsymbol{W}$, a new vector field, called their *commutator* $[\boldsymbol{V}, \boldsymbol{W}]$, is defined by

$$[\boldsymbol{V}, \boldsymbol{W}](f) = \boldsymbol{V}[\boldsymbol{W}(f)] - \boldsymbol{W}[\boldsymbol{V}(f)]. \tag{2.15}$$

The commutator for any two coordinate basis vector fields vanishes. If f and g are any two smooth functions, it can be seen that $[\boldsymbol{V}, \boldsymbol{W}](f+g) = [\boldsymbol{V}, \boldsymbol{W}](f) + [\boldsymbol{V}, \boldsymbol{W}](g)$ and that $[\boldsymbol{V}, \boldsymbol{W}](\alpha f) = \alpha[\boldsymbol{V}, \boldsymbol{W}](f)$ for any $\alpha \in R$. It can be shown that

$$[\boldsymbol{V}, \boldsymbol{W}](fg) = f[\boldsymbol{V}, \boldsymbol{W}](g) + g[\boldsymbol{V}, \boldsymbol{W}](f), \tag{2.16}$$

which is the product property. By expanding in a coordinate basis it is seen that the commutator $[\boldsymbol{V}, \boldsymbol{W}]$ will be a smooth vector field if and only if both $\boldsymbol{V}$ and $\boldsymbol{W}$ are smooth. Note that $[\boldsymbol{V}, \boldsymbol{V}] = 0$ and that $[\boldsymbol{V}, \boldsymbol{W}] = -[\boldsymbol{W}, \boldsymbol{V}]$. Furthermore, the commutator is linear in each of its arguments with respect to addition, that is

$$[\boldsymbol{V}_1 + \boldsymbol{V}_2, \boldsymbol{W}] = [\boldsymbol{V}_1, \boldsymbol{W}] + [\boldsymbol{V}_2, \boldsymbol{W}]. \tag{2.17}$$

Any smooth function f on M defines a one-form df, called the *differential* of f, by the rule

$$\langle df, \boldsymbol{V} \rangle \equiv \boldsymbol{V} f. \tag{2.18}$$

Therefore, in a coordinate basis,

$$\langle df, \boldsymbol{V} \rangle = V^a \frac{\partial f}{\partial x^a}. \tag{2.19}$$

The local coordinate functions $(x^1, \ldots, x^n)$ are used to define a set of one-forms $(dx^1, \ldots, dx^n)$, which is a basis dual to the coordinate basis because

$$\left\langle dx^a, \frac{\partial}{\partial x^b} \right\rangle = \frac{\partial x^a}{\partial x^b} = \delta^a_b. \tag{2.20}$$

Also,

$$df = \left\langle df, \frac{\partial}{\partial x^a} \right\rangle dx^a = \frac{\partial f}{\partial x^a} dx^a, \tag{2.21}$$

which is the usual definition of the differential df.

If f is a non-constant function, the surfaces $f = \text{const.}$ define an $(n-1)$-dimensional submanifold of M. Consider the set of all the vectors $\boldsymbol{V} \in T_p$ such that

$$\langle df, \boldsymbol{V} \rangle = \boldsymbol{V} f = 0, \tag{2.22}$$

then the vectors $\boldsymbol{V}$ are tangent to curves in the $f = \text{const.}$ submanifold, through p. Therefore, the differential df is normal to the surface $f = \text{const.}$ at p.

2.1.3 Topological structure

A C^∞-maximal atlas on a spacetime manifold M induces a natural *topology on* M, given by the companion Euclidian space by requiring each ϕ_α to be a homeomorphism. Therefore, the open sets in M are pre-images of open sets in R^n and their unions. Then, the collection $\{u_\alpha\}$ provides a basis for the spacetime topology, where R^n has its canonical topology, defined by the metric

$$d(x, y) = \left[(x_1 - y_1)^2 + \cdots + (x_n - y_n)^2\right]^{1/2} \tag{2.23}$$

for any $x, y \in R^n$.

Several topological regularity conditions that are assumed for a physically reasonable spacetime manifold are now given. The spacetime is assumed to be *Hausdorff*, that is, given p and q with $p \neq q$ in M, there are disjoint open sets u_α and u_β in M such that $p \in u_\alpha$ and $q \in u_\beta$. Physically interesting spacetime examples such as the Schwarzschild geometry and Robertson–Walker models are Hausdorff. This is a reasonable requirement on a spacetime which ensures the uniqueness of limits of convergent sequences, and incorporates the intuitive notion of distinct spacetime events.

Next, the spacetime M has *no boundary.* A boundary represents, in a sense, the 'edge' of the universe, which is not detected by any astronomical observations. Mathematically, it is common to have manifolds without a boundary. For example, for a two-sphere S^2 in R^3, no point in S^2 is a boundary point in the induced topology as implied by the natural topology on R^3, because all neighborhoods of any $p \in S^2$ are contained within S^2 in this induced topology. Assume M to be *connected*, that is, $M = X_1 \cup X_2$, with X_1 and X_2 being two open sets and $X_1 \cap X_2 = \phi$ is not possible. This is because disconnected components of the universe cannot interact by means of any signals, and the observations are confined to the connected component where the observer is situated. But, M could be either simply connected or multiply connected. For further discussion on multiply connected spacetimes and the notion of a wormhole in the Schwarzschild geometry, see Wheeler (1962, 1964) and Misner, Thorne, and Wheeler (1973). Such wormholes are like 'handles' in the multiply connected topology of space and can connect widely separated regions in space.

It is known, however, that such wormholes are not stable and collapse as soon as created, unless the violation of the energy condition in an averaged sense is allowed, thus implying negative energy fields (see for example, Deutsch and Candelas, 1980; Lee, 1983; Morris, Thorne, and Yurtsever, 1988). Therefore, a wormhole may be stabilized only by shifting the energy of vacuum to be negative by some quantum processes. The process of topology change could also give rise to a multiply connected spacetime. It is not clear if the topology of space could change while it evolves in time, and if so, what physical agencies cause it. A topology change can affect the structure of spacetime severely to cause naked singularities (Joshi and Saraykar, 1987).

A spacetime is assumed to be *non-compact*, because compact spacetimes violate causality and admit closed timelike curves. One could then enter one's own past, which is considered to be highly unphysical. Usually, M is also taken to be paracompact. An atlas $\{u_\alpha, \phi_\alpha\}$ is called *locally finite* if there is an open set containing every $p \in M$ that intersects only a finite number of the sets u_α. A manifold M is called *paracompact* if, for every atlas $\{u_\alpha, \phi_\alpha\}$, there is a locally finite atlas $\{O_\beta, \psi_\beta\}$ with each O_β contained in some u_α. For a further discussion on these topological concepts, see Simmons (1963) and Willard (1970).

For a connected, Hausdorff manifold, the paracompactness property is equivalent to the existence of a countable base for the topology of M. The existence of a Lorentz metric globally on M implies any Hausdorff manifold with a C^r Lorentz metric tensor must be paracompact (Geroch, 1968b).

Let B be the set of all ordered basis $\{\boldsymbol{e}_i\}$ for T_p. If $\{\boldsymbol{e}_i\}$ and $\{\boldsymbol{e}_{j'}\}$ are in B, then

$$\boldsymbol{e}_{j'} = a^i_{j'}\boldsymbol{e}_i. \tag{2.24}$$

If a denotes the matrix $[a^i_j]$, then $\det[a] \neq 0$. An equivalence relation in B is introduced by the condition that $\boldsymbol{e}_i \sim \boldsymbol{e}_{j'}$ if and only if $\det[a] > 0$. Clearly, there are exactly two such equivalence classes that are called the *orientations* of T_p. By an arbitrary choice, one of these classes is called a *positive orientation* and the other a *negative orientation* for T_p. Now, let M be a manifold and $p \in M$. Suppose for all p there is a neighborhood U of p and n continuous linearly independent vector fields $\{\boldsymbol{\xi}_1, \ldots, \boldsymbol{\xi}_n\}$ such that for any $q \in U$, the basis $\{\boldsymbol{\xi}_1(q), \ldots, \boldsymbol{\xi}_n(q)\}$ belongs to the same equivalence class under a transformation of coordinates when one chooses another chart $\boldsymbol{V}$ containing q. Then, under all possible coordinate transformations, the determinant for the transformation matrix of the basis has the same sign, and M is called an *orientable* manifold. This definition could then also be stated as below. An n-dimensional manifold M is called orientable if M admits an atlas $\{U_i, \phi_i\}$ such that, whenever $U_i \cap U_j \neq \emptyset$, then

$$J = \det\left(\frac{\partial x^i}{\partial x^j}\right) > 0, \tag{2.25}$$

where $\{x^i\}$ and $\{x^j\}$ are local coordinates in U_i and U_j respectively. The Möbius strip is an example of a non-orientable manifold.

2.1.4 Tensors

Tensors are geometric objects on a spacetime, and are invariant under the change of coordinates. The stress–energy tensor represents various matter fields existing on a spacetime, such as the electromagnetic field, dust, and so on. On the other hand, the global geometry and curvature of the manifold are described by tensor fields such as the metric tensor and the curvature tensor. In the general theory of relativity, the form of physical laws remains unchanged under the general transformation of coordinates (principle of general covariance). So, physical fields are represented by various tensor fields on a spacetime, and the laws governing these are tensor equations, valid under arbitrary coordinate transformations. In an inertial coordinate frame, these reduce to special relativity laws.

A *tensor* T of type (r, s) at $p \in M$ is a multilinear real-valued functional on the Cartesian product, as given by

$$T^*_p \times \cdots \times T^*_p \times T_p \times \cdots \times T_p \to R, \tag{2.26}$$

where there are r-factors of T_p^* and s-factors of T_p. Therefore, T acts on one-forms and vectors in general to produce a real number.

If T is a tensor of (r, s) type at $p \in M$, then

$$T(\boldsymbol{\omega}_1, \ldots, \boldsymbol{\omega}_r, \boldsymbol{V}_1, \ldots, \boldsymbol{V}_s) = T(\omega_{i_1} \boldsymbol{e}^{i_1}, \ldots, \omega_{i_r} \boldsymbol{e}^{i_r}, V^{j_1} \boldsymbol{e}_{j_1}, \ldots, V^{j_s} \boldsymbol{e}_{j_s}). \tag{2.27}$$

Using multilinearity, the above equation can be written as

$$T^{i_1 \ldots i_r}{}_{j_1 \ldots j_s} \omega_{i_1} \ldots \omega_{i_r} V^{j_1} \ldots V^{j_s}, \tag{2.28}$$

where

$$T^{i_1 \ldots i_r}{}_{j_1 \ldots j_s} \equiv T(\boldsymbol{e}^{i_1}, \ldots, \boldsymbol{e}^{i_r}, \boldsymbol{e}_{j_1}, \ldots, \boldsymbol{e}_{j_s}), \tag{2.29}$$

and $\{\boldsymbol{e}_i\}$ and $\{\boldsymbol{e}^i\}$ are basis vectors at p for T_p and T_p^* respectively.

The space of all tensors of type (r, s) at p is called the *tensor product* $T_s^r(p)$, denoted by,

$$T_s^r(p) = T_p \otimes \cdots \otimes T_p \otimes T_p^* \otimes \cdots \otimes T_p^*, \tag{2.30}$$

with r-factors of T_p and s-factors of T_p^*. The dimension of T_s^r is n^{r+s}, with n being the dimension of the manifold. This is a vector space of all (r, s) tensors over real numbers with the addition of tensors and scalar multiplication defined in a natural manner. In particular, a vector is a tensor of type (1,0) and a one-form is a tensor of type (0,1). Using the basis vectors $\{\boldsymbol{e}_i\}$ and $\{\boldsymbol{e}^i\}$ for the tangent space and cotangent space at p, the set

$$\{\boldsymbol{e}_{i_1} \otimes \cdots \otimes \boldsymbol{e}_{i_r} \otimes \boldsymbol{e}^{j_1} \otimes \cdots \otimes \boldsymbol{e}^{j_s}\} \tag{2.31}$$

forms a basis for the tensor product $T_s^r(p)$ with all indices running from 1 to n. Then, any tensor $T \in T_s^r$ can be expressed as

$$T = T^{i_1 \ldots i_r}{}_{j_1 \ldots j_s} \boldsymbol{e}_{i_1} \otimes \cdots \otimes \boldsymbol{e}_{i_r} \otimes \boldsymbol{e}^{j_1} \otimes \cdots \otimes \boldsymbol{e}^{j_s}, \tag{2.32}$$

with the tensor components $T^{i_1 \ldots i_r}{}_{j_1 \ldots j_s}$ defined as above.

Consider a change of coordinates, giving a change of basis $\{\boldsymbol{e}^i\}$ to $\{\boldsymbol{e}^{i'}\}$ and $\{\boldsymbol{e}_j\}$ to $\{\boldsymbol{e}_{j'}\}$. If a coordinate basis $\{\partial/\partial x^i\}$ for T_p and a corresponding basis $\{dx^i\}$ for the cotangent space T_p^* are chosen, then under a change of coordinates, the components of T in the new coordinates $\{x^{i'}\}$ are

$$T^{i'_1 \ldots i'_r}{}_{j'_1 \ldots j'_s} = T\left(dx^{i'_1}, \ldots, dx^{i'_r}, \frac{\partial}{\partial x^{j'_1}}, \ldots, \frac{\partial}{\partial x^{j'_s}}\right). \tag{2.33}$$

Since $x^{i'}$ are functions of x^i, substituting in the above equation for $\partial/\partial x^{i'}$ and $dx^{i'}$ gives, for the transformed components,

$$T^{i'_1 \dots i'_r}{}_{j'_1 \dots j'_s} = T^{i_1 \dots i_r}{}_{j_1 \dots j_s} \frac{\partial x^{i'_1}}{\partial x^{i_1}} \cdots \frac{\partial x^{i'_r}}{\partial x^{i_r}} \frac{\partial x^{j_1}}{\partial x^{j'_1}} \cdots \frac{\partial x^{j_s}}{\partial x^{j'_s}}. \tag{2.34}$$

Therefore, the components of a vector $\boldsymbol{V}$ and a one-form $\boldsymbol{\omega}$ transform as

$$V^{i'} = \frac{\partial x^{i'}}{\partial x^i} V^i, \quad \omega_{i'} = \frac{\partial x^i}{\partial x^{i'}} \omega_i. \tag{2.35}$$

For a general set of basis vectors, the transformation of components of a tensor can be written similarly.

For a tensor of type (r, s), the *contraction* of T over a contravariant index and a covariant index is defined to be a tensor $C(T)$ of type $(r-1, s-1)$. For example, if one contracts over the first contravariant and covariant indices, this gives

$$C_1^1(T) = T^{ij\dots l}{}_{im\dots n} \boldsymbol{e}_j \otimes \cdots \otimes \boldsymbol{e}_l \otimes \boldsymbol{e}^m \otimes \cdots \otimes \boldsymbol{e}^n. \tag{2.36}$$

Using the equations above for a transformation under the change of basis vectors, it is seen that the contraction $\boldsymbol{C}^1{}_1$ is indcpendent of the basis used and so invariant under a change of coordinates. Similarly, T can be contracted over any pair of contravariant and covariant indices.

In the space of all tensors of type (r, s) at p, the *addition* of two tensors T and T', and multiplication by a real number α, are defined as

$$\begin{aligned}(T + T')(\boldsymbol{\omega}^1, \dots, \boldsymbol{\omega}^r, \boldsymbol{X}_1, \dots, \boldsymbol{X}_s) &= T(\boldsymbol{\omega}^1, \dots, \boldsymbol{\omega}^r, \boldsymbol{X}_1, \dots, \boldsymbol{X}_s) \\ &\quad + T'(\boldsymbol{\omega}^1, \dots, \boldsymbol{\omega}^r, \boldsymbol{X}_1, \dots, \boldsymbol{X}_s), \\ (\alpha T)(\boldsymbol{\omega}^1, \dots, \boldsymbol{\omega}^r, \boldsymbol{X}_1, \dots, \boldsymbol{X}_s) &= \alpha T(\boldsymbol{\omega}^1, \dots, \boldsymbol{\omega}^r, \boldsymbol{X}_1, \dots, \boldsymbol{X}_s). \end{aligned} \tag{2.37}$$

The *outer product* of the two tensors T and S of type (r, s) and (r', s') can now be defined in terms of their components to give a new tensor $T \otimes S$,

$$(T \otimes S)^{i_1 \dots i_{r+r'}}{}_{j_1 \dots j_{s+s'}} = T^{i_1 \dots i_r}{}_{j_1 \dots j_s} S^{i_{r+1} \dots i_{r+r'}}{}_{j_{s+1} \dots j_{s+s'}}. \tag{2.38}$$

This allows new tensors to be constructed out of vectors and dual vectors.

A *tensor field* of type (r, s) on M is an assignment of a tensor of the same type at all p in M. It is C^k-*differentiable* if all the components of T have the same differentiability as functions of the coordinates.

If T is a $(0, 2)$ type tensor, it acts on the pairs of vectors $\boldsymbol{V}, \boldsymbol{W}$ to produce a real number $T(\boldsymbol{V}, \boldsymbol{W}) = T_{ij} V^i W^j$. Then T is called *symmetric* if

$$T(\boldsymbol{V}, \boldsymbol{W}) = T(\boldsymbol{W}, \boldsymbol{V}). \tag{2.39}$$

If $\{e_i\}$ is a basis for the tangent space, this implies $T(e_i, e_j) = T(e_j, e_i)$, that is

$$T_{ij} = T_{ji}. \tag{2.40}$$

Similarly, T is called *antisymmetric* if

$$T_{ij} = -T_{ji}. \tag{2.41}$$

One can formulate this in terms of symmetric and antisymmetric parts of T. For T_{ij}, its *symmetric part* is written as

$$T_{(ij)} = \frac{1}{2!}(T_{ij} + T_{ji}), \tag{2.42}$$

and its *antisymmetric part* is written as

$$T_{[ij]} = \frac{1}{2!}(T_{ij} - T_{ji}). \tag{2.43}$$

Then, T is called *symmetric* if $T_{(ij)} = T_{ij}$, and *antisymmetric* if $T_{[ij]} = T_{ij}$. In general, for a tensor $T_{i_1 \ldots i_r}$ of type $(0, r)$, $T_{(i_1 \ldots i_r)}$ is defined as the sum over all permutations of indices $i_1, \ldots, i_r$ divided by $r!$. Similarly, $T_{[i_1 \ldots i_r]}$ is defined as the alternating sum over all permutations of the indices $i_1, \ldots, i_r$ divided by $r!$. Therefore, for example

$$T^i_{\;[jkl]} = \frac{1}{3!}[T^i_{\;jkl} + T^i_{\;klj} + T^i_{\;ljk} - T^i_{\;kjl} - T^i_{\;lkj} - T^i_{\;jlk}]. \tag{2.44}$$

In general, a tensor of type (r, s) is called symmetric over a collection of indices if it equals its symmetric part over these indices, and antisymmetric tensors are defined in a similar manner.

2.2 The metric tensor

The notion of distance between any two infinitesimally separated points of a spacetime manifold is defined by the metric tensor. These distances locally reduce to those given by special relativity, which have a flat metric with an indefinite signature on the Minkowski spacetime. As special relativity is seen to be valid by experiments, it must hold when confined to local regions in the spacetime, corresponding to measurements of space and time intervals at the laboratory scale. Therefore, the spacetime distances between events need not be positive definite.

This is carried out by assuming the existence of an indefinite *metric tensor field* g defined globally on M as a $(0, 2)$ type, symmetric tensor. Therefore, the metric tensor acts on pairs of vectors to give a number, and is symmetric in its indices. Choosing a coordinate basis,

$$g \equiv g_{ij}\, dx^i \otimes dx^j, \tag{2.45}$$

where $g_{ij} = g(\partial/\partial x^i, \partial/\partial x^j)$. If $\boldsymbol{V}$ and $\boldsymbol{W}$ are any vectors, this gives $g(\boldsymbol{V}, \boldsymbol{W}) = g_{ij} V^i W^j$. This is written conventionally in the form of a distance between two infinitesimally separated points in the spacetime as

$$ds^2 = g_{ij}\, dx^i\, dx^j. \tag{2.46}$$

For a single vector $\boldsymbol{V}$, $g(\boldsymbol{V}, \boldsymbol{V})$ is the magnitude of $\boldsymbol{V}$, which is $g_{ij} V^i V^j$.

The metric tensor is assumed to be *non-degenerate*, that is, there is no non-zero vector $\boldsymbol{V} \neq 0$ such that $g(\boldsymbol{V}, \boldsymbol{W}) = 0$ for all $\boldsymbol{W} \in T_p$. Then, the matrix $[g_{ij}]$ is non-singular, so there must be an inverse matrix g^{ij} such that

$$g^{ij} g_{jk} = \delta^i_k. \tag{2.47}$$

So the tensors g^{ij} and g_{ij} give an *isomorphism* or a unique correspondence between the space of covariant and contravariant vectors as

$$X_i = g_{ij} X^j, \qquad X^i = g^{ij} X_j. \tag{2.48}$$

Similarly, a second rank tensor T can also be written as

$$T^i{}_j = g^{ik} T_{kj}, \qquad T^j{}_i = g^{jk} T_{ki}, \qquad T^{ij} = g^{ik} g^{jl} T_{kl}. \tag{2.49}$$

In particular,

$$g^{ik} g_{km} = g^i{}_m = \delta^i{}_m, \tag{2.50}$$

and the Kronecker delta δ^i_m transforms as components of a tensor, so δ^i_m and g^i_m are identical tensors.

The tensors $T^i{}_j$, $T^j{}_i$, or T^{ij} are treated as representations of the same geometric object because these are uniquely associated. Such an isomorphism between the covariant and contravariant arguments is equivalent to the procedure of 'raising' and 'lowering' of indices as pointed out above. The multilinear map

$$g : T_p \times T_p \to R \tag{2.51}$$

can also be viewed as a linear correspondence from T_p to T^*_p in the sense of the mapping $\boldsymbol{V} \to g(., \boldsymbol{V})$. The non-degeneracy of the metric implies that this map is one–one and onto and so g establishes a one–one correspondence between vectors and dual vectors. The components $V_i = g_{ij} V^j$ are the one-form components uniquely associated with the vector components V^j.

Suppose M is an n-dimensional manifold with g being the metric on it. Then, at any $p \in M$ one can always choose an orthonormal basis $\{\boldsymbol{e}_i\}$ such that the metric components of the g_{ij} values have the diagonal form

$$g_{ij} = \mathrm{diag}(+1, \ldots, +1, -1, \ldots, -1). \tag{2.52}$$

If the metric has the form $g_{ij} = (+1, \ldots, +1)$ then it is called *positive definite.* Then $g(\boldsymbol{X}, \boldsymbol{X}) = 0$ implies $X = 0$. It is called a *Lorentzian metric* if the form is

$$g_{ij} = \mathrm{diag}(+1, \ldots, +1, -1), \tag{2.53}$$

with $(n-1)$ terms being positive. The metric is indefinite in the sense that the magnitude of a non-zero vector could be positive, negative or zero. The vector $\boldsymbol{X} \in T_p$ is called *timelike, null,* or *spacelike,* as defined by

$$g(\boldsymbol{X}, \boldsymbol{X}) < 0, \qquad g(\boldsymbol{X}, \boldsymbol{X}) = 0, \qquad g(\boldsymbol{X}, \boldsymbol{X}) > 0, \tag{2.54}$$

respectively. An indefinite metric divides the vectors in T_p into three disjoint classes, which are the timelike, null, and spacelike vectors. The null vectors form a cone in the tangent space T_p that separates the timelike vectors and the spacelike vectors.

The differentiable manifold of four dimensions, with a globally defined Lorentzian metric tensor is called a *spacetime manifold.* The *signature* of the metric tensor is defined as the number of its positive eigenvalues minus the number of negative eigenvalues. Therefore, a spacetime is a four-dimensional differentiable manifold with a Lorentzian metric globally defined, and which has the signature $+2$.

In the special theory of relativity, the spacetime admits a global coordinate frame covering the entire manifold so that the metric has the form given by (2.53) globally. The metric coefficients are constants on the manifold, which is called the *Minkowski spacetime.* The tangent vector for a particle traveling with a constant velocity less than that of light through a point p in such a spacetime is represented by a timelike vector at p. The particle must travel within the future light cone at p, which satisfies the equation $g(\boldsymbol{X}, \boldsymbol{X}) = 0$. This defines the set of all null vectors at p representing the photon paths. Now, according to special relativity, no material particles and signals travel at a velocity more than that of light. Thus, the metric determines the causal structure of spacetime in the sense that an event p is causally related to another event q if and only if there is a timelike or null signal between p and q. All such events lie on or within the double cone at p, defined by the metric tensor as above.

For the spacetime continuum of general relativity that is non-flat, the metric coefficients are functions of spacetime coordinates and one has to solve for the metric as a solution to the Einstein field equations. As for the existence of a Lorentz metric on a spacetime, any C^r-paracompact manifold admits a C^{r-1} Lorentz metric if and only if it admits a non-vanishing C^{r-1} line element field (an assignment of a pair of equal and opposite vectors $(\boldsymbol{V}, -\boldsymbol{V})$ globally on M) at each of its points (see for example, Hawking and Ellis, 1973). Such a field is always defined for a non-compact manifold and hence a Lorentz metric always exists.

Let (M, g) be a spacetime and γ be a C^1-curve in M. Then γ is a *timelike, null*, or *spacelike curve* if the tangent vector is timelike, null, or spacelike respectively at all its points. A timelike or null curve is also sometimes called a *non-spacelike curve.* The tangent space magnitudes defined by g,

$$\boldsymbol{X} \rightarrow |\, g(\boldsymbol{X}, \boldsymbol{X})\, |^{1/2}, \tag{2.55}$$

are related to the distances on the manifold as below. If $\boldsymbol{X}$ is the tangent vector along γ such that $g(\boldsymbol{X}, \boldsymbol{X})$ has the same sign at all its points, then the *arc length* between $p = \gamma(t_1)$ and $q = \gamma(t_2)$ along the curve is given by

$$L(\gamma) = s = \int_a^b (|\, g(\boldsymbol{X}, \boldsymbol{X})\, |)^{1/2} dt. \tag{2.56}$$

The equations (2.56) and (2.55) are equivalent to the expression $ds^2 = g_{ij}\, dx^i\, dx^j$, which represents the infinitesimal arc length along γ.

2.3 Connection

The notion of parallel transport of a given vector $\boldsymbol{X}$ in Euclidian spaces can be defined by requiring that, in going from a point p to q, both the magnitude and direction of $\boldsymbol{X}$ must not change. If both these for the tangent vector remain unchanged along a curve, it is called a straight line, along which the tangent is parallel transported. In Euclidian space, if a vector is parallel transported from points p to q along two different curves, the result is the same, independent of the path taken. However, this is not necessarily the case for a general affine manifold. For a differentiable manifold, the notion of the parallel transport of vectors is defined by introducing a connection on M.

Let $\boldsymbol{X}$ be a vector field on M, the *derivative operator* $\nabla_{\boldsymbol{X}}$ on M then gives the rate of change of vectors or tensor fields along $\boldsymbol{X}$ at all $p \in M$. If $\boldsymbol{Y}$ is another vector field at p, then the operator $\nabla_{\boldsymbol{X}}$ maps $\boldsymbol{Y}$ into a new vector field $\boldsymbol{Y} \rightarrow \nabla_{\boldsymbol{X}} \boldsymbol{Y}$ such that the following are satisfied:

(1) $\nabla_{\boldsymbol{X}}(\alpha \boldsymbol{Y} + \beta \boldsymbol{Z}) = \alpha \nabla_{\boldsymbol{X}} \boldsymbol{Y} + \beta \nabla_{\boldsymbol{X}} \boldsymbol{Z}$ for all $\alpha, \beta \in R$;
(2) $\nabla_{f\boldsymbol{X}+g\boldsymbol{Y}} \boldsymbol{Z} = f \nabla_{\boldsymbol{X}} \boldsymbol{Z} + g \nabla_{\boldsymbol{Y}} \boldsymbol{Z}$;
(3) $\nabla_{\boldsymbol{X}}(f\boldsymbol{Y}) = f \nabla_{\boldsymbol{X}} \boldsymbol{Y} + \boldsymbol{Y} \boldsymbol{X}(f)$.

A *connection* ∇ at a point $p \in M$ is a rule that assigns, to each vector field $\boldsymbol{X}$ at p, a differential operator $\nabla_{\boldsymbol{X}}$ that maps an arbitrary C^r vector field $\boldsymbol{Y}$ at p into a vector field $\nabla_{\boldsymbol{X}}\boldsymbol{Y}$, such that (1), (2), and (3) are satisfied (for a further discussion, see Hicks, 1965). The *covariant derivative of* $\boldsymbol{Y}, \nabla\boldsymbol{Y}$, is defined as a type $(1,1)$ tensor field that gives a vector $\nabla_{\boldsymbol{X}}\boldsymbol{Y}$ when contracted with the vector $\boldsymbol{X}$. In such a case, condition (3) above implies

$$\nabla(f\boldsymbol{Y}) = df \otimes \boldsymbol{Y} + f\nabla\boldsymbol{Y}. \tag{2.57}$$

A C^r *connection* ∇ on a C^k manifold ($k \geq r+2$) is a rule assigning a connection ∇ to each $p \in M$ such that if $\boldsymbol{Y}$ is a C^{r+1} vector field, then $\nabla\boldsymbol{Y}$ is a C^r tensor field of type (1,1),

$$\nabla\boldsymbol{Y} = Y^i{}_{;j}\boldsymbol{e}^j \otimes \boldsymbol{e}_i. \tag{2.58}$$

Here, $Y^i{}_{;j}$ is often called the *covariant derivative* of the vector Y^i, completely defined by the n^3 *connection coefficients* Γ^i_{jk}, given by choosing the vector fields $\boldsymbol{X}$ and $\boldsymbol{Y}$ to be the basis vector fields

$$\nabla_{\boldsymbol{e}_j}\boldsymbol{e}_k \equiv \Gamma^i_{jk}\boldsymbol{e}_i. \tag{2.59}$$

It is easy to see that this is equivalent to the condition

$$\langle \boldsymbol{e}^i, \nabla_{\boldsymbol{e}_j}\boldsymbol{e}_k \rangle = \Gamma^i{}_{jk}. \tag{2.60}$$

Therefore, in a coordinate basis,

$$\left\langle dx^i, \nabla_{\partial/\partial x^j}\left(\frac{\partial}{\partial x^k}\right)\right\rangle = \Gamma^i_{jk}. \tag{2.61}$$

Consider now the vector $\nabla_{\boldsymbol{X}}\boldsymbol{Y}$. Defining

$$\nabla_{\partial/\partial x^i}\boldsymbol{Y} \equiv \nabla_i\boldsymbol{Y}, \tag{2.62}$$

using the rules defining the connection given above, and the relation

$$\boldsymbol{X}(f) = X^i\frac{\partial}{\partial x^i}(f) = X^i\frac{\partial f}{\partial x^i}, \tag{2.63}$$

one obtains

$$\nabla_{\boldsymbol{X}}\boldsymbol{Y} = X^i\left(\frac{\partial Y^k}{\partial x^i} + \Gamma^k{}_{ij}Y^j\right)\left(\frac{\partial}{\partial x^k}\right). \tag{2.64}$$

Comparing this with (2.61)

$$\nabla_{\boldsymbol{X}}\boldsymbol{Y} = Y^k{}_{;i}X^i\left(\frac{\partial}{\partial x^k}\right), \tag{2.65}$$

where

$$Y^k{}_{;i} \equiv \frac{\partial Y^k}{\partial x^i} + \Gamma^k{}_{ij} Y^j. \tag{2.66}$$

It can be seen that the components of the vector $\nabla_{\boldsymbol{X}} \boldsymbol{Y}$ are given as $Y^k{}_{;i} X^i$, and

$$Y^i{}_{,j} \equiv \frac{\partial Y^i}{\partial x^j} \tag{2.67}$$

can be defined.

Then, taking the transformation of the coordinates $\{x^i\} \to \{x^{i'}\}$ when the basis vectors transform as $\boldsymbol{e}_i \to \boldsymbol{e}_{i'}$, it can be seen that $Y^i{}_{,j}$ does not transform like the components of a tensor. Similarly, consider the connection coefficients in the new coordinate system,

$$\Gamma^{k'}{}_{i'j'} = \langle \boldsymbol{e}^{k'}, \nabla_{\boldsymbol{e}_{i'}} \boldsymbol{e}_{j'} \rangle. \tag{2.68}$$

Transforming the dashed vectors to the original coordinate system and using the conditions (2) and (3) above gives, in a coordinate basis,

$$\Gamma^{k'}{}_{i'j'} = \frac{\partial x^{k'}}{\partial x^k} \left(\frac{\partial x^i}{\partial x^{i'}} \frac{\partial x^j}{\partial x^{j'}} \Gamma^k{}_{ij} + \frac{\partial^2 x^k}{\partial x^{i'} \partial x^{j'}} \right). \tag{2.69}$$

It follows that, because of the presence of the second derivative terms above, the coefficients $\Gamma^i{}_{jk}$ also do not transform like the components of a tensor. Consider, however,

$$\nabla_{\boldsymbol{X}} \boldsymbol{Y} = (Y^i{}_{;j} X^j) \left(\frac{\partial}{\partial x^i} \right) = (Y^{i'}{}_{;j'} X^{j'}) \left(\frac{\partial}{\partial x^{i'}} \right), \tag{2.70}$$

which implies

$$Y^i{}_{;j} X^j = Y^{i'}{}_{;j'} \frac{\partial x^{j'}}{\partial x^j} \frac{\partial x^i}{\partial x^{i'}} X^j. \tag{2.71}$$

Since the above is true for an arbitrary vector X^j, $Y^i{}_{;j}$ are components of a tensor.

Further, if $\Gamma^i{}_{jk}$ and $\Gamma^{\bar{i}}{}_{jk}$ are components of two different connections on M, then using the coordinate transformations, it can be seen that the quantities

$$C^i{}_{jk} = \Gamma^{\bar{i}}{}_{jk} - \Gamma^i{}_{jk} \tag{2.72}$$

are components of a tensor.

For a connection ∇ on M, the *torsion tensor* T is defined by

$$T(\boldsymbol{X}, \boldsymbol{Y}) = \nabla_{\boldsymbol{X}} \boldsymbol{Y} - \nabla_{\boldsymbol{Y}} \boldsymbol{X} - [\boldsymbol{X}, \boldsymbol{Y}]. \tag{2.73}$$

Writing the components,

$$T(\boldsymbol{X}, \boldsymbol{Y}) = (\Gamma^i{}_{jk} - \Gamma^i{}_{kj}) X^j Y^k \boldsymbol{e}_i. \tag{2.74}$$

This is a type $(1, 2)$ tensor with components

$$T^i{}_{jk} = \Gamma^i{}_{jk} - \Gamma^i{}_{kj}. \tag{2.75}$$

A connection is called *symmetric* when the torsion tensor vanishes,

$$\Gamma^i{}_{jk} = \Gamma^i{}_{kj}, \tag{2.76}$$

or $[\boldsymbol{X}, \boldsymbol{Y}] = \nabla_{\boldsymbol{X}} \boldsymbol{Y} - \nabla_{\boldsymbol{Y}} \boldsymbol{X}$. Symmetric connections are always worked with in this book, and the torsion tensor is assumed to be vanishing.

The notion of connection is generalized to arbitrary tensor fields to obtain a tensor $\nabla_{\boldsymbol{X}} T$ of type (r, s) for any given tensor T of type (r, s) by assuming first that ∇ is linear and obeys the Leibnitz rule. That is,

$$\nabla_{\boldsymbol{X}}(\alpha S + \beta T) = \alpha \nabla_{\boldsymbol{X}} S + \beta \nabla_{\boldsymbol{X}} T, \qquad \alpha, \beta \in R, \tag{2.77}$$

and

$$\nabla_{\boldsymbol{X}}(S \otimes T) = \nabla_{\boldsymbol{X}} S \otimes T + S \otimes \nabla_{\boldsymbol{X}} T \tag{2.78}$$

for any vector field $\boldsymbol{X}$ and tensor fields S and T. Furthermore, ∇ must agree with the usual notion of a directional derivative, that is,

$$\nabla_{\boldsymbol{X}} f = \langle df, \boldsymbol{X} \rangle = \boldsymbol{X} f = X^i \frac{\partial f}{\partial x^i}. \tag{2.79}$$

Finally, ∇ must commute with contractions, that is,

$$(\nabla_a T)^{i_1 \ldots l \ldots i_r}{}_{j_1 \ldots l \ldots j_s} = \nabla_a T^{i_1 \ldots l \ldots i_r}{}_{j_1 \ldots l \ldots j_s}. \tag{2.80}$$

As shown earlier,

$$\nabla_{\boldsymbol{X}} T = T^{i_1 \ldots i_r}{}_{j_1 \ldots j_s; a} \boldsymbol{X}^a e_{i_1} \otimes \cdots \otimes \boldsymbol{e}_{i_r} \otimes \boldsymbol{e}^{j_1} \otimes \cdots \otimes \boldsymbol{e}^{j_s}, \tag{2.81}$$

with

$$\nabla_{\boldsymbol{X}} T^{i_1 \ldots i_r}{}_{j_1 \ldots j_s} = T^{i_1 \ldots i_r}{}_{j_1 \ldots j_s; a} X^a. \tag{2.82}$$

Now, by considering the expansion for $\nabla_i(\boldsymbol{e}_j \otimes \boldsymbol{e}^k)$ it is seen that

$$\nabla_a \boldsymbol{e}^i = -\Gamma^i{}_{ac} \boldsymbol{e}^c, \tag{2.83}$$

and if $\boldsymbol{\omega}$ is a one-form then

$$\nabla_{\boldsymbol{e}_j} \boldsymbol{\omega} = \omega_{k;j} \boldsymbol{e}^k, \tag{2.84}$$

with

$$\omega_{k;j} \equiv \frac{\partial \omega_k}{\partial x^j} - \Gamma^i{}_{jk}\omega_i. \tag{2.85}$$

In general, the covariant derivative of a tensor T can be written as

$$T^{i_1\ldots i_r}{}_{j_1\ldots j_s;a} = \frac{\partial T^{i_1\ldots i_r}{}_{j_1\ldots j_s}}{\partial x^a} + \sum_m \Gamma^{i_m}{}_{he} T^{i_1\ldots e\ldots i_r}{}_{j_1\ldots j_s}$$
$$- \sum_n \Gamma^e{}_{hj_n} T^{i_1\ldots i_r}{}_{j_1\ldots e\ldots j_s}. \tag{2.86}$$

Given a Lorentzian metric on M, the condition $\nabla_{\boldsymbol{X}} g = 0$ defines a unique torsion-free connection on M. Then,

$$(\nabla_{\boldsymbol{X}} g)_{ij} = g_{ij;k} X^k = 0, \tag{2.87}$$

which implies that

$$g_{ij;k} = 0. \tag{2.88}$$

The parallel transport of vectors preserve the scalar product defined by the metric tensor g, and the connection coefficients $\Gamma^i{}_{jk}$ are determined in terms of the first derivatives of the metric components. Since all the information on spacetime structure is supposed to be contained in the ten metric functions g_{ij}, this is to be expected. One way to see this is the following. Using (2.86), the covariant derivative of the metric can be written as

$$g_{ij;k} = \frac{\partial g_{ij}}{\partial x^k} - \Gamma^m{}_{il} g_{mj} - \Gamma^m{}_{jl} g_{mi}. \tag{2.89}$$

Now, using the condition $g_{ij;k} = 0$ and defining

$$g_{mj}\,\Gamma^m{}_{il} = \Gamma_{jil}, \tag{2.90}$$

the above equation can be written as

$$g_{ij,k} \equiv \frac{\partial g_{ij}}{\partial x^k} = \Gamma_{jil} + \Gamma_{ijl}. \tag{2.91}$$

Using the above equations and the symmetry property of the connection,

$$\Gamma_{jil} = \tfrac{1}{2}(g_{ji,m} - g_{mj,i} + g_{im,j}). \tag{2.92}$$

This can also be seen by specializing to the frame of free fall. In such a case, all the connection coefficients vanish and the metric is locally that of special relativity. Then, $g_{ij} = \eta_{ij}$ and the partial derivatives of g_{ij} vanish. Therefore, from the above equation for $g_{ij;k}$, one obtains $g_{ij;k} = 0$. As this is a tensor equation, it must hold in all frames in general, and one can again proceed as earlier.

2.4 Non-spacelike geodesics

In Euclidian spaces, the line of shortest distance between any two points is the straight line joining them, along which the tangent does not change in direction or magnitude. That is, the tangent is parallel transported. Now, let $\gamma(t) : R \to M$ be a C^1-curve in M. If T is a $C^r (r \geq 0)$ tensor field on M, then the *covariant derivative of* T *along* $\gamma(t)$ is defined as

$$\frac{DT}{\partial t} = T^{i\ldots l}{}_{k\ldots m;h} X^h, \tag{2.93}$$

where $\boldsymbol{X}$ is the tangent to γ. Then, γ is called a *geodesic* if its tangent vector is parallel transported along it. That is, if $\boldsymbol{X}$ is the tangent vector field along γ, then $\nabla_{\boldsymbol{X}}\boldsymbol{X}$ is proportional to $\boldsymbol{X}$. Therefore, there exists a function f such that

$$\nabla_{\boldsymbol{X}}\boldsymbol{X} = f\boldsymbol{X}. \tag{2.94}$$

Writing the components, this implies $(X^i{}_{;j}X^j)\boldsymbol{e}_i = fX^i\boldsymbol{e}_i$ always holds, and so $X^i{}_{;j}X^j = fX^i$ along the geodesic curve. But, it is always possible to reduce f to zero by a suitable choice of the curve parameter t along γ, so the equation for the geodesic is written as

$$X^i{}_{;j}X^j = 0, \tag{2.95}$$

where the X^i values are components of the tangent vector to the geodesic. The parameter t is called an *affine parameter* along γ, which is an *affinely parametrized geodesic.* If $\{x^i\}$ denotes a local coordinate system, the components X^i are written as $X^i = dx^i/dt$ and the equation for geodesics is

$$\frac{d^2x^i}{dt^2} + \Gamma^i{}_{jk}\frac{dx^j}{dt}\frac{dx^k}{dt} = 0. \tag{2.96}$$

The affine parameter along the geodesic is determined up to an additive and multiplicative constant. Thus, if t is an affine parameter, then so is $t' = at+b$ and $X^i{}_{;j}X^j = 0$. Here $b \neq 0$ gives a new choice of the initial point $\gamma(0)$ and $a \neq 0$ implies a renormalization of the vector $\boldsymbol{X}$.

A geodesic in (M, g) is called *timelike, spacelike*, or *null* if its tangent vector is timelike, spacelike, or null respectively. Here, the timelike or null geodesics that represent the paths of particles or photons in the spacetime are mainly considered. Since the tangent to a geodesic is parallel transported, a timelike or null geodesic remains the same always and it cannot become spacelike. In a Riemannian manifold with a positive definite metric, such geodesics give the curves of shortest distance between its points. However, in a spacetime with a Lorentzian metric, the non-spacelike geodesics maximize the distance between the points, as defined by (2.56). If there is a timelike geodesic between the points p and q, there is no shortest distance geodesic between them because, by introducing null geodesic pieces, one could always join these points by curves of arbitrary small lengths. On the other hand, any maximal length curve between p and q must necessarily be a timelike geodesic.

The geodesic equations above are n equations in n variables x^i with $i = 1, \dots, n$. Thus, the existence theorems for differential equations ensure that, given x^i and dx^i/dt, that is, given any initial point p and the value of the tangent vector X^i, a unique geodesic through p with this value of tangent exists. This can be used to define the *exponential map* $E_p : T_p \to M$ from the tangent space at p into the spacetime. Under this map, any given tangent vector X^i in T_p is mapped to a point in M, a unit affine parameter distance away along the unique geodesic determined by p and X^i. It is clear that the exponential map may not be defined on all of T_p because all the geodesics in M passing through p may not extend to all the values of the affine parameter. In such a case, M is called *geodesically incomplete*. On the other hand, if the exponential map is defined on all of T_p for all points p, then M is called *geodesically complete*. Then, all geodesics in M extend for all values of their affine parameter. Also, the map E_p may not be one–one, because the geodesics might cross each other. However, it can be shown (Bishop and Critendon, 1964) that for a sufficiently small neighborhood N_p of p there is a neighborhood of thc origin in T_p which is diffeomorphically mapped onto N_p by the exponential map which is one–one and well-defined on this neighborhood. In such a case, the exponential map can be used to define the *normal coordinates* on the neighborhood N_p of p. Since T_p is an n-dimensional vector space equivalent to R^n, the coordinates of any $r \in N_p$ can be chosen to be the n-coordinates of the vector $\boldsymbol{X}_p$ that is mapped onto it. This coordinate system has the property that the geodesics are mapped into straight lines and the connection coefficients vanish at p. Therefore, this coordinate system turns out to be quite convenient for calculations at the point p. The neighborhood N_p can have a further property that any two points in it can be joined by a unique geodesic contained totally within N_p. Such a neighborhood of p is called a *convex normal neighborhood.*

The geodesic equations are derived from the parallel transport required for the tangent vector. If one requires that the curve must extremize the length, namely that $\delta l = 0$, and works out

$$\delta l = \delta \int_a^b \left| g_{ab} \frac{dx^a}{dt}\frac{dx^b}{dt} \right|^{1/2} dt = 0, \tag{2.97}$$

using the variational methods, it turns out that the resulting equations are precisely the geodesic equations (2.96). So, the geodesics extremise the lengths of curves between any two spacetime points. If the events p and q are timelike related, and if there is a maximum length timelike curve from p

to q, that curve must be a timelike geodesic. Therefore, in many situations, the useful way to work out geodesic equations is to choose the Lagrangian as

$$L = \frac{1}{2} g_{ab} \dot{x}^a \dot{x}^b, \tag{2.98}$$

and to write the Lagrange equations, which are the equations of the spacetime geodesics. Then, by comparison with the geodesic equations, it is also possible to evaluate the quantities $\Gamma^i{}_{jk}$ for the spacetime.

In the Minkowski spacetime, the surface $t = 0$ is a three-dimensional surface with the time direction always normal to it. Any other surface of $t =$ const. is also a spacelike surface in the same way. In general, let S be an $(n-1)$-dimensional manifold. If there is a C^∞-map $\phi : S \to M$ that is locally one–one (there is a neighborhood N for every $p \in S$ such that ϕ restricted to N is one–one) and ϕ^{-1} is also C^∞ as defined on $\phi(N)$, then $\phi(S)$ is called an *immersed submanifold* of M. If ϕ is globally one–one, then $\phi(S)$ is called an *embedded submanifold of* M. It may also be required that ϕ be a homeomorphism with the induced topology on $\phi(S)$ from M. Lower dimensional embedded submanifolds in M represent well-behaved surfaces in the spacetime.

A *hypersurface* S of any n-dimensional manifold M is defined as an $(n-1)$-dimensional embedded submanifold of M, and V_p is denoted by the $(n-1)$-dimensional subspace of T_p of the vectors tangent to S at any $p \in S$. It follows that a vector $\boldsymbol{n} \in T_p$ exists that is unique up to the scale, and that is orthogonal to all the vectors in V_p. This is called the *normal* to S at p. If the magnitude of $\boldsymbol{n}$ is either positive or negative at all points of S without changing the sign, then $\boldsymbol{n}$ could be normalized so that $g_{ab}n^a n^b = \pm 1$. If $g_{ab}n^a n^b = -1$, then the normal vector is timelike everywhere and S is called a *spacelike hypersurface.* If the normal is spacelike everywhere with a positive magnitude, S is a *timelike hypersurface*, and S is a *null hypersurface* if the normal n^a is null at S.

The timelike geodesics could be used to define the *synchronous coordinate system* in the neighborhood of a spacelike hypersurface in the spacetime as below. Let ∇_a be the metric connection that satisfies $\nabla_a g_{bc} = g_{bc;a} = 0$. Let S be a spacelike hypersurface, then, for every $p \in S$, let γ be the unique timelike geodesic with a tangent n^a, that is, the congruence of these curves at points of S is orthogonal to S. Then, in the neighborhood of that portion of S, the coordinates $q \to (x^1, \dots, x^{n-1}, t)$ are assigned for any point q in the future of p along γ, where t is the parameter along γ and $x^1, \dots, x^{n-1}$ are the spatial coordinates of p. In particular, if the geodesics in the congruence are parametrized by the proper time t with the magnitude of the tangent given by -1 along γ, then the spacelike surfaces are given as $\{t = \text{const.}\}$ surfaces. Note that S can be labeled as the $\{t = 0\}$ spacelike surface.

The synchronous coordinates have the important property that when the congruence $\{\gamma\}$ is orthogonal to S_0, it will also be orthogonal to subsequent surfaces S_t given by $t = \text{const}$. Clearly, $\{\gamma\}$ is orthogonal to S_0 by construction. To see that this holds for any S_t for t within the domain of construction, let X^a be any basis vector for the tangent space at a point of S_t. Then,

$$n^b \nabla_b (n_a X^a) = (n^b \nabla_b n_a) X^a + n^b n_a \nabla_b X^a. \tag{2.99}$$

Since n^b is a tangent to the geodesic, the above equals $n_a n^b \nabla_b X^a$. But, $\boldsymbol{X}$ and $\boldsymbol{n}$ are coordinate vectors, implying $\nabla_{\boldsymbol{n}} \boldsymbol{X} = \nabla_{\boldsymbol{X}} \boldsymbol{n}$, that is,

$$n^a \nabla_a X^b = X^a \nabla_a n^b. \tag{2.100}$$

This implies

$$n^b \nabla_b (n_a X^a) = n_a X^b \nabla_b n^a = \tfrac{1}{2} X^b \nabla_b (n^a n_a) = 0, \tag{2.101}$$

because $n^a n_a = -1$. Therefore, $n_a X^a = 0$ in the future of S_0 in the domain of the validity of the synchronous coordinate system.

2.5 Spacetime curvature

The curvature for a spacetime is measured by the non-commutation of the tangent vectors when these are parallel transported along different curves to arrive at the same spacetime point. This is measured by the *Riemann curvature tensor*, which is defined as a type $(1,3)$ tensor, $R : T_p^* \times T_p \times T_p \times T_p \to R$. In a coordinate basis, the Riemann tensor can be written as

$$R = R^i_{jkl} \boldsymbol{e}_i \otimes \boldsymbol{e}^j \otimes \boldsymbol{e}^k \otimes \boldsymbol{e}^l. \tag{2.102}$$

If the vector $R(\boldsymbol{X}, \boldsymbol{Y})\boldsymbol{Z}$ is defined as

$$R(\boldsymbol{X}, \boldsymbol{Y})\boldsymbol{Z} \equiv \nabla_{\boldsymbol{X}} (\nabla_{\boldsymbol{Y}} \boldsymbol{Z}) - \nabla_{\boldsymbol{Y}} (\nabla_{\boldsymbol{X}} \boldsymbol{Z}) - \nabla_{[\boldsymbol{X}, \boldsymbol{Y}]} \boldsymbol{Z}, \tag{2.103}$$

then the components of the Riemann tensor are given by

$$R^i{}_{jkl} = \langle \boldsymbol{e}^i, \boldsymbol{R}(\boldsymbol{e}_k, \boldsymbol{e}_l) \boldsymbol{e}_j \rangle. \tag{2.104}$$

Working out the components gives

$$R(\boldsymbol{X}, \boldsymbol{Y})\boldsymbol{Z} = R^i{}_{jkl} \frac{\partial}{\partial x^i} X^k Y^l Z^j. \tag{2.105}$$

Now, in order to evaluate (2.104), note that

$$[\nabla_{\boldsymbol{X}} (\nabla_{\boldsymbol{Y}} \boldsymbol{Z})]^i = \nabla_{\boldsymbol{X}} (Z^i{}_{;j} Y^j) = Z^i{}_{;jk} Y^j X^k + Z^i{}_{;j} Y^j{}_{;k} X^k. \tag{2.106}$$

Similarly,

$$[\nabla_{\boldsymbol{Y}}(\nabla_{\boldsymbol{X}}\boldsymbol{Z})]^i = Z^i{}_{;jk}X^jY^k + Z^i{}_{;j}X^j{}_{;k}Y^k. \tag{2.107}$$

Finally,

$$\begin{aligned} -\nabla_{[\boldsymbol{X},\boldsymbol{Y}]}\boldsymbol{Z} &= -\nabla_{(Y^i{}_{;j}X^j - X^i{}_{;j}Y^j)(\partial/\partial x^i)}\boldsymbol{Z} \\ &= -Z^k{}_l Y^l{}_{;j}X^j + Z^k{}_l X^l{}_{;j}Y^j. \end{aligned} \tag{2.108}$$

Combining the above equations,

$$R(\boldsymbol{X},\boldsymbol{Y})\boldsymbol{Z} = (Z^i{}_{;lk} - Z^i{}_{;kl})X^kY^l. \tag{2.109}$$

Comparing (2.109) and (2.105),

$$Z^i{}_{;lk} - Z^i{}_{;kl} = R^i{}_{jkl}Z^j, \tag{2.110}$$

which is the same as

$$\nabla_k\nabla_l Z^i - \nabla_l\nabla_k Z^i = R^i{}_{jkl}Z^j. \tag{2.111}$$

The above equation could also be taken as the defining equation for the components of the curvature tensor. As shown by the left-hand side of (2.111), the Riemann curvature tensor provides the measure of non-commutation of a tangent vector when parallel transported along different curves to arrive at the same spacetime point.

In place of the vectors $\boldsymbol{X}$,$\boldsymbol{Y}$, and $\boldsymbol{Z}$ the basis vectors $\boldsymbol{e}_i$ can now be chosen. Then,

$$\begin{aligned} \nabla_{\boldsymbol{e}_j}\nabla_{\boldsymbol{e}_k}\boldsymbol{e}_l &= \nabla_{\boldsymbol{e}_j}(\Gamma^a{}_{kl}\boldsymbol{e}_a) \\ &= \boldsymbol{e}_j(\Gamma^a{}_{kl})\boldsymbol{e}_a + \Gamma^a{}_{kl}\Gamma^h{}_{ja}\boldsymbol{e}_h. \end{aligned} \tag{2.112}$$

Consider the definition of the components of the Riemann tensor as given by (2.111). In particular, if a coordinate basis is chosen, then $[\boldsymbol{e}_i, \boldsymbol{e}_j] = 0$ and

$$R^i{}_{jkl} = \langle \boldsymbol{e}^i, \nabla_{\boldsymbol{e}_k}\nabla_{\boldsymbol{e}_l}\boldsymbol{e}_j\rangle - \langle \boldsymbol{e}^i, \nabla_{\boldsymbol{e}_l}\nabla_{\boldsymbol{e}_k}\boldsymbol{e}_j\rangle. \tag{2.113}$$

Then, using (2.61) and a coordinate basis, the coordinate components of the Riemann curvature tensor can be given in terms of the coordinate components of the connection as

$$R^i{}_{jkl} = \frac{\partial\Gamma^i{}_{lj}}{\partial x^k} - \frac{\partial\Gamma^i{}_{kj}}{\partial x^l} + \Gamma^i{}_{ka}\Gamma^a{}_{lj} - \Gamma^i{}_{la}\Gamma^a{}_{kj}. \tag{2.114}$$

As pointed out earlier, given the metric tensor g on M, there is a unique, torsion-free connection on M defined by the condition $\nabla_{\boldsymbol{X}}g = 0$, which is equivalent to the vanishing covariant derivative of the metric tensor, $g_{ij;k} = 0$.

Then, parallel transport of vectors preserves the scalar product defined by g and $g(\boldsymbol{V}, \boldsymbol{V}) = \text{const.}$ along a geodesic γ, where $\boldsymbol{V}$ is the tangent to γ. Then,

$$\begin{aligned} \nabla_{\boldsymbol{X}}(g(\boldsymbol{Y}, \boldsymbol{Z})) &= \boldsymbol{X} g(\boldsymbol{Y}, \boldsymbol{Z}) \\ &= \nabla_{\boldsymbol{X}}(g_{ij} Y^i Z^j) \\ &= g(\nabla_{\boldsymbol{X}} \boldsymbol{Y}, \boldsymbol{Z}) + g(\nabla_{\boldsymbol{X}} \boldsymbol{Z}, \boldsymbol{Y}). \end{aligned} \tag{2.115}$$

Evaluating $\boldsymbol{Y}(g(\boldsymbol{Z}, \boldsymbol{X}))$ and $\boldsymbol{Z}(g(\boldsymbol{X}, \boldsymbol{Y}))$ and adding the first and subtracting the second from (2.115) gives

$$\begin{aligned} g(\boldsymbol{Z}, \nabla_{\boldsymbol{X}} \boldsymbol{Y}) =& \tfrac{1}{2}[\boldsymbol{X} g(\boldsymbol{Y}, \boldsymbol{Z}) + \boldsymbol{Y} g(\boldsymbol{Z}, \boldsymbol{X}) - \boldsymbol{Z} g(\boldsymbol{X}, \boldsymbol{Y}) + g(\boldsymbol{Y}, [\boldsymbol{Z}, \boldsymbol{X}]) \\ &+ g(\boldsymbol{Z}, [\boldsymbol{X}, \boldsymbol{Y}]) - g(\boldsymbol{X}, [\boldsymbol{Y}, \boldsymbol{Z}])]. \end{aligned} \tag{2.116}$$

Choosing the basis vectors $\boldsymbol{e}_i$ in place of the vectors $\boldsymbol{X}$, $\boldsymbol{Y}$, and $\boldsymbol{Z}$ in (2.116) gives the connection coefficients in terms of the derivatives of g_{ij} and the Lie derivatives of the basis vectors

$$g(\boldsymbol{e}_i, \nabla_{\boldsymbol{e}_j} \boldsymbol{e}_k) = g_{im} \Gamma^m{}_{jk} = \Gamma_{ijk}. \tag{2.117}$$

Choosing a coordinate basis with $[\boldsymbol{e}_i, \boldsymbol{e}_j] = 0$ gives the usual *Christoffel symbols*

$$\Gamma_{ijk} = \frac{1}{2}\left(\frac{\partial g_{ij}}{\partial x^k} + \frac{\partial g_{ik}}{\partial x^j} - \frac{\partial g_{jk}}{\partial x^i}\right). \tag{2.118}$$

It follows then that the Riemann tensor components are expressed in terms of the metric tensor and its second derivatives when the connection defined by the metric is used. From now on, 'the connection' means this unique connection as defined by the metric tensor.

The expression, as given by (2.114) and earlier definitions, implies that the Riemann tensor has the symmetry given by

$$R^i{}_{jkl} = -R^i{}_{jlk}, \tag{2.119}$$

which is equivalent to $R^i{}_{j(kl)} = 0$. Furthermore, the curvature tensor obeys the *cyclic identity* $R^i{}_{[jkl]} = 0$, which can be written as

$$R^i{}_{jkl} + R^i{}_{klj} + R^i{}_{ljk} = 0. \tag{2.120}$$

The covariant derivatives of the Riemann tensor satisfy the *Bianchi identities* given by $R^i{}_{j[kl;a]} = 0$, which is the same as

$$R^i{}_{jkl;a} + R^i{}_{jla;k} + R^i{}_{jak;l} = 0. \tag{2.121}$$

A straightforward proof would involve writing down each term above explicitly, substituting from (2.114), and then taking a summation. There are certain additional symmetries that are valid when the connection is the one induced by the metric. In this case,

$$\Gamma_{ijk} = g_{il}\Gamma^{l}{}_{jk}, \quad R_{ijkl} = g_{ia}R^{a}{}_{jkl}, \quad \Gamma^{l}{}_{jk} = g^{li}\Gamma_{ijk}. \tag{2.122}$$

The Riemann tensor R_{ijkl} defined by the metric has the symmetry

$$R_{ijkl} = -R_{jikl}, \tag{2.123}$$

which means $R_{(ij)kl} = 0$. Also, in this case the Riemann tensor is symmetric in the pairs of the first two and last two indices,

$$R_{ijkl} = R_{klij}. \tag{2.124}$$

The spacetime (M, g) is said to have a *flat connection* if and only if $R^{i}{}_{jkl} = 0$, that is, all the components of the Riemann tensor must be vanishing. This is the necessary and sufficient condition for a vector at a point p to remain unaltered after parallel transport along an arbitrary closed curve through p. This is subject to the condition that all such curves can be shrunk to zero, or the spacetime has to be simply connected. In general, the parallel transport of vectors does not hold in a spacetime manifold in the sense that given a connection, if a given vector is parallel transported along two different spacetime curves to arrive at the same point, the resultant vector will be different in general. However, when all the components of the Riemann tensor vanish, it can be shown that whenever a vector is transported from one point to the other in the spacetime, the result is independent of the path taken. In such a case, the connection is also said to be *integrable* and a necessary and sufficient condition for this to happen is the vanishing of all the components of the Riemann tensor.

When a symmetric connection is integrable, the manifold is called *flat*. Furthermore, in the case of the connection being the metric connection, the vanishing of all the Riemann tensor components provides a necessary and sufficient condition for the spacetime metric to be *flat*, that is, a global coordinate system in M exists such that the metric reduces to the diagonal form with values ± 1 everywhere.

The *Ricci tensor* is defined as a type $(0, 2)$ tensor, obtained by contracting the Riemann tensor,

$$R_{jl} = R^{i}{}_{jil}. \tag{2.125}$$

As a consequence of the symmetries discussed above, it follows that the Ricci tensor is symmetric, and

$$R^{i}{}_{ikl} = 0. \tag{2.126}$$

A further contraction of the Ricci tensor gives the *curvature scalar* R, which is defined as

$$R = g^{ij} R_{ij}. \tag{2.127}$$

The quantity R has the property that it depends only on the values of g_{ij} and on their derivatives only up to the second order. Furthermore, it is linear in the second derivatives of the metric components. The total number of independent scalars that could be constructed from the metric and its derivatives up to the second order is 14.

As a consequence of the various symmetries described above, the total number of independent components of R_{ijkl} reduces to 20 when the dimension of the manifold is four. When the dimension is three, R_{ijkl} has six independent components essentially given by R_{ij}, and when the dimension is two, there is only one independent component, which is R.

Another important tensor that can be constructed from R_{ijkl} is the *Weyl tensor*, which is also sometimes called the *Weyl conformal tensor*,

$$C_{ijkl} = R_{ijkl} + \left\{ g_{i[l} R_{k]j} + g_{j[k} R_{l]i} \right\} + \tfrac{1}{3} R g_{i[k} g_{l]j}. \tag{2.128}$$

The symmetry properties of the Weyl tensor follow from those of the Riemann tensor discussed above; it possesses the same symmetries as the Riemann tensor. Also, it can be verified that the following identically vanishes,

$$g^{ik} C_{ijkl} = 0. \tag{2.129}$$

The Weyl tensor is that part of the curvature tensor for which all contractions vanish for any pair of contracted indices,

$$C^{i}{}_{jil} = 0. \tag{2.130}$$

If the Weyl tensor vanishes throughout the spacetime with $C_{ijkl} = 0$ at all points, then it can be shown that the metric g_{ij} must be *conformally flat*. This means that a conformal function $\Omega(x^i)$, $0 < \Omega < \infty$, exists such that

$$g_{ij} = \Omega^2 \eta_{ij}, \tag{2.131}$$

where η_{ij} is the flat Minkowskian metric. The Weyl tensor is conformally invariant in the sense that under a conformal transformation $g_{ij} \rightarrow \overline{g}_{ij} = \Omega^2 g_{ij}$,

$$\overline{C^{i}_{jkl}} = C^{i}_{jkl}. \tag{2.132}$$

It is possible to show that a necessary and sufficient condition for the spacetime metric to be conformally flat is that the Weyl tensor must vanish everywhere.

The *geodesic deviation equation*, which is also called the *Jacobi equation*, is now derived. This characterizes the coming together, or moving away, of spacetime geodesics from each other as a result of the spacetime curvature. Consider a smooth one-parameter family of affinely parametrized non-spacelike geodesics, characterized by the parameters (t, v), where t is the affine parameter along a geodesic and $v =$ const. characterizes different geodesics in the family with $t, v \in R$. Such non-spacelike geodesics span a two-dimensional submanifold on which t and v could be chosen as coordinates. The vectors $\boldsymbol{T} = \partial/\partial t$ and $\boldsymbol{V} = \partial/\partial v$ are then coordinate vectors for which $[\boldsymbol{T}, \boldsymbol{V}] = 0$. Then, since the torsion tensor is vanishing,

$$\nabla_{\boldsymbol{T}} \boldsymbol{V} = \nabla_{\boldsymbol{V}} \boldsymbol{T}, \tag{2.133}$$

which implies $T^i \nabla_i \boldsymbol{V} = V^i \nabla_i \boldsymbol{T}$. Because $\boldsymbol{T}$ is a tangent to the geodesics, $T^i \nabla_i T^j = 0$. Now, define the operator D by $D \equiv T^i \nabla_i$. Then,

$$DV^j = V^i \nabla_i T^j. \tag{2.134}$$

Taking another derivative,

$$\begin{aligned} D^2 V^j &= DV^i \nabla_i T^j + V^i D(\nabla_i T^j) \\ &= (T^k \nabla_k V^i)(\nabla_i T^j) + V^i T^l \nabla_l \nabla_i T^j. \end{aligned} \tag{2.135}$$

However, by the definition of the Riemann curvature tensor,

$$\nabla_l \nabla_i T^j - \nabla_i \nabla_l T^j = R^j{}_{kli} T^k. \tag{2.136}$$

Substituting this into (2.135),

$$\begin{aligned} D^2 V^j &= (V^k \nabla_k T^i)(\nabla_i T^j) + \nabla_i \nabla_l T^j V^i T^l + R^j{}_{kli} T^k V^i T^l \\ &= V^k((\nabla_k T^i)(\nabla_i T^j) + (\nabla_k \nabla_l T^j) T^l) + R^j{}_{kli} T^k V^i T^l \\ &= V^k(\nabla_k (T^i \nabla_i T^j)) + R^j{}_{kli} T^k V^i T^l \\ &= R^j{}_{kli} T^k V^i T^l. \end{aligned} \tag{2.137}$$

The equation

$$D^2 V^j = -R^j{}_{kil} T^k V^i T^l \tag{2.138}$$

is called the Jacobi equation, or the equation of geodesic deviation. It is clear from the above that $D^2 V^j = 0$ if and only if all the components of the Riemann tensor vanish. On the other hand, whenever some components are non-zero, then the neighboring non-spacelike geodesics will necessarily accelerate towards or away from each other.

2.6 The Einstein equations

General relativity is a theory of gravity defined on a spacetime manifold, where the force of gravitation is described in terms of the curvature of the spacetime. These curvatures are in turn generated by the matter fields existing in the spacetime, as governed by the Einstein equations. The Einstein equations involve second derivatives of the metric tensor. So, it is assumed that the metric components are at least C^2-differentiable functions of the coordinates. All pairs (M', g') that are diffeomorphic to (M, g) are regarded as equivalent and (M, g), which represents this entire equivalence class of spacetimes with equivalent physical properties, is studied.

The local causality and local conservation of the energy and momentum are accepted as the basic physical postulates for the spacetimes in general relativity. The basic mathematical criterion while formulating general relativity has been that the matter distribution determines the geometry of the spacetime universe in terms of the Riemann curvature tensor. Next, the motion of any test particles in such a gravitational field is always independent of its own mass and composition. This is the *principle of equivalence*, which has now been verified to a great degree of accuracy to show that any two objects with different masses and different compositions always arrive at the same time on the surface of the earth when they have left from the same height. A logical consequence of this is that any reference frame uniformly accelerated with respect to an inertial frame of the special relativity is locally identical to a frame at rest in a gravitational field. Finally, in general relativity, one postulates the *principle of general covariance*, namely that all physical laws are expressed as tensor equations so that they are valid in a general frame of reference, and are invariant under arbitrary coordinate transformations. When restricted to the frame of free fall, these must produce the laws of special relativistic physics.

All matter fields on the spacetime, such as electromagnetic fields, dust, perfect fluids or scalar fields, are assumed to be represented by a second rank tensor T^{ij}, called the *energy–momentum tensor*, in the sense that T^{ij} vanishes on any open region in the spacetime if and only if all the matter fields vanished there. Such matter fields then obey tensor equations on the spacetime and the derivatives involved will only be the covariant derivatives with respect to the unique connection defined by the metric tensor. This is because for any other connection defined on M, its difference from the metric connection, which is a tensor again as shown earlier, could always be regarded as another physical field on M.

Such a stress–energy tensor T^{ij} then describes the matter fields on the manifold. For example, for *dust*, which is the matter distribution composed of non-interacting material particles, the field is characterized by the proper density ρ_0 of the flow and the four velocities of the particles given by $dx^i/d\tau$, where τ is the proper time along the timelike trajectory describing the particle worldline. The simplest second rank tensor constructed from these two quantities is

$$T^{ij} = \rho_0 u^i u^j. \tag{2.139}$$

The component T^{00} of this energy–momentum tensor is

$$T^{00} = \rho_0 \frac{dx^0}{d\tau}\frac{dx^0}{d\tau}. \tag{2.140}$$

In a special relativistic frame, this can be interpreted as the relativistic energy density of matter. It can also be shown that requiring this tensor to have zero divergence in such a frame gives the conservation of the energy and momentum.

A *perfect fluid* is characterized by an additional scalar quantity, the pressure $p = p(x^i)$. In the limit as the pressure vanishes, this must reduce to the dust form of matter. Furthermore, one also demands the conservation laws in a special relativistic frame, and that these should reduce to the classical equations of continuity and the Navier–Stokes equations in the appropriate limits. Then this energy–momentum tensor is given in a general frame as

$$T^{ij} = (\rho + p)u^i u^j + p g^{ij}, \tag{2.141}$$

which can be taken as the definition of a perfect fluid in general relativity.

In general, the energy–momentum tensors of various fields can be constructed by using a variational principle where there is a proposed Lagrangian, and the change in action is considered due to the change in the metric.

For an arbitrary frame, and for other matter fields such as the electromagnetic field, or a charged scalar field, the principle of local conservation of energy and momentum states that

$$T^{ij}{}_{;j} = 0. \tag{2.142}$$

The above equation for the stress–energy tensor contains much information on the matter fields in a spacetime. For example, if the spacetime contains a Killing vector K^i then the above equation could be integrated to give a conservation law. The conserved vector in such a case is defined as $P^i = T^{ij}K_j$, and $P^i{}_{;i} = 0$ as a consequence of (2.142) and the *Killing equation*

$$K_{i;j} + K_{j;i} = 0. \tag{2.143}$$

Then, the integration of $P^i{}_{;i}$ over a compact region implies that the total flux over a closed surface of the energy–momentum is zero in the direction of the Killing vector (Hawking and Ellis, 1973). Even when the spacetime does not admit a Killing vector, given any point p a Riemannian normal coordinate system at p could be set up so that the metric components have the Minkowskian values and the connection coefficients $\Gamma^{i}{}_{jk}$ vanish at p. A small enough neighborhood of p could then be chosen so that the values of g_{ij} and $\Gamma^{i}{}_{jk}$ differ by an arbitrarily small amount from the values at p. Using this fact, it could be shown that isolated test particles should move along timelike geodesics (Fock, 1939; Dixon, 1970).

Furthermore, all matter fields are assumed to obey the postulate of *local causality*, that is, the equations governing the matter fields are such that for any $p \in M$, there is an open neighborhood U of p in which a signal can be sent between any two points of U if and only if there is a non-spacelike curve joining these points. This principle is valid in special relativity and is also accepted in general relativity.

The above principles effectively imply that it is the spacetime metric, and the quantities derived from it, that must appear in the equations for physical quantities, and that these equations must reduce to the flat spacetime equations when the metric is Minkowskian. This is the basic content of the general theory of relativity, where the spacetime manifold is now allowed to have topologies other than R^4, and the metric g_{ij} can be non-flat globally. In general relativity, the matter fields expressed by the stress–energy tensor are related to the non-flat nature of the spacetime by means of the Einstein equations, which are the basic equations satisfied by the spacetime metric. In Einstein's theory, one does not discuss the physical interaction of matter fields in a fixed background metric prescribed in advance. Actually, the g_{ij} values are treated as dynamical variables that depend on the matter content of the spacetime and that are to be solved from the Einstein equations.

The Newtonian theory gives an important indicator towards obtaining this relationship between the matter content and the spacetime geometry, where the gravitational field is described by a potential ϕ. The tidal acceleration between nearby particles is given in terms of the separation between them and the second derivatives of ϕ. In a curved spacetime manifold, such tidal accelerations are described by the Jacobi equation (2.138), in terms of the Riemann curvature tensor. Furthermore, the Poisson equation

$$\nabla^2\phi = 4\pi\rho \tag{2.144}$$

must be recovered in the Newtonian limit. Both in special and general theories of relativity, the matter content is described by the stress–energy tensor T_{ij}, and the mass–energy density ρ corresponds to the quantity $T_{ij}V^iV^j$. Therefore, each side of Poisson's equation corresponds to the Riemann tensor as expressed in the Jacobi equation and $T_{ij}V^iV^j$ respectively. Another important indicator for the comparison is provided by the Bianchi identities.

Contracting i with k in (2.121) gives

$$\nabla_a R_{jl} + \nabla_i R^i{}_{jla} - \nabla_l R_{ja} = 0. \tag{2.145}$$

Then, raising j and contracting with a,

$$\nabla_a R^a{}_l + \nabla_i R^i{}_l - \nabla_l R = 0. \tag{2.146}$$

This then gives $2\nabla_i R^i{}_l - \nabla_l R = 0$. Therefore,

$$\nabla^i G_{ij} = 0, \tag{2.147}$$

where G_{il} is the *Einstein tensor* defined by

$$G_{il} \equiv R_{il} - \tfrac{1}{2} g_{il} R. \tag{2.148}$$

From the above considerations, one can make a comparison given by $R_{ij} = 4\pi T_{ij}$ as field equations for general relativity. This does not, however, work because the contracted Bianchi identities imply $\nabla_l R = 0$, so the trace $T =$ const. throughout the spacetime. This is an unphysical restriction on the matter content.

Using the above indicators, Einstein proposed the field equations

$$G_{ij} = R_{ij} - \tfrac{1}{2} R g_{ij} = 8\pi T_{ij}. \tag{2.149}$$

Then, the contracted Bianchi identities actually imply the local conservation of energy and momentum through the Einstein equations. Taking the trace of the above,

$$R = -8\pi T. \tag{2.150}$$

Substituting this back in the above gives an alternative form of the Einstein equations,

$$R_{ij} = 8\pi (T_{ij} - \tfrac{1}{2} T g_{ij}). \tag{2.151}$$

The definition of the Ricci tensor suggests that the Einstein equations depend on the derivatives of the metric up to the second order. These equations are highly non-linear in g_{ij}, however, they are linear in the second derivatives of g_{ij}. In fact, the quantities R_{ij} and Rg_{ij} are the only second-rank symmetric tensors that are linear in the second derivatives of the metric, and involve only up to the second derivatives of g_{ij}. Actually, the Einstein equations are a coupled system of non-linear second order partial differential equations for g_{ij}. This makes the task of solving these extremely difficult. Several symmetry assumptions on the spacetime generally need to be imposed in order to work out the metric components as a solution to the Einstein equations. Some solutions that are useful in the context of gravitational collapse are discussed later.

Given the T^{ij} values, the field equations may be viewed as a set of differential equations to determine the gravitational potential values, g^{ij}, that fix the resulting geometry. A particularly important case here is that of vacuum solutions when $T^{ij} = 0$. On the other hand, one could arbitrarily specify the ten metric potentials and then compute the Einstein tensor G_{ij}. Then, the field equations determine the energy–momentum tensor T_{ij}. However, in that case, the resulting T^{ij} values turn out to be unphysical most of the time in that the energy conditions ensuring the positivity of mass–energy density may be violated. Such a violation of the energy conditions is rejected on physical grounds as all observed classical fields obey positivity of energy density, which is closely connected with the physical features of gravitation theory.

In general, the field equations are ten equations, connecting a total of twenty quantities, which are the ten components of g_{ij} and the other ten components of T_{ij}. Therefore, the field equations are the conditions placing constraints on the simultaneous choice of these twenty quantities. If part of the gravitational potentials and the matter contents are determined from physical conditions, then such conditions can fully determine the matter and geometry. In particular, if the vacuum equations are considered,

$$G_{ij} = R_{ij} - \tfrac{1}{2}Rg_{ij} = 0, \tag{2.152}$$

then there are ten equations to determine the ten quantities g_{ij}. However, the Bianchi identities

$$\nabla_j G^{ij} = 0 \tag{2.153}$$

place four differential constraints on these equations, which are not all independent. Therefore, there is an indeterminacy in that there are fewer equations than unknowns to be determined. Furthermore, there is an intrinsic gauge freedom available in the general theory of relativity that does not allow a complete determination of the metric potentials. This is given by the coordinate freedom that allows a transformation from one set of coordinates x^i to any other set of coordinates $x^{i'}$. However, this coordinate freedom could be used to impose conditions on the metric components. For example, choosing the normal coordinates gives $g_{00} = 1$ and $g_{0\alpha} = 0, \alpha = 1, 2, 3$ in this coordinate system. This leaves six other components to be determined from the field equations. The issue is closely linked to the Cauchy problem in general relativity where the basic problem is that, given initial data on a regular spacelike hypersurface, one would like to determine its unique evolution in the future or past.

The Einstein equations can admit a cosmological constant. Note that the most general second rank tensor which can be constructed out of G_{ij} and g_{ij}, which is divergence free and involves the derivatives of the metric tensor up to second order only, is the linear combination $G_{ij} + \Lambda g_{ij}$ (Lovelock, 1972), where Λ is a constant. Therefore, addition of such a constant multiple of g_{ij} to the Einstein tensor preserves all the required properties discussed above. Einstein historically introduced the *cosmological term* Λ in the equations in order to generate static cosmological solutions, and wrote the equations as

$$R_{ij} - \frac{1}{2}Rg_{ij} + \Lambda g_{ij} = 8\pi T_{ij}. \tag{2.154}$$

It is seen that for an empty spacetime with $T_{ij} = 0$, the Einstein equations then become

$$R_{ij} = -\Lambda g_{ij}. \tag{2.155}$$

If $\Lambda \neq 0$, then one does not obtain the Newtonian theory in the limit of slow motions and weak fields. However, if the magnitude of Λ is very small, then such departures will be quite negligible and approximate agreement with the Newtonian theory is obtained.

2.7 Exact solutions

Owing to their complexity and the fact that they are a highly non-linear system of partial differential equations, sufficiently general solutions of the Einstein equations are difficult to find. Almost all known exact solutions assume a high degree of symmetry, such as the spherical or axial symmetry, the existence of various Killing vector fields on the spacetime, or similar other conditions. These represent special and idealized situations. (For a detailed discussion on exact solutions with various symmetries see Stephani *et al.*, 2003).

Such spacetime examples, however, do provide a good idea of what is possible within the framework of general relativity. They also give an indication of certain important features that may be present in a situation that is much more general than the actual solution being studied. An example of this is the singularity theorems in general relativity. Some exact solutions, such as the Schwarschild and Friedmann–Robertson–Walker models, exhibited the occurrence of a spacetime singularity where physical quantities were blowing up. Earlier, it was thought that this was a mathematical artifact due to the symmetry of these solutions, and singularity would not be present in sufficiently general models once the symmetries are relaxed. It turned out, however, that the singularity was a fairly general feature of general relativity and would occur under rather general physical assumptions, as shown by the singularity theorems. Some of the solutions also have interesting applications. For example, the Minkowski spacetime is both the geometry of special relativity and locally that of any general relativistic model, and the Schwarzschild spacetime is used to test physical predictions of general relativity.

Some of the most interesting and physically relevant solutions to the Einstein equations are spherically symmetric; these include the Friedman–Robertson–Walker models in cosmology, and the Schwarzschild solution that can model an isolated star. In order to understand the possible final fate of a gravitationally collapsing massive star, we discuss the spherical symmetry in some detail. Such a symmetry represents a high degree of idealization of the physical situation, but the advantage is that one can solve it many times analytically to get exact results. It is also possible that several salient physical features of the situation are preserved when departures from spherical symmetry are taken into account.

In fact, the basic motivation for the theory and the idea of a blackhole come from the case of a spherically symmetric homogeneous dust cloud collapse. Independently of the interior solution, the metric exterior to such a spherical body must be the Schwarzschild spacetime, as implied by *Birkhoff's theorem* (Birkhoff, 1923), which states that the only vacuum, spherically symmetric gravity field must be static. The spherically symmetric and asymptotically flat spacetimes such as the Schwarzschild geometry are useful to model the spacetime outside the sun and stars, and could be used to obtain conclusions relevant for the experimental verification of general relativity. Such solutions can also possibly represent the outcome of a complete gravitational collapse of a massive star. Similarly, the Vaidya geometry can model the field outside a radiating star. The Schwarzschild and Vaidya geometries have useful applications, and are particularly relevant to the gravitational collapse scenarios studied here. Some of the properties of these exact solutions of the Einstein equations that are referred to in later chapters are reviewed below.

2.7.1 Minkowski spacetime

The Minkowski spacetime mathematically is the manifold $M = R^4$ with the Lorentzian metric

$$ds^2 = -dt^2 + dx^2 + dy^2 + dz^2, \tag{2.156}$$

with $-\infty < t, x, y, z < \infty$ giving the range of coordinates. This is a flat spacetime with all components of the Riemann tensor $R^i{}_{jkl} = 0$, and hence it is the simplest empty spacetime solution to the Einstein equations $G_{ij} = 8\pi T_{ij} = 0$, which underlies the physics of special relativity. The vector $\partial/\partial t$ provides a time orientation here. If the spherical polar coordinates (t, r, θ, ϕ) given by $x = r\sin\theta\sin\phi, y = r\sin\theta\cos\phi$, and $z = r\cos\theta$ are used, the above metric becomes

$$ds^2 = -dt^2 + dr^2 + r^2(d\theta^2 + \sin^2\theta\, d\phi^2). \tag{2.157}$$

The range of the coordinates r, θ, ϕ are $0 < r < \infty, 0 < \theta < \pi$, and $0 < \phi < 2\pi$. Two such coordinate neighborhoods are needed to cover all of the Minkowski spacetime.

In coordinates (t, x, y, z) it is obvious that the geometry is flat, because all the metric components are constants, so all the connection coefficients are vanishing. In other coordinate systems, such as spherical coordinates (t, r, θ, ϕ), the connection coefficients $\Gamma^i{}_{jk}$ do not all vanish (for example, $\Gamma^1{}_{22} = r$); however, all the Riemann curvature tensor components do still vanish.

The *Lorentz transformations* on the Minkowski spacetime are defined as the set of metric preserving isometries that are linear and homogeneous. Physically, these represent the change of reference frame from one to another inertial observer, given by the coordinate change

$$x^i \to x^{i'} = L^i{}_j x^j. \tag{2.158}$$

In addition to the metric preserving property, the above implies det $L^i{}_j = \pm 1$, so the matrix is non-singular. If $\det L^i{}_j = +1$ and $L^0{}_0 \geq 1$, then the Lorentz transformations preserve the orientations in both space and time. The set of all Lorentz transformations, form a group where the identity map is $\delta^i{}_j$, and the inverse is the inverse matrix. The Lorentz group is a subgroup of the Poincaré group of transformations, which are general inhomogeneous mappings that leave the Minkowskian metric invariant. It consists of a Lorentz transformation together with an arbitrary translation in space and time. This is a ten-parameter group, consisting of six Lorentz parameters and four translation parameters, and physically it is a mapping of one inertial frame into another in a general position in the spacetime.

The geodesics of Minkowski spacetime are the straight lines of the underlying Euclidian geometry. Given an event in M, the lines at 45° to the time axis through that event give null geodesics in M. They form the boundary of the chronological future or past $I^{\pm}(p)$ of an event p, which is the set of all events that lie in the future (past) of p, and which contain all possible timelike particle trajectories through p including timelike geodesics. The causal future $J^+(p)$ is the closure of $I^+(p)$ in Minkowski space, which includes all the events in M that are either timelike or null related to p by means of future directed non-spacelike curves from p. The causal past of an event is dually defined. The spacelike hypersurfaces $t =$ const. in the Minkowski spacetime give a family of Cauchy surfaces that covers all of M. A Cauchy surface is a spacelike hypersurface such that all inextendible non-spacelike curves in M meet this surface once and only once. However, all spacelike hypersurfaces in M need not be Cauchy surfaces. For example, the family given by $-t^2 + x^2 + y^2 + z^2 = A =$ const. with $A < 0$ are inextendible spacelike surfaces that are not Cauchy surfaces. All these surfaces are fully contained inside the chronological past or the future of the origin and there are timelike geodesics outside this past or future cone that do not meet any of these surfaces.

2.7.2 The ideal points boundary

In general, the spacetime boundary consists of points at infinity and the spacetime singularities. To understand global properties and the structure of the infinity of the Minkowski spacetime, the procedure of Penrose (1968) and the general procedure of Geroch, Kronheimer, and Penrose (1972) can be used. An arbitrary event p in the Minkowski spacetime is uniquely determined either by its chronological future $I^+(p)$ or past $I^-(p)$. If a future directed non-spacelike curve γ has a future end point at p, $I^-(\gamma) = I^-(p)$. (By definition, $I^-(\gamma)$ is the union of all $I^-(q)$ with q being any point on γ.) On the other hand, if γ is a future inextendible curve without any future end point, the set $I^-(\gamma)$ determines a *point at infinity* of M. (A future or past inextendible curve, in the context of Minkowski spacetime, is a trajectory which goes off to infinity in the future or past without stopping anywhere.) Two such curves γ_1 and γ_2 determine the same ideal point, or a point at infinity, if $I^-(\gamma_1) = I^-(\gamma_2)$.

Such a procedure defines the future ideal points. Past ideal points are defined dually using past inextendible non-spacelike curves. In the case of Minkowski spacetime, there are future directed inextendible curves γ that are timelike and have the same past, which is the entire spacetime M, that is, $I^-(\gamma) = M$. Hence, all such timelike curves determine a single future ideal point i^+, called the *future timelike infinity.* The past timelike infinity i^- is similarly defined. If γ is chosen to be a future endless null geodesic or a null curve, it is possible to have a situation where $I^-(\gamma)$ is not the entire Minkowski spacetime. Certain timelike curves also have this property. For example, consider the past of the timelike hyperbola

$$t = \sinh\lambda, \quad x = \cosh\lambda, \quad y, z = 0, \quad -\infty < \lambda < \infty. \tag{2.159}$$

Then, $I^-(\gamma)$ lies completely to the past of the null hypersurface $x = t$. It can be shown in general (Geroch, Kronheimer, and Penrose, 1972) that, for a non-spacelike curve γ, if $I^-(\gamma) \neq M$ and if γ is future endless, then a null hypersurface S_γ exists, the half-space below which coincides with $I^-(\gamma)$.

If the collection of ideal points can be denoted by $\mathcal{I}^+$, then there is a one–one and onto correspondence between the points of $\mathcal{I}^+$ and such null hypersurfaces. Any such null hypersurface is determined by the value of the time t at which it intersects the time axis, and by the direction of the null vector at the point of intersection. Since the set of all possible light ray directions at any point is equivalent to the two-sphere S^2, it follows that $\mathcal{I}^+$ is a three-dimensional manifold with topology $S^2 \times R$.

The three-dimensional null hypersurfaces $\mathcal{I}^+$ and $\mathcal{I}^-$ are called the future and past *null infinities* respectively for the Minkowski spacetime. A general spacetime would also admit such a boundary construction under certain conditions, such as being asymptotically flat and empty. It can be shown for the Minkowski spacetime that all complete null hypersurfaces are flat and so are like the surfaces $\{x = t\}$, in which case the topological structure of the null infinity is clearly $\mathcal{I}^+ = S^2 \times R$. However, the topological structure of null infinities for a general spacetime need not be the same.

2.7.3 Conformal compactification

A differential structure and a metric on the future null infinity $\mathcal{I}^+$ can be introduced. A convenient way to attach the ideal points boundary $\mathcal{I}^+$ to M is to use a suitable conformal factor Ω to obtain a transform of the original spacetime metric η_{ij},

$$g_{ij} = \Omega^2 \eta_{ij}, \quad \Omega > 0. \tag{2.160}$$

This leaves the causal structure of M invariant, because the null geodesic paths of the metric η_{ij} and the unphysical metric g_{ij} are the same under a conformal mapping, up to a reparametrization. Therefore, the past of any non-spacelike curve γ is unchanged and there is a natural correspondence between ideal points in both the spacetimes. Since light cones are unaltered by a conformal transformation, the boundary attachment obtained in this manner is coordinate independent.

In the Minkowski spacetime, the *advanced* and *retarded* null coordinates can be introduced, where

$$v = t + r, \quad u = t - r, \tag{2.161}$$

which gives a reference frame, based on null cones, which is most suitable to analyze the radiation fields.

Under this transformation of coordinates, the metric becomes

$$ds^2 = -du\, dv + \tfrac{1}{4}(u - v)^2(d\theta^2 + \sin^2\theta\, d\phi^2), \tag{2.162}$$

with $-\infty < v < \infty$ and $-\infty < u < \infty$. Now, the information at future null infinity corresponds to taking the limit as $v \to \infty$, which amounts to moving in the future along $u =$ const. light cones. Similarly, past null infinity corresponds to $u \to \infty$. This procedure could be made precise in a coordinate independent way. The Minkowski spacetime M can be compactified by means of a conformal transformation of the metric as given by

$$\Omega^2 = (1+v^2)^{-1}(1+u^2)^{-1}, \tag{2.163}$$

and then by adding the closure to add the null infinities. New coordinates p and q can be introduced by

$$v = \tan p, \quad u = \tan q. \tag{2.164}$$

Then, the corresponding ranges for p and q are

$$-\frac{\pi}{2} < p < \frac{\pi}{2}, \quad -\frac{\pi}{2} < q < \frac{\pi}{2}, \tag{2.165}$$

and the metric $\bar{g}_{ij}$ on the unphysical spacetime $\overline{M}$, after the conformal transformation, is given by

$$d\bar{s}^2 = -dp\,dq + \sin^2(p-q)(d\theta^2 + \sin^2\theta\,d\phi^2). \tag{2.166}$$

It is possible to see now that the metric above, with the coordinate ranges as given above, is a manifold embedded as a part of the Einstein static universe. To see this, let

$$T = p + q, \quad R = p - q, \tag{2.167}$$

and then (2.166) becomes, in (T, R, θ, ϕ) coordinates,

$$d\bar{s}^2 = -dT^2 + dR^2 + \sin^2 R(d\theta^2 + \sin^2\theta\,d\phi^2), \tag{2.168}$$

with the coordinate ranges given by

$$-\pi < T + R < \pi, \quad -\pi < T - R < \pi. \tag{2.169}$$

This is precisely the natural Lorentz metric on $S^3 \times R$, which is the Einstein static universe, except that the coordinate ranges are now restricted by (2.169). In this picture, the future null infinity $\mathcal{I}^+$ is given by $T = \pi - R$ for $0 < R < \pi$, and the past null infinity is given by $T = -\pi + R$ for $0 < R < \pi$. (For a further discussion see, for example, Wald, 1984.) The structure of infinity for any spherically symmetric spacetime can be depicted by a similar *Penrose diagram.* As mentioned above, $\mathcal{I}^+$ here is topologically $S^2 \times R$.

2.7.4 Schwarzschild solution

The Schwarzschild solution gives the geometry exterior to a spherically symmetric massive body such as a star and has been used extensively to verify experimentally the predictions of the general theory of relativity. This is an empty exterior solution where the Ricci tensor vanishes and is matched at the boundary to an interior solution inside the body. In (t, r, θ, ϕ) coordinates, the metric can be given as

$$ds^2 = -\left(1 - \frac{2m}{r}\right) dt^2 + \left(1 - \frac{2m}{r}\right)^{-1} dr^2 + r^2 d\Omega^2, \tag{2.170}$$

where $d\Omega^2 = d\theta^2 + \sin^2\theta\, d\phi^2$. The coordinate t is timelike and r, θ, ϕ are spacelike coordinates. The radial coordinate r has the property that the two-sphere given by $t =$ const., $r =$ const. has the two-metric given by

$$ds^2 = r^2(d\theta^2 + \sin^2\theta\, d\phi^2), \tag{2.171}$$

so the area of any such two-sphere would be $4\pi r^2$.

The coordinate r is restricted to $r > 2m$ because the above metric has an apparent singularity at $r = 2m$. The coordinate t has the range $-\infty < t < \infty$. The solution is obtained by solving the vacuum Einstein equations for a spherically symmetric spacetime, where the quantity m is the constant of integration, with its value determined by using the weak field Newtonian limit of general relativity. If Φ is the Newtonian gravitational potential, then in non-relativistic units,

$$g_{00} \simeq 1 + \frac{2\Phi}{c^2} = 1 - \frac{2GM}{c^2 r}, \tag{2.172}$$

where G is the Newtonian constant of gravity, c is the velocity of light, and M is the point mass at the origin which gives rise to the Newtonian potential Φ. This determines the constant of integration m in the Schwarzschild solution as

$$m = \frac{GM}{c^2}. \tag{2.173}$$

Therefore, this is interpreted as a solution describing the gravitational field of a point particle with mass m (in relativistic units $G = c = 1$) situated at the center.

The Schwarzschild metric is static in the sense that $\partial/\partial t$ is a timelike Killing vector that is a gradient. The metric components g_{ij} here are independent of time. Also, there are no mixed terms in the metric that involve both space and time. Therefore, there is no rotation inherent in the spacetime and the metric is stationary. For a detailed discussion on stationary and static solutions of the field equations, see Stephani *et al.* (2003). In the present case, using the Birkhoff theorem, a spherically symmetric vacuum solution of Einstein equations must be necessarily static. An important implication of this theorem is that even when a spherically symmetric star undergoes pulsations or changes in shape, while maintaining the spherical symmetry, it cannot radiate any disturbances in the exterior, such as gravitational waves. Therefore, any spherically symmetric solution of Einstein equations with $R_{ij} = 0$ is necessarily the Schwarzschild solution. So, the Schwarzschild exterior can be used to describe the outside metric for several situations such as a spherically symmetric star that is either static, or that undergoes radial pulsations, or a radial spherically symmetric gravitational collapse.

The spherical symmetry of the Schwarzschild spacetime is seen in that the metric components g_{00} and g_{11} are functions of r alone and not of θ and ϕ, as implied by the angular part of the metric. Specifically, the isometry group of M contains a subgroup which is isomorphic to the group $SO(3)$, and the orbits of this subgroup are two-dimensional spheres. These isometries are interpreted as rotations and so the metric remains invariant under rotations in general for any spherically symmetric spacetime. The parameter m serves here as the source of the gravitational field, and setting $m = 0$ gives the flat Minkowski spacetime. As pointed out above, the comparison with Newtonian theory shows that m is to be treated as the gravitational mass of the body producing the field, as measured from infinity. The spacetime here is asymptotically flat in the sense that as r tends to infinity, the flat spacetime metric is recovered, and the gravitational field diminishes to zero.

Generally, the Schwarzschild metric is taken to represent the outside metric for a star, with coordinate $r > r_0$, where r_0 gives the boundary of the star. The metric inside $r < r_0$ is a different interior metric determined by the matter distribution T_{ij} inside the star, and it is matched at the boundary $r = r_0$ with the Schwarzschild solution. However, in the case of a complete gravitational collapse, when all the mass collapses at $r = 0$, it is necessary to consider the metric above as an empty spacetime solution for all the values of r. Clearly, the metric has singularities at $r = 0$ and $r = 2m$, and hence it represents only one of the patches $0 < r < 2m$ or $2m < r < \infty$. If one confines to the manifold given by the latter range of values of r, it is necessary to determine if the spacetime is extendible, that is, if a bigger spacetime (M', g') exists with M embedded in M' and $g = g'$ on M. That this should be possible is indicated by the fact that even though the form of the metric is singular at $r = 2m$, the curvature scalars are all well-behaved here, and so this could be merely a singularity due to an inappropriate choice of coordinates. A decision on whether a given spacetime manifold is maximal or not can be made by looking at the geodesics. In a maximal manifold, it would be required that all the geodesics be extended in both the directions to an infinite value of their affine parameter, or they must terminate at an intrinsic singularity of the spacetime that is not removable. On the other hand, if this metric is taken to describe the patch $0 < r < 2m$ of the spacetime, then it is seen that as r tends to zero, the curvature scalar

$$R^{ijkl}R_{ijkl} = \frac{m^2}{r^6} \tag{2.174}$$

diverges, so the point $r = 0$ is a real spacetime singularity. It is not possible to extend the spacetime across this singularity in a continuous manner. A maximal extension of the Schwarzschild manifold, which covers both the patches above, was obtained by Kruskal (1960) and Szekeres (1960).

The vacuum Schwarzschild geometry exterior to the collapsing matter arises for a dust cloud collapse. In the case of homogeneous collapse, a Schwarzschild blackhole is necessarily produced in a complete gravitational collapse, and the event horizon fully covers the resulting spacetime singularity of infinite curvature and density. In other words, as the collapse evolves, the event horizon develops earlier than the formation of the singularity. This situation is significant for the cosmic censorship hypothesis and blackhole formation scenarios. The interior metric in this case is exactly that of a closed Friedmann model in a time reversed sense.

In the extended Schwarzschild manifold, $r = 2m$ is a null hypersurface and each point on this surface is a two-sphere of area $16\pi m^2$. Note that the metric component $g_{00} = (1 - 2m/r)$ is positive for $r > 2m$; however, $g_{00} < 0$ for $r < 2m$. Therefore, it is no longer possible to use t as a time coordinate in that range, as the coordinates t and r reverse their roles there, and spacetime is no longer static. Therefore, the $r = 2m$ surface is also called the *static limit*. The vector $\partial/\partial t$ with components $\xi^i = \delta^i{}_0 = (1, 0, 0, 0)$ gives the time translation, leaving the g_{ij} unchanged as it does not involve the time coordinate. Therefore, ξ is a Killing vector that leaves the spacetime geometry unchanged. One obtains $\boldsymbol{\xi}^2 = g_{ij}\xi^i\xi^j = g_{00}$, and for the Schwarzschild metric $\boldsymbol{\xi}^2$ vanishes on $r = 2m$. Hence, at the static limit the timelike Killing vector becomes null.

The Schwarzschild geometry is an illustration of the basic principle that Einstein used to formulate gravitation theory, namely that matter tells the spacetime how to curve. To see this, consider the Schwarzschild solution in a spacelike surface $t = \text{const.}$ and in the equatorial plane $\theta = \pi/2$. The metric of this two-dimensional curved surface is given by

$$ds^2 = \frac{dr^2}{(1 - 2m/r)} + r^2\, d\phi^2. \tag{2.175}$$

The geometry of such a curved surface can be visualized as embedded in the ordinary Euclidian space. Here, the region $0 < r < r_\text{b}$ is to be considered as filled by the matter that represents the spherical star with a boundary at $r = r_\text{b}$, and the curved surface would then represent the geometry outside such a star.

Consider a static observer along a Killing direction, for whom the four-velocities are $u^i = \xi^i/|\boldsymbol{\xi}|$. Suppose a static source with four-velocity u_1^i emits a photon with four-momentum p^i (so $p_{i;\,j}p^j = 0$ with a suitable parametrization) and is observed by a static observer with four-velocity u_2^i. Now, take the directional derivative of $\xi_i p^i$ along the geodesic tangent p^i,

$$(\xi_i p^i)_{;\,j} p^j = \xi_{i;\,j} p^i p^j + \xi_i p^i{}_{;\,j} p^j = 0. \tag{2.176}$$

The first term vanishes because ξ^i is a Killing vector and the second term vanishes because of the geodesic equation. Therefore, the ratio of energies measured at these two points by static observers is given by

$$\frac{E_1}{E_2} = \frac{\left(u^i p_i\right)_1{}^{1/2}}{\left(u^i p_i\right)_2{}^{1/2}}. \tag{2.177}$$

Using $u^i = \xi^i/|\boldsymbol{\xi}|$ and $\xi^i p_i =$ const. along the geodesic,

$$\frac{E_1}{E_2} = \frac{\left(\xi^i \xi_i\right)_1{}^{1/2}}{\left(\xi^i \xi_i\right)_2{}^{1/2}}. \tag{2.178}$$

Since $\boldsymbol{\xi}^2 = g_{00}$, this is the *gravitational red-shift* formula for a static source and observer in terms of the metric components. It is now seen that if the observer remains at a finite radius but the source approaches $r = 2m$, the red-shift tends to infinity. Therefore, as a particle falls into the blackhole, approaching $r = 2m$, the light rays emitted by it are infinitely red-shifted as observed by a distant static observer in the outside spacetime.

2.7.5 Homogeneous collapse and the blackhole

Consider a spherically symmetric massive star, collapsing gravitationally when it has exhausted its internal nuclear fuel. In order to investigate the final state for such a collapse, the dynamical interior for such a star needs to be considered. As such, there is no unique interior solution available to represent this situation, which basically depends on the properties of the matter, the equation of state obeyed, and the physical processes taking place within the stellar interior. However, assuming the matter to be pressureless and homogeneous dust allows the problem to be solved analytically, which provides many important insights (Datt, 1938; Oppenheimer and Snyder, 1939). In that case, the energy–momentum tensor is given by $T^{ij} = \rho u^i u^j$ and the Einstein equations for the spherically symmetric form of the metric given above have to be solved. Solving the Einstein equations determines the metric potentials completely and the interior geometry of the star, which is a collapsing dust ball. This is the same line element as that of the closed homogeneous and isotropic Friedmann models given by

$$ds^2 = -dt^2 + R^2(t)\left[\frac{dr^2}{1-r^2} + r^2 d\Omega^2\right], \tag{2.179}$$

where $d\Omega^2$ represents the metric on a two-sphere. The geometry outside the star is a vacuum and is then the Schwarzschild spacetime as implied by the Birkhoff theorem. It can be shown that the interior geometry of the dust cloud is matched at the boundary of the star $r = r_{\rm b}$ with the exterior Schwarzschild spacetime.

When the collapse is complete, in the case of an empty and asymptotically flat exterior, the spacetime settles to a vacuum Schwarzschild geometry. Apart from predicting small observable departures from the Newtonian gravity, the Schwarzschild solution is important for the theory of gravitational collapse, as sufficiently massive stars unable to support themselves against the pull of their own gravity at the end of their life cycle could finally settle to that geometry. The final state of a spherically symmetric homogeneous dust collapse is necessarily a Schwarzschild configuration that contains a spacetime singularity hidden within an event horizon. This gives rise to a blackhole in the spacetime, which is a region from which no causal signals can reach a faraway observer. This scenario forms the basis of much of the theory and applications of blackhole physics.

If the star has internal pressures and is radiating, which would be a physically realistic scenario, then it can be matched to an exterior Vaidya, or a generalized Vaidya solution, as is discussed later. To understand the vacuum case, if m is the total mass of the star, then the structure of the resulting configuration can be discussed by considering the radial null geodesics in this metric, defined by $ds^2 = 0$ and $\dot{\theta} = \dot{\phi} = 0$. Taking the positive sign solutions since the interest here is in the outgoing null geodesics, these are given by

$$\frac{dt}{dr} = \frac{r}{r-2m}. \tag{2.180}$$

Integrating this gives

$$t = r + 2m \ln \mid r - 2m \mid + \text{const.} \tag{2.181}$$

In the region $r > 2m$ we have $dr/dt > 0$ and hence r increases with increasing t. Thus the above describes the congruence of outgoing radial null geodesics. The in going null trajectories are given by the negative sign solutions, which are given by

$$t = -r - 2m \ln \mid r - 2m \mid + \text{const.} \tag{2.182}$$

In the region below $r = 2m$, the coordinates r and t change their spacelike and timelike nature and hence the light cones tip over. As a result, no observer in the region $r < 2m$ can remain at a constant value of r, but they must move inwards to fall within the intrinsic curvature singularity at $r = 0$. Each point in the (t, r) plane represents a two-sphere of area $4\pi r^2$.

This gives an idea of the phenomena happening in the region $r < 2m$, namely that any material particle or photon here must fall in the spacetime singularity, and that it cannot escape to larger values of r to communicate with external observers in the spacetime. Hence, this is called the blackhole region in the spacetime. However, the Schwarzschild picture gives the impression that from the outside $r > 2m$, no photons or particles could fall in this blackhole and they will take infinite time to reach the surface $r = 2m$. It turns out that this is actually a coordinate defect arising due to the coordinate singularity at $r = 2m$, as can be seen by going to the Eddington–Finkelstein coordinates, where the idea is to choose a new time coordinate such that the in going null geodesics become straight lines in the spacetime. It is clear from the Schwarzschild consideration above that the appropriate change may be given by

$$t \to t + 2m \ln(r - 2m), \tag{2.183}$$

for the $r > 2m$ region. The solution with such a coordinate change is now regular at $r = 2m$ and the coordinate range is now $0 < r < \infty$. This is called an analytic extension of the Schwarzschild solution. A time-reversed solution is obtained if a different time coordinate is introduced,

$$t \to t - 2m \ln(r - 2m), \tag{2.184}$$

in which case the outgoing null geodesics are straight lines. A simpler way to write the metric in the new coordinate system is to introduce the advanced null coordinate v defined by

$$v = t + r + 2m \ln(r - 2m). \tag{2.185}$$

The metric then has the form

$$ds^2 = -\left(1 - \frac{2m}{r}\right) dv^2 + 2dv\, dr + r^2(d\theta^2 + \sin^2\theta\, d\phi^2). \tag{2.186}$$

The radially infalling null geodesics are now given by $v =$ const. It can be seen that at $r = 2m$, the radially outgoing photons stay at a constant value of r, and below this surface they must also fall to the singularity at $r = 0$. Further, radially infalling material particles must also fall to the spacetime singularity within a finite amount of proper time, as measured along their trajectory.

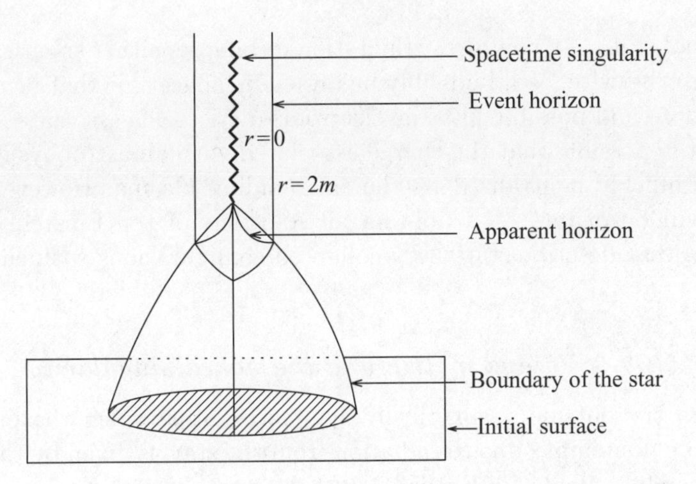

Fig. 2.3 Homogeneous dust cloud collapse. The trapped surfaces form when the star enters a $r = 2m$ radius. The event horizon forms prior to the singularity, creating a blackhole as the collapse endstate.

The basic features of such a collapsing spherically symmetric homogeneous dust cloud configuration are summarized in Fig. 2.3. The gravitational collapse initiates when the star surface is outside its Schwarzschild radius $r = 2m$, and a light ray emitted from the surface of the star can escape to infinity. However, once the star has collapsed below $r = 2m$, a *blackhole*, that is a region of no escape, develops in the spacetime, which is bounded by the event horizon at $r = 2m$. Any point in this empty region below the surface $r = 2m$ represents a *trapped surface* (which is a two-dimensional sphere in spacetime) in that both the outgoing and in going families of null geodesics emitted from this point converge, and hence no light ray comes out of this region bounded by $r = 2m$. Then, the collapse to an infinite density and curvature singularity at $r = 0$ becomes inevitable in a finite proper time, as measured by an observer on the surface of the star. In this case, the blackhole region in the resulting vacuum Schwarzschild spacetime is given by $0 < r < 2m$ and the outer boundary of this region, $r = 2m$, is called the *event horizon*. On the event horizon, only the radial outwards photons stay where they are, but all the rest of the photons move inwards. No information from this blackhole region can propagate outside the $r = 2m$ boundary to any outside observer. In the Schwarzschild geometry, for a source situated outside $r = 2m$, part of the photon trajectories emitted with decreasing r values will go towards the blackhole and fall into the singularity. All the other null geodesics will escape to infinity and they intersect the future null infinity. If a source is located below $r = 2m$, no null geodesic would come out of the blackhole and they necessarily end up in the singularity in the future.

The final state of a complete gravitational collapse, either spherically symmetric or otherwise, could possibly be a vacuum spacetime that incorporates the rotation, and possibly also the electromagnetic fields associated with the object. It is possible that the charge associated with an astrophysical object could be quickly neutralized by the surrounding plasma. However, in any case it will be of interest to obtain all solutions of the Einstein–Maxwell equations that describe stationary collapsed configurations with charge.

2.7.6 Vaidya metric and the naked singularity

The geometry outside a spherically symmetric star, when the exterior is taken to be non-empty due to radiation from the star, is given by the Vaidya metric (Vaidya, 1943, 1951, 1953). Just as astrophysical bodies are found to be rotating, they radiate energy in the form of electromagnetic radiation. The Schwarzschild solution does not describe this as it corresponds to an empty exterior given by $T_{ij} = 0$. In the case of a normal star, the effect of radiation on the overall exterior spacetime could be negligible, and effects such as rotation, magnetic fields, and so on, are considered as small perturbations from spherical symmetry. However, the radiation effects would be important during the later stages of gravitational collapse when the star could be throwing away considerable mass as radiation, or when abundant neutrinos are radiated from a collapsing supernova core (see for example, Kahana, Baron, and Cooperstein, 1984). Such a non-static distribution as the radiating star would then be surrounded by an ever-expanding zone of radiation. This radiating system, together with its radiation, could be treated as forming an isolated object in an otherwise empty, asymptotically flat universe. Beyond the zone of pure radiation, the spacetime can be described by the empty Schwarzschild solution.

Here, a spherically symmetric solution to the Einstein equations $G_{ij} = 8\pi T_{ij}$ is sought with a geometrical optics type stress–energy tensor for the radiation of the form

$$T_{ij} = \sigma k_i k_j, \tag{2.187}$$

where k_i is a null vector radially directed outwards. The metric is best given in the null coordinates (u, r, θ, ϕ) as

$$ds^2 = -\left(1 - \frac{2m(u)}{r}\right) du^2 - 2du\, dr + r^2 d\Omega^2, \tag{2.188}$$

with $m(u)$ being an arbitrary non-increasing function of the retarded time u. The above gives the Vaidya metric in the radiation zone, which is to be matched by the interior metric of the radiating body at the boundary of the star, and is matched by the Schwarzschild metric in the exterior of the radiation zone.

The form for the energy–momentum tensor σ is defined to be the energy density of the radiation as measured locally by an observer with a four-velocity vector v^i. Thus, σ is the energy flux as well as the energy density measured in this frame,

$$\sigma \equiv T_{ij} v^i v^j, \tag{2.189}$$

with $v^i v_i = -1$. Working out the connection coefficients, the Ricci tensor in null coordinates is given by

$$R_{ij} = -\frac{2}{r^2}\frac{dm(u)}{du}\delta^0{}_i\delta^0{}_j. \tag{2.190}$$

This implies that the Ricci scalar $R^i{}_i = R = 0$, and hence the Einstein equations give

$$T_{ij} = -\frac{1}{4\pi r^2}\frac{dm(u)}{du}\delta^0{}_i\delta^0{}_j, \tag{2.191}$$

which is the energy–momentum tensor of a radiating field in the geometric optics form. From (2.189) and (2.191),

$$\sigma = -\frac{1}{4\pi r^2}\frac{dm(u)}{du}, \tag{2.192}$$

which is the expression for the energy density of radiation.

In the case when $m(u) =$ const., the relationship of the null coordinates used here with the Schwarzschild coordinates (t, r, θ, ϕ) is not difficult to see. In such a case, the transformation given by Finkelstein (1958) can be used to diagonalize (2.188),

$$u = T - r - 2m\log(r - 2m), \tag{2.193}$$

which gives the Schwarzschild metric in the (T, r, θ, ϕ) coordinate system.

The energy flux from the star, as seen by an outside observer, was computed by Lindquist, Schwartz, and Misner (1965) by considering radially moving observers only. As pointed out, σ is the energy flux measured in a local frame, and if $U \equiv v^r = dr/d\tau$ for a radially moving observer with $v^i v_i = -1, v^\theta = 0$ and $v^\phi = 0$, then from (2.191) and (2.192), the energy density q is

$$q = -\frac{1}{4\pi r^2}\frac{dm}{du}(\gamma + U)^{-2}. \tag{2.194}$$

Since q must be positive as it is energy density, it follows from the above that $dm/du \leq 0$. For an observer at rest at infinity, the total luminosity is given by

$$L_\infty(u) = \lim_{r\to\infty, U=0} 4\pi r^2 q = -\frac{dm}{du}, \tag{2.195}$$

that is, it is the negative rate of change of mass of the radiating body.

The surface $r = 2m(u)$ has many interesting properties, as pointed out by Lindquist, Schwartz, and Misner (1965). Unlike the Schwarzschild case, where $r = 2m$ is a null hypersurface, for the Vaidya radiating star metric, this is a spacelike hypersurface. The induced metric on this hypersurface is given by

$$ds^2|_{r=2m(u)} = -2\left(\frac{dm}{du}\right)du^2 + r^2\, d\Omega^2. \tag{2.196}$$

This induced metric has the signature $(+,+,+)$ since $dm/du < 0$. As a result, the position of light cones on this surface is such that for all timelike vectors in the forward light cone at all points on this surface, $dr/du > 0$. Therefore, no timelike trajectory from the outside region $r > 2m(u)$ can come and cross this surface to enter inside the $r < 2m(u)$ region. The solution is of type D in the Petrov classification of spacetimes and possesses a normal shear-free congruence with a non-zero expansion.

The Vaidya solution has been used extensively in the context of the cosmic censorship hypothesis, and to study gravitational collapse endstates (see for example, Papapetrou, 1985; Joshi, 1993, for details). In that case, imploding radiation shells are considered, rather than the outgoing case considered here. Then, the function m is taken to be non-decreasing and the advanced null coordinate $t + r$ is used. For the case of a linear mass function $m(u) = \lambda u$, an interesting causality and horizon structure arises, depending on the rate of collapse $\dot{m}(u)$, that is, as decided by the magnitude of λ. In this case, the model is self-similar in that it admits a homothetic Killing vector. When the collapse is fast enough, the horizon forms well before the singularity to fully cover it, and a blackhole forms as the collapse endstate. However, for a slow enough collapse when λ is smaller than a critical value, the horizon and trapped surface formation is delayed and a naked singularity develops as the collapse final state (see Fig. 2.4).

The above scenario may be of physical interest if, in the very late stages of collapse, the star converts itself into some kind of an imploding radiation ball. On the other hand, it can be argued that real collapsing stars may not have the form of the energy–momentum tensor of a radiation fluid in the very late stages of their collapse. In any case, not much is known today on the actual equation of state for the collapsing star in the very late stages of its gravitational collapse. The above consideration really shows what is actually possible within the framework of general relativistic gravitational collapse, subject to various physical reasonability and regularity conditions, such as the validity of energy conditions and evolution from regular initial data.

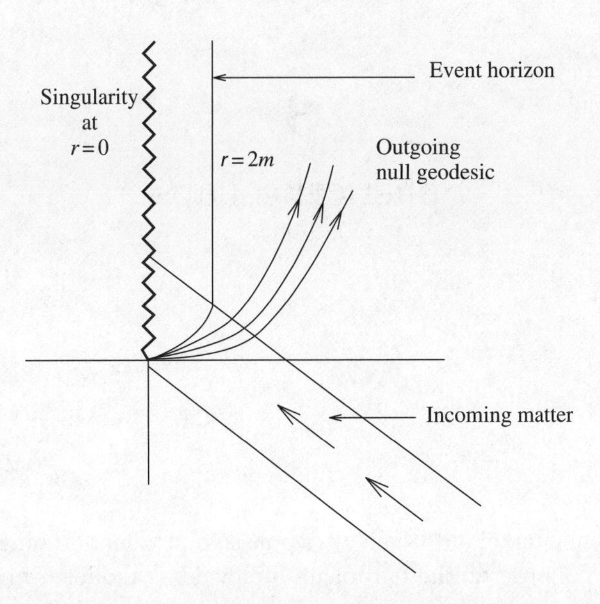

Fig. 2.4 Collapse of radiation shells. A thick shell of radiation arrives at the center in an otherwise empty spacetime. The mass function is $m(u) = \lambda u$ and for a slow enough collapse, that is, for λ below a critical value, a naked singularity forms at the center $r = 0$ from which non-spacelike curves escape to infinity.

In fact, the homogeneous dust collapse is also a rather special situation, as discussed in the next chapter. A physically realistic density profile has to be inhomogeneous, typically higher at the center for an astrophysical object such as a star, decreasing as one moves away from the center. In such a case, it is seen that the structure of the trapped surfaces' geometry changes radically as soon as inhomogeneities are included into the collapse considerations. In the case of a homogeneous collapse, the trapped surfaces start forming early during the evolution of the collapse, and the spacetime singularity forms later. This results in an event horizon structure that fully covers the singularity, thus producing a blackhole in the spacetime. However, when the density is higher at the center, the trapping is delayed and trapped surfaces form only at the epoch of the onset of the singularity. Such a singularity is then visible in principle to external observers, and thus a naked singularity develops as the collapse endstate, rather than a blackhole.

3
Spherical collapse

A collapsing matter cloud, such as a massive star undergoing a continual gravitational collapse at the end of its life cycle, is modeled in general relativity as a dynamical spacetime geometry with a suitable energy–momentum tensor. The time evolution here is governed by the Einstein equations. The cloud has a boundary, with its interior collapsing continually as time evolves, and, at the boundary the interior spacetime is matched to a suitable exterior geometry so as to complete the full model of gravitational collapse. The physical situation considered here is that of the force of gravity being so overwhelming that no final, stable configuration, such as a neutron star or white dwarf, is possible as a collapse endstate, and a continual collapse inevitably proceeds. If the initial mass of the collapsing star is sufficiently high, then such a situation is realized.

In such a scenario, the classical theory leads the collapse to the formation of a spacetime singularity, as predicted by the singularity theorems of general relativity. The spacetime singularity is a region close to where the densities, spacetime curvatures, and all other physical quantities grow without bounds. At the singularity itself these are infinite, and hence, strictly speaking, the singularity is not part of the spacetime and is regarded as the boundary of the spacetime manifold. Eventually, as one moves closer to the singularity, the quantum gravity effects may dominate, which could resolve the classical singularity. Even if the final singularity is removed due to quantum effects, the formation of superdense regions of extreme gravity, as signaled by the occurrence of a spacetime singularity, are really important physically. The curvatures and physical conditions are extreme in the vicinity of such regions that come into being as a result of continual gravitational collapse, which follows from the physical process of the evolution of a massive star at the end of its life cycle. Both strong gravity, as well as quantum effects, should come into their own in such regions.

The singularity theorems predicting the existence and formation of such ultra-dense regions also allow these to be either visible to external observers, or otherwise. This aspect is actually determined by the causal structure of the spacetime in the very late stages of the dynamical collapse, as governed by the Einstein equations. If the causal structure covers the ultra-dense regions within the event horizons of gravity, a blackhole develops as the final state of collapse. On the other hand, when trapped surfaces are delayed in the continual collapse, a naked singularity forms, which is nothing but a visible ultra-dense region that may communicate physical effects to outside observers in the universe. The theoretical and observational properties of these objects would typically be quite different from each other, and so it is of much interest to gain an insight into how each of these phases come about as endstates of a dynamically developing collapse that is governed by gravity.

The purpose here is to investigate the gravitational collapse scenarios within the framework of Einstein gravity, and to determine when either a blackhole or a naked singularity will result in a collapse endstate. For the sake of focus and clarity, spherical collapse models are considered here in order to obtain explicit results, and certain more general aspects will be discussed later. The collapsing matter fields are chosen to be of general *type I*, which is a broad class that includes most of the physically reasonable matter fields. It is shown that, given the initial data for matter (in terms of the initial density and pressure profiles at an initial surface $t = t_i$ from which the collapse develops), the rest of the initial functions and classes of solutions of the Einstein equations constructed here exist, such that the spacetime evolution goes to either a blackhole or a naked singularity final state, depending on the nature of the initial data and the evolutions. The validity of the weak energy condition and various regularity conditions are preserved. Assumptions such as self-similarity of spacetime are not imposed here, in order to keep the treatment general.

The matter fields treated here are a general and broad class that include most of the physically reasonable matter such as dust, perfect fluids, massless scalar fields and such others, and any special restrictions are not imposed on the form of the matter. In order to investigate the collapse final states, given the initial data for the matter in terms of the initial density and pressure profiles at an initial surface $t = t_i$, classes of solutions to the Einstein equations are constructed such that the spacetime evolution goes to a final state that is either a blackhole or a naked singularity, depending on the nature of the rest of the free initial data functions and allowed evolutions. The methodology used here allows the genericity and stability aspects related to the occurrence of naked singularities in gravitational collapse to be considered in some detail.

In Section 3.1, the basic framework for a collapse scenario is described. Various regularity conditions for gravitational collapse are discussed in Section 3.2. The dynamical evolution of the collapsing clouds is considered in Section 3.3. The apparent horizon and structure of a trapped surface formation is discussed in Section 3.4, which describes the evolving causal structure of the collapsing cloud that determines the nature of the singularity in terms of being either visible or covered. The exterior spacetime and matching conditions are given in Section 3.5. As an example to illustrate the formalism here, the gravitational collapse of a dust cloud with vanishing pressures is discussed in some detail in Section 3.6. Finally, the issues related to the equation of state and those of the validity of the energy conditions in the treatment here are considered in Section 3.7.

3.1 Basic framework

To consider spherical symmetry, let P be any point at a distance r from the origin O, then the system must be invariant under rotations around O. Such rotations will generate a two-sphere around O, and the line element on it is given by

$$ds^2 = r^2(d\theta^2 + \sin^2\theta\, d\phi^2), \tag{3.1}$$

which is the metric on the two-sphere given by $t = \text{const.}$, $r = \text{const.}$ in a general spherically symmetric spacetime. Furthermore, as the metric must be invariant under the reflections $\theta \to \pi - \theta$ and $\phi \to -\phi$, there are no cross terms in the metric in $d\theta$ and $d\phi$. As the line elements must not change with any change in θ and ϕ, they have to occur in the metric only in the form of the two-metric given above. Then, in the (t, r, θ, ϕ) coordinate system, the spacetime metric has the form

$$ds^2 = -A\, dt^2 + 2B\, dt\, dr + C\, dr^2 + D(d\theta^2 + \sin^2\theta\, d\phi^2). \tag{3.2}$$

Here, the quantities A, B, C, and D are the functions of t and r that are to be determined. Introducing a new radial coordinate $r' = D^{1/2}$ and a new time coordinate t' by requiring that $dt' = F[A'dt - B'dr']$, where $F(t, r')$ is a suitable integrating factor, the line element reduces to

$$ds^2 = -e^{\nu}\, dt^2 + e^{\psi}\, dr^2 + r^2(d\theta^2 + \sin^2\theta\, d\phi^2), \tag{3.3}$$

where the primes have been dropped. Here, $\nu = \nu(r, t)$ and $\psi = \psi(r, t)$, and the quantities $e^{\nu} = 1/F^2 A'$ and $e^{\psi} = C' - B'^2/A'$ appearing in the metric are always positive.

In general, the spherical symmetry of a spacetime can be defined using Killing vectors. There must be three linearly independent spacelike Killing vector fields $\boldsymbol{X}^1, \boldsymbol{X}^2$, and $\boldsymbol{X}^3$ in the spacetime that satisfy the commutator relations

$$[\boldsymbol{X}^1, \boldsymbol{X}^2] = \boldsymbol{X}^3, \quad [\boldsymbol{X}^2, \boldsymbol{X}^3] = \boldsymbol{X}^1, \quad [\boldsymbol{X}^3, \boldsymbol{X}^1] = \boldsymbol{X}^2, \tag{3.4}$$

and their orbits must be closed. Using these properties, the line element above could be derived rigorously for a spherically symmetric spacetime.

A general formalism for a spherically symmetric gravitational collapse including pressure, was developed by Misner and Sharp (1964). The spherically symmetric spacetime is written in comoving coordinates as

$$ds^2 = -e^{2\phi}\, dt^2 + e^{\lambda}\, dr^2 + R^2(t,r)\, d\Omega^2, \tag{3.5}$$

where $d\Omega^2 = d\theta^2 + \sin^2\theta\, d\phi^2$ is the usual metric on the two-sphere, and ϕ and λ are functions of t and r. The stress–energy tensor is that of a perfect fluid given by

$$T_{ij} = (\rho + p) u_i u_j + p g_{ij}. \tag{3.6}$$

The spatial velocities u^i are vanishing here and the spatial coordinates of a given particle remain constant throughout the collapse. A function $m(r,t)$ is introduced by the definition

$$e^{\lambda} = \left(1 + \dot{R}^2 - \frac{2m}{r}\right)^{-1} R'^2, \tag{3.7}$$

where a dash denotes a derivative with respect to r, and for any function f,

$$\dot{f} = e^{-\phi}\left(\frac{\partial f}{\partial t}\right). \tag{3.8}$$

The coordinate t here gives the proper time along the particle world lines. Integrating the conservation equation $T^{ij}{}_{;j} = 0$ and solving the Einstein equations, the Misner–Sharp equations for the spherically symmetric collapse can be written as

$$\dot{m} = -4\pi R^2 p \dot{R}, \tag{3.9}$$

$$\ddot{R} = \left(\frac{1 + \dot{R}^2 - 2m/r}{\rho + p}\right)\left(\frac{\partial p}{\partial R}\right) - \frac{m + 4\pi R^3 p}{R^2}, \tag{3.10}$$

$$\frac{\partial m}{\partial R} = 4\pi R^2 \rho. \tag{3.11}$$

The above equations, when combined with an equation of state relating ρ and p, determine the dynamical evolution of the collapse. However, when the pressure $p \neq 0$, the situation is quite complex and numerical computation (see for example, May and White, 1966) can be opted for in order to get an idea of the evolution of the collapsing system.

The general spherically symmetric metric, in a spacetime of dimension N, can be written in the form

$$ds^2 = -g_{ab}\,(x^0,\,x^1)\,dx^a\,dx^b + R^2(x^0,\,x^1)d\Omega^2_{N-2}, \tag{3.12}$$

where a and b run from 0 to 1, and

$$d\Omega^2_{N-2} = \sum_{i=1}^{N-2}\left[\prod_{j=1}^{i-1}\sin^2(\theta^j)\right](d\theta^i)^2 \tag{3.13}$$

is the metric on an $(N-2)$ sphere with θ^i being the spherical coordinates. From this metric, the components of the Einstein tensor can be obtained as (Rocha and Wang, 2000)

$${}^N G_{ab} = \frac{N-2}{2R^2}\left[g_{ab}\{2R\Box R + (N-3)Q\} - 2RR_{,ab}\right], \tag{3.14}$$

$${}^N G_{22} = -\frac{1}{2}\left[R^2\Re + (N-3)\{2R\Box R + (N-4)Q\}\right], \tag{3.15}$$

$${}^N G_{ii}(i>2) = \left[\prod_{k=1}^{i-2}\sin^2(\theta^k)\right]{}^N G_{22}, \tag{3.16}$$

where

$$Q = 1 + R^{,a}R_{,a}, \quad \Box R = g^{ab}R_{;ab}, \tag{3.17}$$

and

$$\Re = g^{ab}\Re_{ab}. \tag{3.18}$$

Here, $\Re_{ab}$ is the Ricci tensor evaluated from the two-metric g_{ab}, and $\Re$ is the scalar curvature also evaluated from the two-metric g_{ab}.

The spacetime geometry within the spherically symmetric collapsing cloud is described by comoving coordinates $(t,\,r,\,\theta^i)$, which are specified below. The matter field is chosen to be of general type I, which is a broad class including most of the physically reasonable matter forms, including dust, perfect fluids, massless scalar fields, and such others. This class of matter is specified by the requirement that the energy momentum tensor for the matter admits one timelike and three spacelike eigenvectors (Hawking and Ellis, 1973; Stephani *et al.*, 2003).

The coordinates $(t,\,r,\,\theta^i)$ are then chosen to be those along these eigenvectors, which makes the coordinate system *comoving*, that is, the coordinate system moves with the matter. The freedom of coordinate transformations of the form $t' = f(t,r)$ and $r' = g(t,r)$ can be used to make the g_{tr} term in (3.12) and the radial velocity of the matter vanish. In this case, the general metric in the comoving coordinates $(t,\, r,\, \theta^i)$ must have three general arbitrary functions of t and r, and this can be written in the form (Landau and Lifshitz, 1975)

$$ds^2 = -e^{2\nu(t,r)}dt^2 + e^{2\psi(t,r)}dr^2 + R^2(t,r)d\Omega^2_{N-2}. \tag{3.19}$$

In this comoving frame, the energy–momentum tensor for any matter field that is type I is given in a diagonal form

$$T^t{}_t = -\rho(t,r), \;\; T^r{}_r = p_r(t,r), \;\; T^{\theta^i}{}_{\theta^i} = p_\theta(t,r). \tag{3.20}$$

The quantities ρ, p_r, and p_θ are the energy density, and radial and tangential pressures ascribed to the matter field respectively.

The matter cloud can be chosen to have a compact support at an initial spacelike surface of $t = t_i$, with $0 < r < r_\mathrm{b}$, where r_b would denote the boundary of the cloud. Outside this boundary, the interior solution has to be matched through suitable junction conditions, with another suitable spacetime metric to complete the full spacetime geometry.

The matter fields are taken to satisfy the *weak energy condition*, that is, the energy density measured by any local timelike observer is non-negative. This ensures the physical reasonability for the collapsing matter fields considered here. Another energy condition frequently used is the *dominant energy condition*, which demands that, for any timelike observer, the local energy flow is non-spacelike. These two conditions are frequently regarded as the main and important energy conditions that are physically reasonable (Hawking and Ellis, 1973; Wald, 1984). All classical observed matter fields satisfy these conditions. For the energy conditions to be satisfied, for any timelike vector V^i,

$$T_{ik}V^iV^k \geq 0 \tag{3.21}$$

and $T^{ik}V_k$ must be non-spacelike. For the energy–momentum tensor (3.20), these amount to the conditions

$$\rho \geq 0, \; \rho + p_r \geq 0, \; \rho + p_\theta \geq 0, \tag{3.22}$$

$$|p_r| \leq \rho, \;\; |p_\theta| \leq \rho. \tag{3.23}$$

Now, with the above metric, the following quantities can be evaluated

$$R^{,a}R_{,a} = \dot{R}^2 e^{-2\nu} - R'^2 e^{-2\psi}, \tag{3.24}$$

$$R = e^{-2\nu}[\ddot{R} + \dot{R}(-\dot{\nu} + \dot{\psi})] - e^{-2\psi}[R'' + R'(\nu' - \psi')], \tag{3.25}$$

and

$$R_{,ab} = \left(\ddot{R} - \dot{\nu}\dot{R} - \frac{e^{2\nu}\nu' R'}{e^{2\psi}}\right)\delta^0_a\delta^0_b + \left(R'' - \psi' R' - \frac{e^{2\psi}\dot{\psi}\dot{R}}{e^{2\nu}}\right)\delta^1_a\delta^1_b + \mathcal{Q}_{ab}, \tag{3.26}$$

where

$$\mathcal{Q}_{ab} = \left(\dot{R}' - \nu'\dot{R} - \dot{\psi}R'\right)\left(\delta^0_a\delta^1_b + \delta^1_a\delta^0_b\right). \tag{3.27}$$

Using the above equations, the Einstein equations $G_{ik} = T_{ik}$ now take the form (in the units $8\pi G = c = 1$)

$$\rho = \frac{(N-2)F'}{2R^{N-2}R'},\ p_r = -\frac{(N-2)\dot{F}}{2R^{N-2}\dot{R}}, \tag{3.28}$$

$$\nu' = \frac{(N-2)(p_\theta - p_r)}{\rho + p_r}\frac{R'}{R} - \frac{p_r'}{\rho + p_r}, \tag{3.29}$$

$$-2\dot{R}' + R'\frac{\dot{G}}{G} + \dot{R}\frac{H'}{H} = 0, \tag{3.30}$$

$$G - H = 1 - \frac{F}{R^{N-3}}, \tag{3.31}$$

where

$$G(t,r) = e^{-2\psi}(R')^2,\ \ H(t,r) = e^{-2\nu}(\dot{R})^2. \tag{3.32}$$

The arbitrary function $F = F(t,r)$ here has the interpretation of the mass function for the cloud, which gives the total mass in a shell of comoving radius r, at an epoch t. The energy condition $\rho \geq 0$ implies $F \geq 0$ and $F' \geq 0$. Since the area radius vanishes at the center of the cloud, from (3.28), it is evident that in order to preserve the regularity of density and pressures at any non-singular epoch t, $F(t,0) = 0$, that is the mass function vanishes at the center of the cloud.

As seen from (3.28), there is a density singularity in the spacetime at $R = 0$ and at $R' = 0$. However, the latter ones are due to shell-crossings (Yodzis, Seifert, and Muller zum Hagen, 1973, 1974), which basically indicate the breakdown of the coordinate system used. These are not generally regarded as genuine spacetime singularities, because they can be possibly removed from the spacetime to extend the manifold (Clarke, 1986, 1993). Hence, only the shell-focusing singularity at $R = 0$, which is a genuine physical singularity where all matter shells collapse to a zero physical radius, will be considered here. This will be discussed in more detail later.

Note that, in general, for a general matter field with non-vanishing pressures as considered here, there are a variety of dynamical time evolutions possible from the given matter density and pressure profiles as prescribed on an initial surface (that are called matter initial data here), from which the collapse evolves. In particular, even if the cloud commences gravitational collapse at the initial surface $t = t_i$, there can be classes of solutions of the Einstein equations where the evolution is such that a bounce is possible at a later stage for the cloud. Here, mainly the continually collapsing class of models are considered, because interest is in the physical situation that corresponds to the case when the mass of the collapsing cloud, or the star, is so high that on exhausting its nuclear fuel, the star must undergo a continual gravitational collapse, completing it in a finite time. Therefore, the continual collapse condition is included as a part of the framework here. In such a case, trapped surfaces form as the collapse evolves and a spacetime singularity necessarily develops as the collapse endstate. The conditions when the final singularity is necessarily covered within an event horizon of gravity (as hypothesized by the cosmic censorship) or when it will be naked with ultra-strong gravity regions being visible to faraway observers then need to be found. In the case of a bounce or dispersal, no singularity needs to form in the spacetime, a situation that is not considered here.

The scaling freedom available for the radial coordinate r can be used to write

$$R = r \tag{3.33}$$

at the initial epoch $t = t_i$. It is interesting to note that, if one had wished to scale the radial coordinate at the initial epoch as $R(t_i, r) = r^\beta$, with β being any constant, then the *only* possible allowed value for β would be unity. This is because, in the other case, either R' would blow up at the center $r = 0$ (which is not allowed by regularity conditions, as the Einstein equations require the metric functions to be at least C^2), or R' would go to zero at the center, causing a shell-crossing singularity (which is avoided by construction) that violates the regularity of the initial data.

A function $v(t, r)$ is now introduced (Joshi and Dwivedi, 1999; Joshi and Goswami, 2004) as defined by

$$v(t, r) \equiv R/r. \tag{3.34}$$

Then, $R(t, r) = rv(t, r)$, and

$$v(t_i, r) = 1, \quad v(t_\mathrm{s}(r), r) = 0, \quad \dot{v} < 0. \tag{3.35}$$

The time $t = t_\mathrm{s}(r)$, that is $v = 0$, corresponds to the shell-focusing singularity at $R = 0$, which is a genuine spacetime singularity where all the matter shells collapse to a vanishing physical radius. The condition $\dot{v} < 0$ corresponds here to a continual collapse of the cloud. The description of the shell-focusing singularity at $R = 0$ in terms of the function $v(t, r)$ as above has several advantages. The physical radius goes to zero at the shell-focusing singularity, but also $R = 0$ at the regular center of the cloud at $r = 0$. This is to be distinguished from the genuine singularity at the collapse endstate, by the fact, for example, that the density and other physical quantities including the curvature scalars all remain finite at the regular center $r = 0$ of the cloud, even though $R = 0$ holds there. This is achieved, as shown below, by a suitable behavior of the mass function, which should go to a vanishing value sufficiently fast in the limit of approach to the regular center where (even though R goes to zero) the density must remain finite. On the other hand, when the function $v(t, r)$ is used, note that at $t = t_i$, $v = 1$ on the entire initial surface, and then as the collapse evolves, the function v continuously decreases to become zero only at the singularity $t_s(r)$. This means that $v = 0$ uniquely corresponds to the genuine spacetime singularity at $R = 0$.

From the point of view of dynamic evolution of the initial data prescribed at the initial epoch $t = t_i$, there are five arbitrary functions of the comoving shell-radius r, as given by

$$\nu(t_i, r) = \nu_0(r), \quad \psi(t_i, r) = \psi_0(r), \quad R(t_i, r) = r,$$
$$\rho(t_i, r) = \rho_0(r), \quad p_r(t_i, r) = p_{r_0}(r), \quad p_\theta(t_i, r) = p_{\theta_0}. \tag{3.36}$$

Note that not all the initial data functions above are mutually independent, as, from (3.29),

$$\nu_0(r) = \int_0^r \left(\frac{(N-2)(p_{\theta_0} - p_{r_0})}{r(\rho_0 + p_{r_0})} - \frac{p'_{r_0}}{\rho_0 + p_{r_0}} \right) dr. \tag{3.37}$$

Therefore, apart from the matter initial data, which describe the initial density and pressure profiles at the initial epoch $t = t_i$, the rest of the initial data that are free is $\psi_0(r)$, which essentially describes the velocities of the collapsing matter shells as discussed later.

Note that in the above, when the pressures are taken to be vanishing, the scenario reduces to a dust collapse. In this case, the only two free initial data functions are the initial density and the velocity profile for the cloud. As discussed later, in this case, for any given initial density profile $\rho(r)$ for the matter cloud, there are classes of initial velocity profiles $\psi(r)$ that take the final collapse outcome to either a blackhole, or a naked singularity, depending on the choice of initial velocities. This is subject to the validity of the energy conditions and regularity conditions for gravitational collapse. The Oppenheimer–Snyder homogeneous dust collapse corresponds here to a special choice of the initial density profile that is a constant function, and another special choice of the initial velocity function. In fact, if the initial density profile is chosen to be constant, but the initial velocities are allowed to take different functional forms, then an initially constant and homogeneous density can also later inhomogenize, and result in a final evolution that is a naked singularity. If the pressures are allowed to be non-vanishing, then the choices in evolution are much wider, as shown below.

3.2 Regularity conditions

Whilst the basic equations for a general spacetime of dimension N (Goswami and Joshi, 2006) have been given above, for the rest of this chapter the usual spacetime dimensions of four ($N = 4$) shall be used for the sake of clarity and transparency (Joshi and Dwivedi, 1999; Joshi and Goswami, 2004). Some of the useful and interesting differences arising when there is a higher spacetime dimension will be pointed out as necessary.

The regularity of the initial data needs to be ensured in order to make sure that the gravitational collapse initiates from regular and physically reasonable initial conditions. The initial pressures must thus be taken to have physically reasonable behavior at the center. Considering that the total force at the center of the collapsing cloud should be zero, the gradients of initial pressures vanish at the center. Also, regularity of the initial data requires that $p_{r_0}(0) - p_{\theta_0}(0) = 0$, that is, at the center the difference between the radial and tangential pressure vanishes. The metric functions have to be C^2-differentiable everywhere as per the requirements of the Einstein equations. As seen from the equation for ν' above, such a condition is implied by the requirement that ν' does not blow up at the regular center. This means that the matter should behave like a perfect fluid at the center of the cloud with the net force vanishing there.

Note that these regularity conditions do not exclude the collapse models with either a purely tangential pressure ($p_r = 0$), or purely radial pressure collapse models ($p_\theta = 0$), which will be referred to later in this chapter. This is because in these cases, the above condition implies that $p_\theta \to 0$ or $p_r \to 0$ respectively, close to the center, where the matter closely approximates dust. Note that the regularity conditions above give a sufficient criterion for the regularity of the metric function $\nu_0(r)$ at any non-singular initial epoch. It follows from (3.37) that at the center of the cloud both ν_0 and ν_0' go to zero. Hence, $\nu_0(r)$ has the form

$$\nu_0(r) = r^2 g(r), \tag{3.38}$$

where $g(r)$ is an arbitrary function which is at least C^1 for $r = 0$, and is at least a C^2-function for $r > 0$, as the Einstein equations demand the metric functions to be at least C^2 everywhere. Another regularity condition frequently used in collapse considerations is that there should be no trapped surfaces at the initial spacelike surface from which the collapse begins.

Therefore, there are five total field equations with seven unknowns, ρ, p_r, p_θ, ψ, ν, R, and F, which are three matter variables, three metric functions and the mass function. This gives the freedom of choice of two free functions. Selection of these free functions, subject to the weak energy condition and the given regular initial data for collapse at the initial surface, determines the matter distribution and metric of the spacetime throughout, and thus leads to a particular time evolution of the initial matter and velocity distributions with which the collapse began. As will be shown, it turns out that, given the matter initial profiles in terms of ρ_0, p_{r_0}, and p_{θ_0}, there are the rest of the initial data at $t = t_i$, and the classes of solutions to the Einstein equations that are explicitly constructed here, which give either a blackhole or a naked singularity as the endstate of collapse. This outcome depends on the nature of the rest of the initial functions, and the classes of the dynamical evolutions, as allowed by the Einstein equations.

An important point to be noted here is that, in the description above no mention has been made so far of the equations of state that the matter must obey. Typically, these are of the form, $p_r = p_r(\rho)$ and $p_\theta = p_\theta(\rho)$. If these are specified, then there is no freedom left, and there are seven equations for seven variables. If this were to be incorporated right away, the only way to proceed to find the collapse endstate would be to *assume* a specific equation of state that the matter must satisfy. Then, one has to examine the collapse problem accordingly, and examine the nature of the final singularity resulting from the dynamical evolution, as governed by the Einstein equations. There have been many collapse studies in the past using this approach, such as for example, for the dust equation of state for perfect fluids, and others. The limitation of such an approach, however, is that there is very little existing knowledge on what a realistic equation of state should be that the matter has to satisfy at the extreme high densities that a continual collapse realizes in its advanced stages. For example, even for neutron star densities that are relatively low compared with those of the later stages of a continual collapse, there is a great deal of uncertainty on the equation of state for such neutron matter. As a result, the neutron star mass limits are uncertain. Therefore, specific or special assumptions used on the equation of state may turn out to be physically unrealistic or restrictive, and untenable in the final stages of the collapse. In fact, diametrically opposite views exist on the possible equation of state in the very late stages of the collapse. For example, while there are many arguments suggesting that pressures must play an important role in the later stages of the collapse, the other view is that in such late stages the matter must necessarily be dust-like (see for example, Hagedorn, 1968; Penrose, 1974a, 1974b).

Under this situation, the approach taken here is that no specific or particular equation of state is *assumed* to begin with. All further considerations are carried out in a general manner, in terms of the allowed initial matter profiles and the allowed dynamical evolutions of the Einstein equations, towards determining the blackhole and naked singularity endstates for collapse.

Subsequently, in Section 3.7, the role that the equation of state plays towards further fine tuning the blackhole and naked singularity outcomes as collapse endstates is discussed. The advantage of such an approach here is that, first all the collapse equations are written in general, and only then are different subcases distinguished, depending on the corresponding equation of state under consideration. As shall be shown, various important subcases and equations of state such as dust, perfect fluids, and others are included as special cases of the treatment given here.

3.3 Collapsing matter clouds

Now the evolution of the matter cloud is considered. The method followed is outlined below. In the case of a blackhole developing as a collapse endstate, the spacetime singularity is necessarily hidden behind the event horizon of gravity. In the case of a naked singularity developing, there are families of future directed non-spacelike trajectories that terminate in the past at the singularity, and which can, in principle, communicate information to faraway observers in the spacetime. The existence of such families confirms the naked singularity formation, as opposed to a blackhole collapse endstate. The final singularity produced by the collapsing matter is studied here, and it is shown that the tangent to the singularity curve at the central singularity at $r = 0$ is related to the radially outgoing null geodesics from the singularity, if there are any. By determining the nature of the singularity curve, and its relation to the initial data and the classes of collapse evolutions, it is possible to deduce whether the trapped surface formation in the collapse takes place earlier than the singularity formation epoch, or is delayed. It is this causal structure of the trapped surface region, and the apparent horizon (that is the boundary of the trapped region), forming during the collapse that determines the possible emergence, or otherwise, of non-spacelike curves from the singularity. This settles the final outcome of the collapse in terms of either a blackhole or naked singularity formation.

It is necessary to clarify here the meaning of the occurrence of a naked singularity developing as the end point of the collapse. At times, the non-existence of trapped surfaces until the formation of the singularity in the collapse is taken as the signature that the singularity is naked. This has a reference generally to some slicing of the spacetime, in which the occurrence, or otherwise, of the trapped surfaces is examined (see for example, Shapiro and Teukolsky, 1991, 1992). At times, however, a slicing dependent definition of naked singularity may not give an accurate picture of what is happening in the spacetime (Wald and Iyer, 1991). What is meant by the development of a naked singularity in the collapse is that families of future directed non-spacelike curves, which in the past terminate at the singularity,

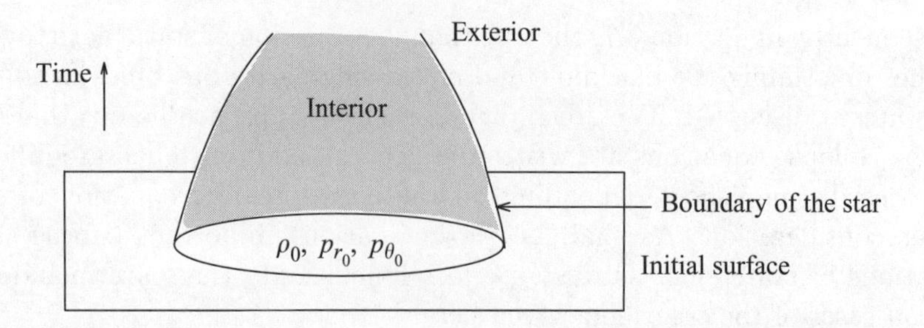

Fig. 3.1 The collapse evolves from regular initial data from an initial spacelike hypersurface. The cloud has a compact support on the spacelike slice and the interior collapse spacetime is matched to a suitable exterior metric at the boundary.

exist. No such families exist, originating from the singularity, when the end product of collapse is a blackhole. This is a definition that is independent of the coordinates used and the slicing chosen. In the case of a blackhole developing, the resultant spacetime singularity will be hidden inside an event horizon of gravity, remaining unseen by external observers. On the other hand, if the collapse ends in a naked visible singularity, there is a causal connection between the region of singularity and faraway observers, thus enabling, in principle, a communication from the super-dense regions close to the singularity to faraway observers.

Given the matter initial profiles in terms of the initial functions as given by $\rho_0(r), p_{r_0}(r), p_{\theta_0}(r)$ at the initial epoch $t = t_i$ from which the collapse commences, the purpose now is to construct and examine possible evolutions (classes of solutions to the Einstein equations) of such a matter cloud from the given initial data, to investigate its final states (see Fig. 3.1).

While constructing the classes of solutions that give the collapse evolutions, given the matter initial data at $t = t_i$, as much generality as possible is preserved. The mass function $F(t, r)$ for the collapsing cloud must have the following general form:

$$F(t, r) = r^3 \mathcal{M}(r, v), \tag{3.39}$$

where $\mathcal{M} > 0$ is at least a C^1-function of r for $r = 0$, and at least a C^2-function for $r > 0$. It is to be noted that F must have this general form, which follows from the regularity and finiteness of the density profile at the initial epoch $t = t_i$, and at all other later regular epochs before the cloud collapses to the final singularity at $R = 0$. This requires, from the Einstein equation for energy density, that F must behave as r^3 close to the regular center. Hence, note that since $\mathcal{M}$ is a general (at least C^2) function, (3.39) is not really any ansatz or a special choice, but quite a generic class of the mass profiles for the collapsing cloud, consistent with and allowed by the regularity conditions. Therefore, no special choice of F is made, but it is allowed to be a general function as given by the Einstein equation (3.28), while constructing the classes of solutions that give collapse evolutions from the given regular initial data.

Then, (3.28) gives

$$\rho(r,v) = \frac{3\mathcal{M} + r\left[\mathcal{M}_{,r} + \mathcal{M}_{,v}v'\right]}{v^2(v + rv')}, \tag{3.40}$$

and

$$p_r(r,v) = -\frac{\mathcal{M}_{,v}}{v^2}. \tag{3.41}$$

The regular density distribution at the initial epoch is given by

$$\rho_0(r) = 3\mathcal{M}(r,1) + r\mathcal{M}(r,1)_{,r}. \tag{3.42}$$

It is evident that, in general, as $v \to 0$, $\rho \to \infty$ and $p_r \to \infty$. That is, both the density and radial pressure blow up at the shell-focusing singularity.

It is shown that, given any regular initial density and pressure profiles for the matter cloud from which the collapse develops, energy profiles or velocity functions for the collapsing matter shells and classes of dynamical evolutions as determined by the Einstein equations always exist, so that the collapse endstate would be either a naked singularity or a blackhole, depending on the nature of the allowed choices. Therefore, given the matter initial data at the initial surface $t = t_i$, these evolutions take the collapse to end either as a blackhole or naked singularity, depending on the choice of the class, subject to the regularity and energy conditions.

To see this, classes of solutions to the Einstein equations need to be constructed. A suitably differentiable function $A(r,v)$ can be defined as

$$\nu'(r,v) = A(r,v)_{,v}R'. \tag{3.43}$$

That is, $A(r,v)_{,v} \equiv \nu'/R'$, and since at $t = t_i$ one has $R = r$, this gives $[A(r,v)_{,v}]_{v=1} = \nu_0'(r)$. The main interest here is in studying the shell-focusing singularity at $R = 0$, which is the physical singularity where all the matter shells collapse to zero physical radius. Therefore, assume that there are no shell-crossing singularities in the spacetime where $R' = 0$, and that the function $A(r,v)$ is well-defined. From (3.38), the form of $\nu(t,r)$ can be generalized and chosen as the class of functions given by

$$\nu(t,r) = r^2 g_1(r,v), \tag{3.44}$$

where $g_1(r,v)$ is a function that is suitably differentiable and $g_1(r,1) = g(r)$. It then follows that $A(r,v)$ has the form

$$A(r,v) = rg_2(r,v). \tag{3.45}$$

The class of solutions considered here is with no shell-crossing singularities developing as the collapse evolves, that is $R' > 0$. This is because, as mentioned above, such singularities are generally weak and can possibly be removed from the spacetime as they are typically gravitationally not strong, and because spacetime extensions have been constructed in certain cases (Clarke, 1993). In contrast, in several physically reasonable collapse models including dust and perfect fluids, the singularity at $R = 0$ turns out to be gravitationally strong and is a powerful curvature singularity. Under this situation, one is interested here only in examining the nature of the shell-focusing singularities at $R = 0$, which are genuine curvature singularities arising at the termination of collapse, where the physical radii for all collapsing shells vanish, and the spacetime necessarily terminates without extension.

Specifically, $R' > 0$ implies that $v + rv' > 0$. Since v is necessarily positive throughout the collapse, it follows that this will be satisfied always whenever v' is greater than or equal to zero. Even when it is negative, the condition that the magnitude of rv' should be less than that of v is sufficient to ensure that there will be no shell-crosses. Later in this section, an expression for the quantity v', in terms of the initial data and the other free evolutions as allowed by the Einstein equations is derived. So, it follows that the condition for avoidance of shell-crossings can be specifically stated in terms of the behavior of these functions. In particular, it turns out that whenever the singularity curve $t_s(r)$ (which corresponds to $R = 0$) is increasing at the center (or when it decreases at a sufficiently slow rate) with a slope greater than or equal to zero at the origin, the shell-crossing singularities are avoided, at least in the vicinity of the regular center $r = 0$. Then, a ball of finite radius around the regular center that contains no shell-crossings until the final singularity formation at $R = 0$ exists.

Coming to the dynamical collapse evolutions, using (3.43) in (3.30), a class of solutions of the Einstein equations can be given as

$$G(r, v) = b(r)e^{2rA(r,v)}. \tag{3.46}$$

Here, $b(r)$ is another arbitrary function of the shell radius r. By using the regularity condition on the function $\dot{v}$ at the center of the cloud,

$$b(r) = 1 + r^2 b_0(r), \tag{3.47}$$

where $b_0(r)$ is the energy distribution function for the shells. Using (3.43) in (3.29),

$$p_\theta = RA_{,v}(\rho + p_r) + 2p_r + \frac{Rp_r'}{R'}. \tag{3.48}$$

It can be seen that in general, both the density and radial pressure blow up at the singularity, so the above equation implies that the tangential pressure would also typically blow up at the singularity.

Now, using (3.39), (3.43), and (3.46) in (3.31),

$$R^{1/2}\dot{R} = -e^{\nu(r,v)}\sqrt{(1+r^2b_0)Re^{2r\,A(r,v)} - R + r^3\mathcal{M}}. \tag{3.49}$$

The negative sign on the right-hand side of the above equation corresponds to a collapse scenario where $\dot{R} < 0$. Defining a function $h(r,v)$ as

$$h(r,v) = \frac{e^{2r\,A(r,v)} - 1}{r^2} = 2A(r,v) + \mathcal{O}(r^2), \tag{3.50}$$

(3.31) becomes

$$v^{1/2}v = -\sqrt{e^{(rA+v)}vb_0 + e^{2v}\,(vh + \mathcal{M})}. \tag{3.51}$$

Integrating the above equation with respect to v,

$$t(v,r) = \int_v^1 \frac{v^{1/2}dv}{\sqrt{e^{(rA+\nu)}vb_0 + e^{2\nu}\,(vh + \mathcal{M})}}. \tag{3.52}$$

Note that the variable r is treated as a constant in the above equation. The above equation gives the time taken for a shell labeled r to reach a particular epoch v from the initial value $v = 1$. Expanding the function $t(v,r)$ around the center of the cloud, provided the functions within the integral above are sufficiently regular, gives

$$t(v,r) = t(v,0) + r\mathcal{X}(v) + \mathcal{O}(r^2), \tag{3.53}$$

where the function $\mathcal{X}(v)$ is given as

$$\mathcal{X}(v) = -\frac{1}{2}\int_v^1 dv \frac{v^{1/2}(vb_1 + h_1 v + \mathcal{M}_1(v))}{(vb_{00} + vh_0 + \mathcal{M}_0(v))^{3/2}}, \tag{3.54}$$

where

$$\begin{aligned} b_{00} &= b_0(0), \quad \mathcal{M}_0(v) = \mathcal{M}(0,v), \quad h_0 = h(0,v), \\ b_1 &= b_0'(0), \quad \mathcal{M}_1(v) = \mathcal{M}_{,r}(0,v), \quad h_1 = h_{,r}(0,v). \end{aligned} \tag{3.55}$$

Hence, it can be seen that the time taken for a shell labeled r to reach the spacetime singularity at $R = 0$ (which is the *singularity curve*) is given as

$$t_s(r) = \int_0^1 \frac{v^{1/2}dv}{\sqrt{e^{(rA+\nu)}vb_0 + e^{2\nu}\,(vh + \mathcal{M})}}. \tag{3.56}$$

As it is the continual collapse to be considered here, only those classes of solutions where $t_s(r)$ is finite and sufficiently regular are focused on. This means that the cloud collapses in a finite amount of time. In the physical situation of a continual collapse of a massive matter cloud in a finite amount of time, the function $t_s(r)$ has to be finite by definition. As for regularity, to check the existence conditions for a well-defined, continuous and C^2-differentiable singularity curve, a function $Q(r, v)$ can be defined as

$$Q(r,v) = \frac{v^{(N-3)/2}}{\sqrt{e^{(rA+\nu)}v^{(N-3)}b_0 + e^{2\nu}\left(v^{(N-3)}h + \mathcal{M}\right)}}. \tag{3.57}$$

Consider now the following functions,

$$\phi_1(r) = \int_0^1 Q(r,v)_{,r}\, dv, \quad \phi_2(r) = \int_0^1 Q(r,v)_{,rr}\, dv. \tag{3.58}$$

Let $\mathcal{A}$ be the rectangular area in the (r, v) plane defined by the lines

$$r = 0, \quad r = \epsilon, \quad v = 0, \quad v = 1. \tag{3.59}$$

Now, if the following conditions:

(1) $Q(r, v)$ is a continuous function of r and v in $\mathcal{A}$;
(2) $Q(r, v)_{,r}$ and $Q(r, v)_{,rr}$ are continuous functions of r and v in $\mathcal{A}$;
(3) the integrals $\phi_1(r)$ and $\phi_2(r)$ converge uniformly in $\mathcal{A}$;

are satisfied, then

$$\phi_1(r) = \frac{d}{dr}[t_s(r)], \quad \phi_2(r) = \frac{d^2}{dr^2}[t_s(r)], \tag{3.60}$$

and this implies that the singularity curve $t_s(r)$ would be a well-defined C^2-function near the center.

Several well-studied collapse models such as dust collapse models and others, satisfy these or stronger conditions, and so the singularity curve is well-defined and expandable. Some examples of such singularity curves that form special cases for the consideration above follow.

In the case of a dust collapse, the expression for the singularity curve is given by (Goswami and Joshi, 2004a, 2004b)

$$t_s(r)_{\text{dust}} = \int_0^1 \frac{v^{1/2}\, dv}{\sqrt{\mathcal{M}(r) + v b_0(r)}}, \tag{3.61}$$

where $\mathcal{M}(r)$ and $b_0(r)$ are well-defined C^2-functions of the comoving coordinate r and are well-defined at $r = 0$. It can then be easily seen that the singularity curve is differentiable at the center with

$$\frac{dt_s(r)}{dr} = -\frac{1}{2}\int_0^1 \frac{v^{1/2}(\mathcal{M}_1 + vb_1)dv}{(\mathcal{M}_0 + vb_{00})^{3/2}}, \tag{3.62}$$

where

$$\begin{aligned} b_{00} &= b_0(0), \quad \mathcal{M}_0 = \mathcal{M}(0), \\ b_1 &= b'(0), \quad \mathcal{M}_1 = \mathcal{M}'(0). \end{aligned} \tag{3.63}$$

Another such example is that of an Einstein cluster (Ec), which describes a non-steady spherically symmetrical system of non-colliding particles moving in such a way that, relative to a suitably moving frame of co-ordinates, their motion is purely transversal. In this case, the singularity curve is given by (Mahajan, Goswami, and Joshi, 2005)

$$t_s(r)_{\mathrm{Ec}} = \int_0^1 \frac{v^{1/2}\sqrt{v^2 + \frac{L(r)^2}{r^2}}\,dv}{e^{\nu}\sqrt{b_0 v^3 - \left(\frac{L(r)^2}{r^4}\right)v + \mathcal{M}(r)\left(v^2 + \frac{L(r)^2}{r^2}\right)}}. \tag{3.64}$$

Here, $L(r)$ is a function of the radial coordinate r only. In the Newtonian limit, this function corresponds to the angular momentum per unit mass of the system. Therefore, the function $L(r)$ is called the *specific angular momentum*, and has the form

$$L(r) \equiv r^2 l(r). \tag{3.65}$$

Again, since $\mathcal{M}(r)$ and $b_0(r)$ and $L(r)$ are well-defined C^2-functions of the coordinate r, it can be seen that the above singularity curve is well-defined and differentiable at $r = 0$.

To generalize this, and to give another explicit example, it can be shown that (Goswami and Joshi, 2006) given any initial data of the form (3.36), classes of dynamical evolutions that give rise to a well-defined and differentiable singularity curve always exist. By the freedom of choice of free functions, the evolution functions $\mathcal{M}(r, v)$ and $A(r, v)$ can be chosen in the following way,

$$\mathcal{M}(r, v) = m(r) - p_r(r)v^3, \;\; A(r, v)_{,v} = \nu_0(r\,v)_{,R}. \tag{3.66}$$

The Einstein equations here imply that, for the above class of evolutions, the radial pressure remains *static*. However, the tangential pressure blows up along with the density at the singularity $v = 0$ and is given by

$$2p_\theta(r, v) = p_r + p_r'(R/R') + \nu_0(rv)_{,R}[\rho(r, v) + p_r]. \tag{3.67}$$

It is now easy to check that the above class of evolutions admits a well-defined and differentiable singularity curve, as both the functions $\mathcal{M}(r,v)$ and $A(r,v)$ are well-defined and are C^2 at $r=0$ and $v=0$.

The point is that, as long as in the construction of classes of the solutions the functions $\mathcal{M}$ and A are taken to be sufficiently regular, then the singularity curve also becomes regular and expandable, at least up to the first order. If the singularity curve is not regular even to that extent, it means that it has no well-defined tangent at the origin, and so the collapse evolution is not regular enough and may not be of physical interest. In any case, the purpose here is not to analyze *all* possible collapse evolutions from a given initial distribution of matter and velocities. The aim really is to show that, given any such regular distribution from which the collapse develops, there exist classes of dynamical evolutions that take the other collapse to either the naked singularity or the other collapse endstates, depending on the choice of allowed functions. It is also shown that well-known classes of collapse models, such as dust, perfect fluids, and others, form special classes of the treatment given here.

Once a singularity curve that is at least C^2 exists, the function can be Taylor expanded near the center as

$$t_s(r) = t_{s_0} + r\mathcal{X}(0) + \mathcal{O}(r^2), \tag{3.68}$$

where t_{s_0} is the time at which the central singularity at $R=0, r=0$ develops, and is given as

$$t_{s_0} = \int_0^1 \frac{v^{1/2}dv}{\sqrt{vb_{00} + vh_0 + \mathcal{M}_0(v)}}. \tag{3.69}$$

From the above equation, it is clear that for t_{s_0} to be defined,

$$vb_{00} + vh_0 + \mathcal{M}_0(v) > 0. \tag{3.70}$$

In other words, a continual collapse in finite time ensures that the above condition holds. Also, from (3.51) and (3.53), for small values of r, along constant v surfaces,

$$v^{1/2}v' = \sqrt{(vb_{00} + vh_0 + \mathcal{M}_0(v))}\mathcal{X}(v) + \mathcal{O}(r). \tag{3.71}$$

It is now clear that the value of $\mathcal{X}(0)$ depends on the functions $b_0, \mathcal{M}$, and h, which in turn depend on the initial data at $t=t_i$, the dynamical variable v, and the evolution function $A(r,v)$. Therefore, a given set of initial matter distributions and the dynamical profiles, including the energy distribution of the shells, completely determine the tangent at the center to the singularity curve.

3.4 Nature of singularities

The occurrence of spacetime singularities in a general spacetime framework in collapse and cosmology is discussed in some detail in the next chapter. Singularities are the boundary points of the spacetime where the normal differentiability and manifold structures break down. These are the points where the energy density as given by the Einstein equation (3.28) above, or the curvature quantities, such as the scalar polynomials constructed out of the metric tensor and the Riemann tensor, diverge. One example of such a quantity is the Kretschmann scalar $\mathcal{K} = R^{ijkl}R_{ijkl}$, which is given in the dust collapse models by

$$\mathcal{K} = 12\frac{F'^2}{R^4R'^2} - 32\frac{FF'}{R^5R'} + 48\frac{F^2}{R^6}. \tag{3.72}$$

Such singularities are indicated by the existence of incomplete future or past directed non-spacelike geodesics in the spacetime that terminate at the singularity. Then, it is required that the curvature quantities stated above assume unboundedly large values in the limit of approach to the singularity along the non-spacelike geodesics terminating there, in the case of a genuine spacetime singularity. If such a condition is satisfied, then the singularity should be considered to be a physically significant curvature singularity.

It is now possible to examine, given the matter initial data at the initial surface $t = t_i$, how the final fate of collapse is determined in terms of either a blackhole or a naked singularity. If there are families of future directed non-spacelike trajectories reaching faraway observers in spacetime, which terminate in the past at the singularity, then a naked singularity forms as the collapse final state. In the other case when no such families exist and the event horizon forms sufficiently earlier than the singularity to cover it, a blackhole is formed. This is decided by the causal behavior of the trapped surfaces developing in the spacetime during the collapse evolution and the apparent horizon, which is the boundary of the trapped surface region in the spacetime.

In general, the equation of the apparent horizon in a spherically symmetric spacetime is given as

$$g^{ik}\, R_{,i}\, R_{,k} = 0. \tag{3.73}$$

Therefore, at the boundary of the trapped region the vector $R_{,i}$ is null. Substituting (3.19) in (3.73),

$$R'^2\, e^{-2\psi} - \dot{R}^2\, e^{-2\nu} = 0. \tag{3.74}$$

Using (3.31), the equation of apparent horizon can be written as

$$\frac{F}{R^{N-3}} = 1 \tag{3.75}$$

in a general N-dimensional spacetime, which gives the boundary of the trapped surface region of the spacetime. For the usual spacetime with $N = 4$, this becomes $F = R$. If the neighborhood of the center gets trapped prior to the epoch of singularity, then it is covered and a blackhole results, otherwise it could be naked when non-spacelike future directed trajectories escape from it.

Therefore, the important point is to determine if there are any future directed non-spacelike paths emerging from the singularity. To investigate this, and to examine the nature of the central singularity at $R = 0, r = 0$, consider the equation for the outgoing radial null geodesics that is given by

$$\frac{dt}{dr} = e^{\psi-\nu}. \tag{3.76}$$

It would be desirable to examine if there are any families of future directed null geodesics emerging from the singularity, thus causing a naked singularity phase as the collapse endstate. The singularity occurs at $v(t_{\rm s}(r), r) = 0$, that is, at $R(t_{\rm s}(r), r) = 0$. Therefore, if there are any future directed null geodesics terminating in the past at the singularity, $R \to 0$ as $t \to t_{\rm s}$ along these curves. Now, writing (3.76) in terms of the variables ($u = r^\alpha$, R),

$$\frac{dR}{du} = \frac{1}{\alpha} r^{-(\alpha-1)} R' \left[1 + \frac{\dot{R}}{R'} e^{\psi-\nu}\right], \tag{3.77}$$

where α is a positive constant to be fixed later. In order to obtain the expression of the tangent to the null geodesics emerging in the (R, u) plane, a particular value of α is chosen such that the geodesic equation is expressed only in terms of known limits. For example, if $\mathcal{X}(0) \neq 0$, and the functions $\mathcal{M}$ and h are well-defined for $0 \leq r \leq r_{\rm b}$ and $0 \leq v \leq 1$, $\alpha = 5/3$ is chosen. Using (3.31) and considering $\dot{R} < 0$, the null geodesic equation can be obtained in the form

$$\frac{dR}{du} = \frac{3}{5}\left(\frac{R}{u} + \frac{v' v^{1/2}}{\left(\frac{R}{u}\right)^{1/2}}\right)\left(\frac{1 - \frac{F}{R}}{\sqrt{G}\left[\sqrt{G} + \sqrt{H}\right]}\right). \tag{3.78}$$

If the future directed outgoing null geodesics do terminate at the singularity in the past with a definite tangent, then at the singularity, the tangent to the geodesics have $dR/du > 0$ in the (u, R) plane, and they must have a finite value. In the case of a massive singularity in dimensions greater than or equal to four, that is, when $F(t_s(r), r) > 0$ for $r \neq 0$, all singularities for $r > 0$ are necessarily covered since $F/R \to \infty$, and hence $dR/du \to -\infty$. This is when both the pressures p_r and p_θ are positive with the energy conditions are satisfied. Therefore, in such a case only the central singularity at $R = 0, r = 0$ could be naked.

Hence, the central singularity at $r = 0, R = 0$ needs to be examined to determine if it is visible or not, and to investigate if any solutions to the outgoing null geodesics equation exist that terminate in the past at the singularity, and that go to faraway observers in the future. The conditions under which this can happen are to be determined. Also, note that since the singularity curve and the evolution functions are regular, the limit of the functions H, G, and F/R at $r \to 0, t \to t_{s_0}$ can be calculated. From (3.46), as $A(r, v)$ is a well-defined function, $G(t_{s_0}, 0) = 1$. Also, from (3.51) at this point, $H \approx r^2/v$. Calculating this limit on the $t = t_{s_0}$ plane from (3.71), at the point $(t_{s_0}, 0)$, $H = 0$. Hence, from (3.31), $F/R = 0$ in this limit.

Let x_0 now be the tangent to the outgoing null geodesics in the (R, u) plane at the central singularity

$$x_0 = \lim_{t \to t_s} \lim_{r \to 0} \frac{R}{u} = \left. \frac{dR}{du} \right|_{t \to t_s; r \to 0}. \tag{3.79}$$

To find out whether the null geodesic equation admits any solution of x_0 that is positive and finite at the central singularity, the values of H, G, and F/R at $(t_{s_0}, 0)$ can be used in (3.78). Also, (3.71) can be used to get the value of $v'v^{1/2}$ on the $v = 0$ surface at $r = 0$ (that is, on the point $(t_{s_0}, 0)$). Therefore, solving (3.78),

$$x_0^{3/2} = \frac{3}{2}\sqrt{\mathcal{M}_0(0)}\mathcal{X}(0), \tag{3.80}$$

and the equation of the radial null geodesic emerging from the singularity is given by $R = x_0 u$ in the (R, u) plane, or in (t, r) coordinates it is given by

$$t - t_s(0) = x_0 r^{5/3}. \tag{3.81}$$

It now follows that if $\mathcal{X}(0) > 0$, then $x_0 > 0$, and a radially outgoing null geodesic emerges from the singularity, giving rise to a naked central singularity. However, if $\mathcal{X}(0) < 0$, then a blackhole solution exists, as there will be no such trajectories (see Fig. 3.2). If $\mathcal{X}(0) = 0$, then the next higher order non-zero term in the singularity curve equation will have to be taken into account, and a similar analysis has to be carried out by choosing a different value of α.

It can be seen that the above is both a necessary and sufficient condition for an outgoing radial null geodesic emerging from the singularity to exist.

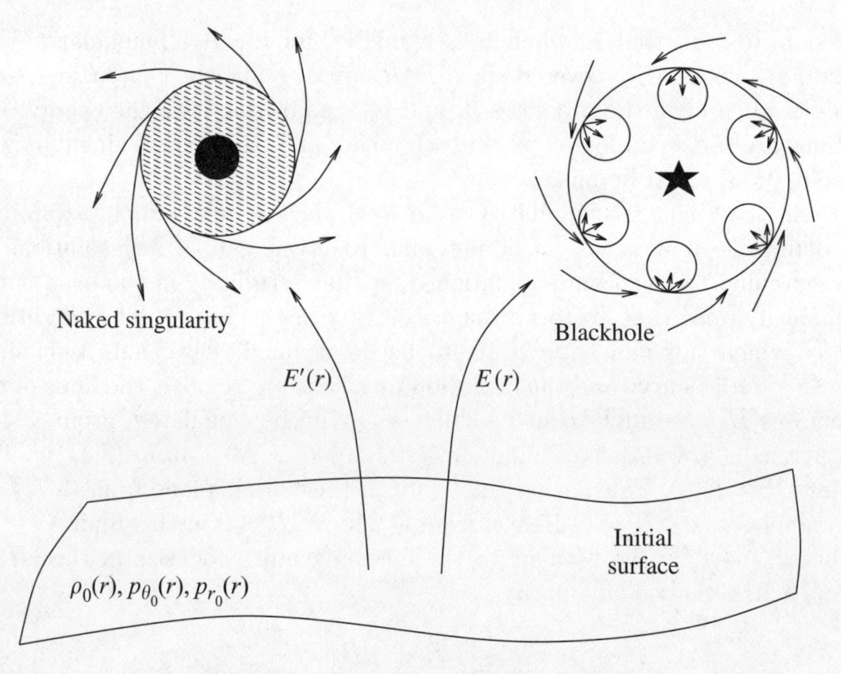

Fig. 3.2 Given the regular matter initial data, the collapse can evolve either to a blackhole or a naked singularity final state, depending on the choice of the rest of the free functions, such as the velocities $E(r)$ of the collapsing shells and the allowed dynamical evolutions as given by the Einstein equations.

Assume that such a geodesic does exist and that in the (R, u) plane it is given by the equation $R = x_0 u$, with $x_0 > 0$. Then, at the central singularity $(R = 0, u = 0)$, the tangent to such a geodesic must be x_0. Also, this tangent must be the root of the equation

$$\frac{dR}{du} - \frac{3}{5}\left(\frac{R}{u} + \frac{v' v^{1/2}}{(R/u)^{1/2}}\right)\left(\frac{1 - F/R}{\sqrt{G}\left[\sqrt{G} + \sqrt{H}\right]}\right) = 0 \tag{3.82}$$

at the point $(R = 0, u = 0)$. This is possible if and only if

$$x_0 = \left[\frac{3}{2}\sqrt{\mathcal{M}_0(0)}\mathcal{X}(0)\right]^{3/2} , \tag{3.83}$$

and for the slope to be defined and positive, $\mathcal{X}(0) \geq 0$.

Next, to see that $\mathcal{X}(0) > 0$ is a sufficient condition for the existence of an outgoing radial null geodesic emerging from the singularity, consider the case that the singularity curve has a positive tangent at the central singularity.

Consider then the curve

$$t - t_s(0) = \left[\frac{3}{2}\sqrt{\mathcal{M}_0(0)}\mathcal{X}(0)\right]^{3/2} r^{5/3}. \tag{3.84}$$

Along this curve $t \to t_{s_0}$ as $r \to 0$. And, as $\mathcal{X}(0) > 0$, this curve is *outgoing* in the sense that t increases as r increases along the curve. The quantity $(-g_{tt}\, dt^2 + g_{rr}\, dr^2)$ along this curve can be calculated in the vicinity of the central singularity. Using (3.71), (3.51), and (3.46), for this curve,

$$\begin{aligned} &- e^{2\nu(t_{s_0},0)} dt^2 + e^{-2rA(t_{s_0},0)} R'(t_{s_0},0)^2 dr^2 \\ &= \frac{5}{3}\left[\frac{3}{2}\sqrt{\mathcal{M}_0(0)}\mathcal{X}(0)\right]^3 \left(r^{2/3} - dr^2 + dr^2\right) = 0. \end{aligned} \tag{3.85}$$

That is, in the vicinity of the central singularity the curve considered above is null. Therefore, for any given positive value of the tangent to the singularity curve at the central singularity, it is always possible to find a *null* and *outgoing* curve terminating in the past at the central singularity, making the singularity naked.

Note that basically it is the geometry of the trapped surfaces and the apparent horizon that decides the visibility, or otherwise, of the spacetime singularity. Different kinds of collapse evolutions lead to different trapped surface configurations, thus leading to visibility, or otherwise, of the final singularity. For example, while in a homogeneous dust collapse, the trapped surfaces and apparent horizon form early enough to cover the singularity; when inhomogeneities are included, the trapping is naturally delayed so as to allow the singularity to be visible.

Hence, some remarks on the nature of the apparent horizon and its relation to the visibility, or otherwise, of the singularity are made below. To find the equation of the apparent horizon near the central singularity, let the time corresponding to a shell labeled by r entering the apparent horizon, in terms of the variable v, be $v_{\rm ah}(r)$. Then, from (3.75), it can easily be seen that $v_{\rm ah}(r)$ is the root of the equation

$$r^2\mathcal{M}(r,v) - v = 0. \tag{3.86}$$

Now, using (3.52), the equation for the apparent horizon in the $(t,\, r)$ plane can be written as

$$t_{\rm ah}(r) = t_s(r) - \int_0^{v_{\rm ah}(r)} \frac{v^{1/2}\, dv}{\sqrt{e^{(rA+\nu)} v b_0 + e^{2\nu}\,(vh + \mathcal{M})}}. \tag{3.87}$$

It is obvious that the necessary condition for the existence of a locally naked singularity is that the apparent horizon curve must be an *increasing* function at the central singularity, in the lowest power of r. If the apparent horizon curve is decreasing at the singularity, the collapse outcome is necessarily a blackhole.

In the above, the functions h and $\mathcal{M}$ are expanded with respect to r around $r = 0$ and the first order terms are considered. At times, however, these are assumed to be expandable with respect to r^2, and it is argued that such smooth functions would be physically more relevant. Such an assumption comes from the analyticity with respect to the local Minkowskian coordinates (see for example, the discussion in Goswami and Joshi, 2004a, 2004b), and it is really the freedom of definition mathematically. The formalism, as discussed above, would also work for such smooth functions, which is a special case of the above discussion.

It can therefore be seen how the initial data, in terms of the free functions available, determine the blackhole and naked singularity phases as the final outcome for the collapse. This is because the quantity $\mathcal{X}(0)$ is determined by these initial and dynamical profiles, as given by (3.54). It is clear, therefore, that given any regular initial density and pressure profiles for the matter cloud from which the collapse develops, velocity profiles can always be chosen so that the endstate of the collapse would be either a naked singularity or a blackhole, and vice versa.

Numerical work on collapse models may provide further insights into these interesting dynamical phenomena, especially when the collapse is non-spherical, which remains a major open problem to be considered. Numerical and some analytical work has been carried out in recent years on spherical scalar field collapse, and also on some perfect fluid models. While the emergence of the null geodesics from the singularity, thus showing it to be naked (or otherwise), has been worked out explicitly here in an analytic manner, the numerical simulations generally discuss the formation, or otherwise, of the trapped surfaces and apparent horizon. Such considerations may possibly break down closer to the epoch of the actual singularity formation. In this case, the actual detection of the blackhole or naked singularity endstates may not be allowed for, whereas important insights into critical phenomena and dispersal have been gained through numerical methods (Choptuik, 1993). Probably, a detailed numerical investigation of the structure of null geodesics in collapse models may provide further insights here.

Note that, in the above consideration, the occurrence of a *locally naked singularity* only, as opposed to that of a *globally naked singularity*, has been deduced. That is, when and in what circumstances the null geodesics escape from the spacetime singularity going out in the future have been shown, but the question of when these actually go out of the boundary of the matter cloud has not been addressed. It is possible, in principle, that the singularity is only locally visible and trajectories do come out, but then they all fall back into the singularity again at a later time, without going out of the boundary

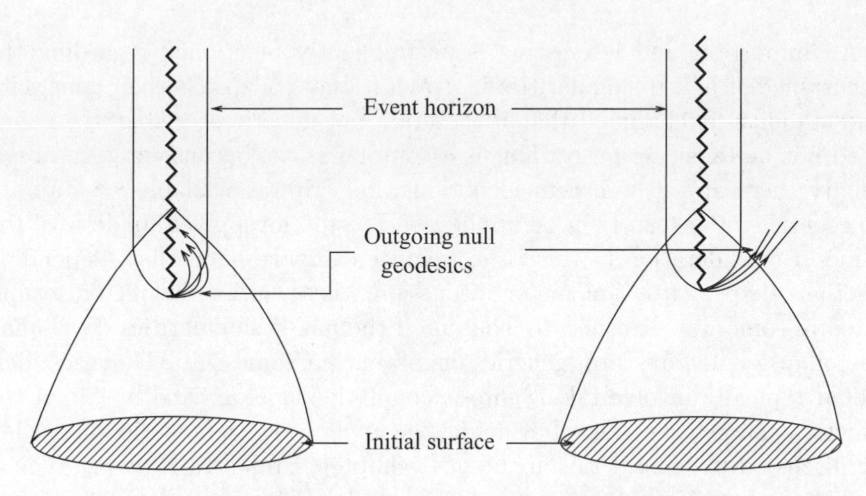

Fig. 3.3 Local versus global visibility. If the singularity is only locally visible, the light rays come out, but they then fall back again at the center without coming out of the cloud. On the other hand, for a globally visible singularity, the outgoing rays reach the boundary of the cloud and can reach faraway external observers.

of the star, thus not allowing the ultra-dense regions to be visible to faraway external observers (see Fig. 3.3).

This issue has still not been studied for the general class of models such as the ones considered here that may be of interest. However, for the case of dust collapse models, this has been studied in some detail (Joshi and Dwivedi, 1993a), and it is shown that, whenever the singularity is locally naked, one can always choose the classes of the mass and energy functions suitably, as one moves away from the center for the larger values of the radial coordinate r, in such a manner that the singularity becomes globally visible. The point is, while the local visibility of the central singularity is basically decided by the conditions near the center, the global visibility depends on the overall behavior of these functions within the matter cloud, away from the center. This behavior can still be freely chosen at larger values of r. In other words, for the dust collapse models, once the singularity is locally visible, there are always classes of mass and energy functions which can be chosen in order to make it globally visible. Another important related point is, as such, there is no scale in the problem, and the size of the collapsing cloud could be quite large. In such a case, even if the singularity is only locally visible, it can still be seen for a long enough time by the observers. Therefore, in principle, a locally naked singularity is also as serious a violation of the cosmic censorship as a globally visible singularity, and there may not be a qualitative difference in the two cases in many situations of physical interest (Penrose, 1979).

An important and interesting issue frequently mentioned regarding the occurrence of naked singularities in gravitational collapse is their genericity and stability. It is argued that if these are not generic or stable, then they need not be taken seriously. This is a complex issue, because in general relativity there are no well-defined notions and criteria available for stability or genericity that can then be applied and tested for a given model. All the same, a consideration of this issue is indeed important in that, depending on the collapse situation under discussion, these notions could be formulated in some way or other, to examine if the naked singularities developing as collapse endstates are 'generic' or 'stable' in some suitable sense. This would typically involve taking into account the topology and metric of the function spaces concerned that define the given collapse scenario. This shall be discussed in some detail in the next chapter. Presently, however, it may be noted that, since the spherical collapse has been formulated here in a general manner for general physically reasonable matter fields, the classes constructed may provide a good arena to explore and test these important issues for the cosmic censorship hypothesis.

A related issue is that of non-spherical collapse. It can be asked if the conclusions available for a spherically symmetric collapse remain the same and stable under possible non-spherical perturbations. Even though some non-spherical collapse models have been discussed and investigated, such as the Szekeres quasi-spherical collapse or some cylindrical collapse models, these may be regarded as somewhat restrictive in nature. It is not clear as yet if this issue can be approached in an analytic manner, and detailed numerical simulations of the collapsing stars could possibly suggest an answer.

Another question that is frequently asked in connection with the occurrence of naked singularities as collapse endstates is how to understand this phenomena physically. A naked singularity signifies the escape of light and particle trajectories from the ultra-dense spacetime regions. However, gravity must become very strong in these regions. In such a case, how can anything escape at all from such a region? Therefore, while a blackhole, which is a region from which not even light would escape, may appear to be the only physically reasonable outcome in such situations, the formation of a naked singularity in the collapse may appear to be counter-intuitive. The point that comes out from such considerations is that the naked singularities are more an artifact of general relativity, rather than that of purely Newtonian physics. Even though the matter density grows higher and higher without bound and blows up closer to a spacetime singularity, which would denote the growth of attractive forces of gravity, there are other important purely general relativistic effects that can delay the formation of trapped surfaces and apparent horizons, which govern the trapping of light.

An interesting effect that does this is the inhomogeneities and related spacetime shear. It is intriguing to find that the physical agencies, such as the spacetime shear and inhomogeneities in matter density distributions within a dynamically collapsing cloud, could naturally delay the formation of trapped surfaces during a gravitational collapse. An explicit example of this will be discussed in Chapter 5. In other words, such physical factors do naturally give rise to naked singularity phases in a collapse, where the formation of the apparent horizon and the trapped surfaces is delayed. Even though the matter densities are arbitrarily large and growing, the shearing effects could distort the trapped surface geometry in such a manner so as to avoid the trapping of light, and facilitate the escape of null rays from such ultra-dense regions. It is thus pointed out how the blackhole and naked singularity endstates arise naturally in a spherical collapse, as governed by the geometry of the trapped surfaces.

3.5 Exterior geometry

To complete the full spacetime model, the interior spacetime of the dynamical collapse needs to be matched to a suitable exterior geometry. As modeling the collapse of astrophysical objects (such as massive stars) is of interest here, the collapsing cloud is taken to have a compact support at the initial surface, with the boundary of the cloud being at some $r = r_b$. If the pressures at the boundary of the cloud vanish, then it is always possible to match the interior collapsing spacetime with an empty Schwarzschild exterior. However, in all cases in general, the pressures need not vanish at the boundary of the cloud. For example, for a dust cloud collapse, the pressure is zero by assumption, and the collapse can be matched at the boundary to an exterior Schwarzschild solution. But, for a perfect fluid with an equation of state $p = k\rho$, the pressure does not have to vanish at the boundary of the cloud.

Hence in such cases, the general practice is that there would be a boundary layer, to which the internal geometry is matched, and which could in turn be matched to an exterior Schwarzschild geometry. This is also appropriate from an astrophysical perspective, because the exteriors of any realistic stars would not be really completely empty and vacuum, but would always be surrounded by a radiation zone, as well as the matter emissions from the star. Such emissions may be particularly important in gravitational collapse situations, where the massive star may be emitting matter and radiation quite significantly as it collapses.

Hence, outlined below is the procedure to match the interior with a general class of exterior metrics, which is the generalized Vaidya spacetime (Joshi and Dwivedi, 1999; Wang and Wu, 1999) at the boundary hypersurface Σ given by $r = r_b$. For the required matching, the Israel–Darmois matching conditions (Israel, 1966a, 1966b) are used, where the first and second fundamental forms are matched. That is, the metric coefficients and the extrinsic curvature are matched at the boundary of the cloud.

Whereas the procedures used below are standard, the particular case treated here will be described in some detail so as to give the exact picture of the overall collapse scenario emerging. Note that since the matching is for the second fundamental form K_{ij}, there is no surface stress energy, or surface tension at the boundary (see for example, Mazur and Mottola, 2004). The metric just inside Σ is

$$ds_-^2 = -e^{2\nu}dt^2 + e^{2\psi(t,r)}dr^2 + R^2(t,r)d\Omega^2, \tag{3.88}$$

which describes the geometry of the collapsing cloud. The metric in the exterior of Σ is given by

$$ds_+^2 = -\left(1 - \frac{2M(r_\mathrm{v},V)}{r_\mathrm{v}}\right)dV^2 - 2dV\,dr_\mathrm{v} + r_\mathrm{v}^2\,d\Omega^2, \tag{3.89}$$

where V is the retarded outgoing null coordinate and r_v is the Vaidya radius. Matching the area radius at the boundary results in

$$R(r_\mathrm{b},t) = r_\mathrm{v}(V). \tag{3.90}$$

Then, on the hypersurface Σ, the interior and exterior metrics are given by

$$ds_{\Sigma-}^2 = -e^{2\nu(t,r_\mathrm{b})}dt^2 + R^2(t,r_\mathrm{b})^2 d\Omega^2 \tag{3.91}$$

and

$$ds_{\Sigma+}^2 = -\left(1 - \frac{2M(r_\mathrm{v},V)}{r_\mathrm{v}} + 2\frac{dr_\mathrm{v}}{dV}\right)dV^2 + r_\mathrm{v}^2\,d\Omega^2. \tag{3.92}$$

Matching the first fundamental form gives

$$\left(\frac{dV}{dt}\right)_\Sigma = \frac{e^{\nu(t,r_\mathrm{b})}}{\sqrt{1 - \frac{2M(r_\mathrm{v},V)}{r_\mathrm{v}} + 2\frac{dr_\mathrm{v}}{dV}}}, \quad (r_\mathrm{v})_\Sigma = R(t,r_\mathrm{b}). \tag{3.93}$$

Next, to match the second fundamental forms (extrinsic curvatures) for the interior and exterior metrics, note that the normal to the hypersurface Σ, as calculated from the interior metric, is given as

$$n_-^\mathrm{i} = \left[0, e^{-\psi(r_\mathrm{b},t)}, 0, 0\right], \tag{3.94}$$

and the non-vanishing components of the normal as derived from the generalized Vaidya spacetime are

$$n_+^V = -\frac{1}{\sqrt{1 - \dfrac{2M(r_\mathrm{v}, V)}{r_\mathrm{v}} + 2\dfrac{dr_\mathrm{v}}{dV}}}, \tag{3.95}$$

$$n_+^{r_\mathrm{v}} = \frac{1 - \dfrac{2M(r_\mathrm{v}, V)}{r_\mathrm{v}} + \dfrac{dr_\mathrm{v}}{dV}}{\sqrt{1 - \dfrac{2M(r_\mathrm{v}, V)}{r_\mathrm{v}} + 2\dfrac{dr_\mathrm{v}}{dV}}}. \tag{3.96}$$

Here, the extrinsic curvature is defined as

$$K_{ab} = \frac{1}{2}\mathcal{L}_n g_{ab}. \tag{3.97}$$

That is, the second fundamental form is the Lie derivative of the metric with respect to the normal vector $\boldsymbol{n}$. The above equation is equivalent to

$$K_{ab} = \frac{1}{2}\left[g_{ab,c}n^c + g_{cb}n^c_{,a} + g_{ac}n^c_{,b}\right]. \tag{3.98}$$

Now, setting $\left[K^-_{\theta\theta} - K^+_{\theta\theta}\right]_\Sigma = 0$ on the hypersurface Σ,

$$RR'e^{-\psi} = r_\mathrm{v}\frac{1 - \dfrac{2M(r_\mathrm{v}, V)}{r_\mathrm{v}} + \dfrac{dr_\mathrm{v}}{dV}}{\sqrt{1 - \dfrac{2M(r_\mathrm{v}, V)}{r_\mathrm{v}} + 2\dfrac{dr_\mathrm{v}}{dV}}}. \tag{3.99}$$

Simplifying the above equation using (3.93) and the Einstein equations,

$$F(t, r_\mathrm{b}) = 2M(r_\mathrm{v}, V). \tag{3.100}$$

Using the above equation and (3.93),

$$\left(\frac{dV}{dt}\right)_\Sigma = \frac{e^\nu(R'e^{-\psi} + \dot{R}e^{-\nu})}{1 - \frac{F(t, r_\mathrm{b})}{R(t, r_\mathrm{b})}}. \tag{3.101}$$

Finally, setting $[K^-_{\tau\tau} - K^+_{\tau\tau}]_\Sigma = 0$, where τ is the proper time on Σ,

$$M(r_\mathrm{v}, V)_{,r_\mathrm{v}} = \frac{F}{2R} + \frac{Re^{-\nu}}{\sqrt{G}}\sqrt{H}_{,t} + Re^{2\nu}\nu'e^{-\psi}. \tag{3.102}$$

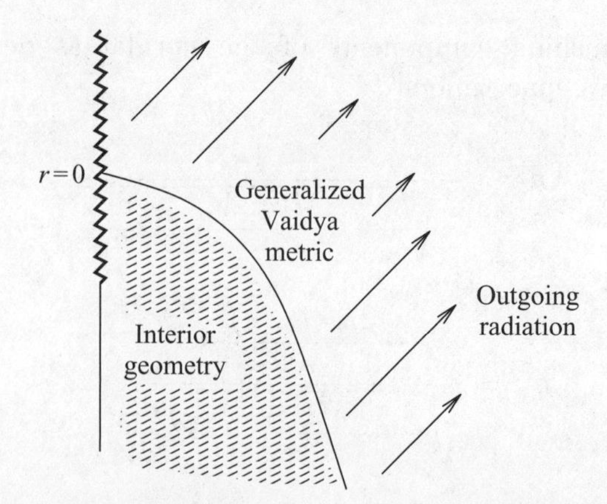

Fig. 3.4 For a general matter field, the interior collapsing metric is matched to an exterior that is a generalized Vaidya geometry.

Any generalized Vaidya mass function $M(v, r_{\mathrm{v}})$ that satisfies (3.102) will then give a unique exterior spacetime with required equations of motion given by other matching conditions, (3.100), (3.101), and (3.90).

To see that the set of all such functions $M(v, r_{\mathrm{v}})$ is non-empty, the examples of a charged Vaidya spacetime $M = M(V) + Q(V)/r_{\mathrm{v}}$, and the anisotropic de Sitter spacetime $M = M(r_{\mathrm{v}})$ are two different solutions of (3.102) (see for example, Joshi and Dwivedi, 1999; Giambo, 2005). This gives two unique exterior spacetimes, both of which are subclasses of the generalized Vaidya metric (see Fig. 3.4).

It is of course also possible to treat this problem from the perspective of only pure general relativity, and the theoretical cosmic censorship aspect. In this case, there is no need to cut off the cloud and match it at the boundary. Suitable fall off conditions can just be imposed for the collapsing matter, so that faraway the metric becomes Minkowskian at the spatial infinity (see for example, Choptuik, 1993, for the case of numerical scalar field collapse models).

3.6 Dust collapse

The dust collapse models have been analyzed extensively towards understanding the final fate of gravitational collapse in Einstein theory, and these can be used to illustrate the ideas above. This class of models have played a very important and fundamental role in gravitation theory because it is, in fact, at the very heart of the blackhole paradigm. As pointed out earlier, after the investigation of Oppenheimer and Snyder (1939), which showed that a blackhole developed as the final state of a continual gravitational collapse of a homogeneous dust cloud, the cosmic censorship hypothesis was introduced by Penrose in 1969 to suggest that this is the generic outcome for a collapse in general relativity for any massive star. This conjecture paved the way for many further developments in blackhole physics. Based on the assumption of censorship, namely that just as in the case of homogeneous dust clouds, all stars that are gravitationally collapsing must end up necessarily as blackholes only, much work was carried out on blackhole theory and its astrophysical applications, while any rigorous formulation and proof of censorship hypothesis is still awaited.

Though any mathematical formulation or proof of censorship is not available as yet, further progress on the better understanding of a dust collapse came only later with work such as the numerical study of Eardley and Smarr (1979), and the analytic treatment of Christodoulou (1984) and Newman (1986). A generic analysis of the inhomogeneous dust collapse with general classes of initial mass and velocity profiles was given by Joshi and Dwivedi (1993a). Gravitational waves in these models were analyzed by Iguchi, Nakao, and Harada (1998).

In the past couple of decades, there have been many investigations on this important class of collapse models, as is discussed below, which has an interesting mathematical structure as well as a definite physical significance. The main reason why such a definite progress for these collapse models could be made was that the full solution of the Einstein equations for the dust class of spacetimes was already given much earlier by Lemaître (1933) and Tolman (1934), and was also discussed by Bondi (1948), which included the homogeneous dust collapse as a special case. These solutions are known as the Tolman–Bondi–Lemaître (TBL) dust collapse models.

As a result of this analysis, it has become clear now that, as soon as one departs from the assumption of the exact homogeneity of the density distribution in space for the collapsing dust cloud, the final collapse outcome is no longer necessarily a blackhole. However, the visible ultra-dense regions or naked singularities could also then develop generically as collapse endstates. The actual outcome in terms of either a blackhole or naked singularity is decided by the nature of the regular initial data, in terms of the initial density and velocity profiles, from which the collapse evolves (see Fig. 3.5, which shows the marginally bound case).

Here, the occurrence and nature of naked singularities for the inhomogeneous gravitational collapse of TBL dust clouds is investigated as a special case of the discussion so far. It is shown that visible ultra-dense regions form at the center of the collapsing cloud in a wide class of these models. The earlier cases considered by Eardley and Smarr (1979), Christodoulou (1984), and Newman (1986) are included as special cases in the general treatment given here. This class also contains self-similar, as well as non-self-similar models.

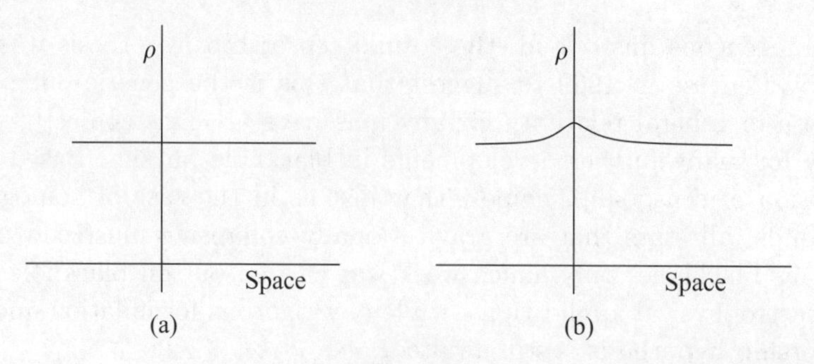

Fig. 3.5 The nature of the initial density profile affects the final outcome of collapse. For a constant density (a) a blackhole develops, but for a profile slightly higher at the center of the cloud (b) the final state is a naked singularity.

It is possible to criticise the dust collapse models in that they incorporate no pressure, or that they are spherically symmetric. While generalizations are discussed above and also later, the important point is that this is the model that has actually inspired the entire blackhole paradigm, as originating from the Oppenheimer–Snyder work, and hence it deserves a serious and complete study. As pointed out, it also answers several important questions regarding the possible endstates of a continual gravitational collapse, and helps to rule out several possible or plausible formulations of cosmic censorship hypothesis that have been tried out earlier. Furthermore, there are also arguments to suggest that dust could possibly be the physically reasonable and realistic equation of state in the very final stages of gravitational collapse (Hagerdorn, 1968; Penrose, 1974a), and that at higher and higher densities the matter may really behave more like the dust equation of state. Again, there is some case for the argument that, eventually in the final stages of collapse, the matter distribution should become almost spherically symmetric (see for example, Nakamura and Sato, 1982). All these factors make a detailed investigation of dust collapse quite interesting and important. Hence, it is clearly useful to examine the inhomogeneous dust collapse as modeled by the TBL spacetimes. Furthermore, a situation analogous to the singularity theorems might develop here where the conclusions derived under the assumption of spherical symmetry are preserved, when small perturbations are taken into account. Thus, spherical symmetry may be a good ansatz to represent certain important classes of gravitational collapse scenarios.

3.6.1 Tolman–Bondi–Lemaître (TBL) spacetimes

The TBL metric represents a general solution to the Einstein equations with a spherically symmetric dust form of matter. In particular, it gives the geometry for the inhomogeneous dust cloud collapse, and in the comoving coordinates, that is, with $u^i = \delta^i_t$, the geometry is

$$ds^2 = -dt^2 + \frac{R'^2}{1+f}\,dr^2 + R^2(d\theta^2 + \sin^2\theta\, d\phi^2). \tag{3.103}$$

The energy–momentum tensor is that of pressureless dust, as given by

$$T^{ij} = \epsilon \delta^i_t \delta^j_t. \tag{3.104}$$

The Einstein equations then give

$$\epsilon = \epsilon(t, r) = \frac{F'}{R^2 R'}, \tag{3.105}$$

where ϵ is the energy density, R is the area radius of the cloud, which is a function of both t and r, and

$$\dot{R}^2 = \frac{F}{R} + f. \tag{3.106}$$

Here, the dot and the prime denote partial derivatives with respect to the parameters t and r respectively.

As the continual gravitational collapse situation is of concern here, it is required that $\dot{R}(t,r) < 0$. The quantities F and f are arbitrary functions of r, as obtained from the integration of the Einstein equations, and, as seen from the above equations, these have the interpretation of the mass and velocity functions respectively. The quantity $4\pi R^2(t,r)$ gives the proper area of the mass shells and the area of such a shell at $r =$ const. goes to zero when $R(t,r) = 0$. Integration of (3.106) gives

$$t - t_0(r) = -\frac{R^{3/2} G(-fR/F)}{\sqrt{F}}, \tag{3.107}$$

where $G(y)$ is a strictly real positive and bounded function that has the range $1 \geq y \geq -\infty$, and is given by

$$G(y) = \left(\frac{\sin^{-1}\sqrt{y}}{y^{3/2}} - \frac{\sqrt{1-y}}{y}\right) \quad \text{for} \quad 1 \geq y > 0, \tag{3.108}$$

$$G(y) = \frac{2}{3} \quad \text{for} \quad y = 0, \tag{3.109}$$

$$G(y) = \left(\frac{-\sinh^{-1}\sqrt{-y}}{(-y)^{3/2}} - \frac{\sqrt{1-y}}{y}\right) \quad \text{for} \quad 0 > y \geq -\infty. \tag{3.110}$$

The function $t_0(r)$ is a constant of integration. So, there are in all three arbitrary functions of r, as given by $f(r)$, $F(r)$, and $t_0(r)$. However, the remaining coordinate freedom left could be used in the choice of scaling r in order to reduce the number of such arbitrary functions to two. Therefore, the area radius R is rescaled using this coordinate freedom, such that

$$R(0,r) = r. \tag{3.111}$$

Then, $t_0(r)$ is evaluated by using (3.111) and (3.107) to give

$$t_0(r) = \frac{r^{3/2} G(-fr/F)}{\sqrt{F}}. \tag{3.112}$$

The time $t = t_0(r)$ corresponds here to the value $R = 0$ where the area of the shell of matter at a constant value of the coordinate r vanishes. It follows that the singularity curve $t = t_0(r)$ corresponds to the time when the matter shells meet the real physical singularity, which corresponds to the entire cloud shrinking to a zero radius. Therefore, the range of the coordinates is given by

$$0 \le r < \infty, \qquad -\infty < t < t_0(r). \tag{3.113}$$

The case considered by Oppenheimer and Snyder (1939), which discussed the homogeneous dust cloud collapse, is a special subcase of the above general dust solution, corresponding to $\epsilon = \epsilon(t)$ only throughout, and with a special value of the velocity function as given by $f(r) = 0$. The above comoving coordinates, tailored to the collapsing matter, which have a somewhat special physical significance, have been used here, and these provide a physical insight into the collapse processes. In the case of a homogeneous density, the interior of the cloud has a special, more simple form, which is the same as the time reversed case of the Friedman solution in cosmology.

It is seen, in general, that when the matter density is allowed to be inhomogeneous and can change as a function of the coordinate radius r, then, unlike the collapsing Friedmann case (where the physical singularity $R = 0$ occurs at a constant epoch of time, say, at $t = t_0$), the singular epoch now is a function of r, as a result of the inhomogeneity in the matter distribution. The Friedmann case could be recovered from the above equations by setting, for example, $t_0(r) = t_0'(r) = 0$. This corresponds to the case of a simultaneous singularity. It follows that, for an inhomogeneous matter distribution, different shells of matter arrive at the singularity at different times as given by $t_{\rm s}(r)$, and a singularity curve rather than a constant singular epoch is obtained, as was the case for a constant density distribution.

The function $f(r)$ above can be used to classify the spacetime as *bound*, *marginally bound*, or *unbound*, depending on the range of its values, which are

$$f(r) < 0, \quad f(r) = 0, \quad f(r) > 0 \tag{3.114}$$

respectively. The function $F(r)$ is interpreted as the weighted mass, as weighted by the factor $\sqrt{1+f}$, within the dust ball B of coordinate radius r that is conserved in the sense

$$m(r) = \frac{F(r)}{2} = \int_B (1+f)^{1/2}\epsilon(t,r)dv = 4\pi \int_0^r \rho(r)r^2 dr, \tag{3.115}$$

where $\epsilon(0,r) = \rho(r)$ is the initial density distribution, from which the gravitational collapse of the cloud develops. For physical reasonableness, the weak energy condition would be assumed throughout the spacetime, that is, $T_{ij}V^iV^j \geq 0$ for all non-spacelike vectors V^i. This implies that the energy density ϵ is positive everywhere ($\epsilon \geq 0$), including the region near the center of the cloud $r = 0$.

The partial derivatives of R, as given by R' and $\dot{R}'$, play an important role in the analysis, and, from the above equations,

$$R' = r^{\alpha-1}\left((\eta-\beta)X + \left(\Theta - \left(\eta - \frac{3}{2}\beta\right)X^{3/2}G(-PX)\right)\left(P + \frac{1}{X}\right)^{1/2}\right)$$
$$\equiv r^{\alpha-1}H(X,r), \tag{3.116}$$

$$\dot{R}' = \frac{\Lambda^{1/2}}{2rX^2}\left(-\beta X^2\left(\frac{1}{X}+P\right)^{1/2} + \Theta - \left(\eta - \frac{3}{2}\beta\right)X^{3/2}G(-PX)\right)$$
$$\equiv \frac{-N(X,r)}{r}, \tag{3.117}$$

where, using the notation given by Joshi and Dwivedi (1993a),

$$X = (R/r^\alpha), \quad \eta = \eta(r) = r\frac{F'}{F}, \quad \beta = \beta(r) = r\frac{f'}{f}, \quad p = p(r) = rf/F, \tag{3.118}$$

$$P = pr^{\alpha-1}, \quad \Lambda = \frac{F}{r^\alpha}, \tag{3.119}$$

and

$$\Theta \equiv \frac{t_0'\sqrt{\Lambda}}{r^{\alpha-1}} = \frac{1+\beta-\eta}{(1+p)^{1/2}\,r^{3(\alpha-1)/2}} + \frac{(\eta-\frac{3}{2}\beta)G(-p)}{r^{3(\alpha-1)/2}}. \tag{3.120}$$

The function $\beta(r)$ is defined to be zero when f is constant and zero. The factor r^α has been introduced here for the sake of convenience in examining the structure of the naked singularity. The exact value of the positive constant $\alpha \geq 1$ is to be determined and chosen later, and it will depend on the different models of the spacetime that allow naked singularities. The functions $H(X,r)$ and $N(X,r)$ are defined by (3.117) and (3.118).

Using the scaling given by $R = r$ as used above, the energy density ϵ on the initial hypersurface $t = 0$ can be written as $\epsilon = F'/r^2$. Since the weak energy conditions are satisfied and the mass function F is a function of r only, it follows that $F' \geq 0$ throughout the spacetime. The energy density can be written as

$$\epsilon = \frac{\eta\Lambda}{R^2 H}. \tag{3.121}$$

Since $F' = \eta\Lambda r^{\alpha-1}$, it follows that $H(X, r) \geq 0$ everywhere and $\eta\Lambda \geq 0$ everywhere as a consequence of the weak energy condition.

In TBL spacetimes singularities occur, as can be seen from the Einstein equations, at the points where $R = 0$, which are the shell-focusing singularities, and also at the points where $R' = 0$. At the points where $R' = 0$, the TBL metric is degenerate and the points $R > 0, F' > 0$, where $R' = 0$, are called the shell-crossing singularities, where the nearby shells of matter cross momentarily (Hellaby and Lake, 1985; Newman, 1986). Such shell-crossings in TBL spacetimes have been analyzed in detail and their nature appears to be fairly well understood. Even though the shell-crossing singularities could be locally naked, the important point is that they have been shown to be gravitationally weak (Newman, 1986). Therefore, it is generally believed that shell-crossing singularities need not be taken seriously as far as the cosmic censorship conjecture is concerned. It was pointed out earlier that the absence of shell-crossings in the spacetime turns out to be related to the condition that the function $t_0(r)$, giving the proper time for the shells to fall into the physical singularity, should be a monotonically increasing function. That is, the shells with increasing r arrive one after the other at the singularity. The dust density and certain components of the curvature blow up near such a singularity. However, the causal structure of the spacetime can be extended through this singularity and the spacetime metric can be defined also in the neighborhood of such a point in a distributional sense (Papapetrou and Hamoui, 1967). In the context of such a situation, it is assumed that there are no shell-crossing singularities in the TBL spacetime, except probably right at the center of the cloud at $r = 0$. Instead, the more serious shell-focusing singularity at $R = 0$ where the entire cloud collapses to a zero physical radius is considered, and a physical spacetime singularity develops.

This does not involve any loss of generality, either for the TBL models, or the general models discussed earlier, as the basic purpose here is to examine the formation and local structure of the shell-focusing singularity at the center of the collapsing cloud. Whereas the existence of shell-crossings will not affect the qualitative nature of these general conclusions, the above assumption allows the calculations to be presented in a more transparent manner. Unlike the shell-crossings, the spacetime metric, however, admits no extension through a shell-focusing singularity occurring at $R = 0$, where all the matter shells reduce to a zero size, and therefore this is a genuine physical singularity. Hence, the occurrence of such shell-focusing singularities at the center of the collapsing dust cloud are investigated here, and their nature and structure for the TBL spacetimes are examined. It is clear that a shell-focusing singularity occurring at $r > 0, R = 0$ is totally spacelike and therefore the discussion would be confined to the singularity at $r = 0$.

The equation for the shell-focusing singularity $R(t_0, r_0) = 0$ is as given earlier, and occurs at $r = r_0$ at the coordinate time $t = t_0$. The singularity is called a *central singularity* if it occurs at $r = 0$. This central shell-focusing singularity can be naked, though gravitationally weak (Newman, 1986), for the class of TBL models considered by Christodoulou (1984), for which the energy density, which is assumed to be positive everywhere and taken to be non-zero at $r = 0$, and the metric functions are even and smooth functions of t and r. Translated in terms of the parameters defined above, this corresponds to the case for which $\eta(0) = 3, \beta(0) = 2$, and $p(r)$ is an even smooth function of r. In terms of the functions $F(r)$ and $f(r)$, this amounts to the conditions

$$F(r) = r^3 \mathcal{F}(r), \quad \infty > \mathcal{F}(0) > 0, \quad 0 < p(r) \leq 1. \tag{3.122}$$

It was, however, pointed out by Waugh and Lake (1988, 1989) and Ori and Piran (1987, 1990) that this class of gravitationally weak naked singularities excludes the self-similar TBL models, where the singularity was shown to be gravitationally strong along the Cauchy horizon, which is a null geodesic coming out of the singularity. It was further pointed out that the strong curvature naked singularity is not necessarily confined to the self-similar models only, but that there are many classes of TBL models that are non-self-similar, and the naked singularity occurring there is gravitationally strong (Dwivedi and Joshi, 1992). In these classes, the collapse terminates in a naked singularity that is gravitationally strong for a wide range of TBL models that are non-self-similar in general, and which include all the self-similar models as a special subclass. In the notation used here, these models are characterized by the conditions $\eta(0) = 1$, with $F(r)$ and $f(r)$ being analytic at $r = 0$.

Rather general differentiability conditions can be required only on the functions $F(r)$ and $f(r)$ in that they are assumed to be at least C^1 at the center $r = 0$, $\infty > \eta(0) > 0$, and $\beta(0)$ is finite. In order to ensure the metric to be C^2, which is the basic requirement in general relativity, these differentiability conditions imposed here on the functions f and F are sufficient. Frequently, much stronger conditions are imposed on these functions and the metric, requiring them to be smooth and analytic. However, these are considered here to be too strong in general. Anyway, the models with stronger conditions become special cases of the above. Furthermore, from the physical reasonableness, one would require $F(0) = 0$ (otherwise there will be a massive singularity present already at $r = 0$), which implies $\eta(0) > 0$.

The condition F being C^1 corresponds to the energy density ϵ not diverging at the center $r = 0$ at all times. Note that the function f, and also its first derivatives (through R') enter the metric potentials. It might actually be argued that the above are more general conditions than should actually be required, as it is customary to assume that the metric is C^2-differentiable (which again ensures the above requirements), so that the metric transformations and other functions connected with the metric are well-defined to carry out regular physics. Hence, the differentiability requirements here may be considered physically reasonable and rather general, and which include practically all the inhomogeneous collapse TBL dust models of interest. In fact, it could be argued that if the metric is not C^2-differentiable, but say only C^1 on the initial surface, it may be considered as being already singular and not defining a regular initial data on an initial spacelike hypersurface.

In order to represent the gravitational collapse scenario, the energy density ϵ is taken to have a compact support on an initial spacelike hypersurface. The TBL spacetime metric above can then be matched at some $r = \text{const.} = r_\text{c}$ to the exterior Schwarzschild field, and

$$ds^2 = -\left(1 - \frac{2M}{r_\text{s}}\right) dT^2 + \frac{dr_\text{s}^2}{1 - 2M/r_\text{s}} + r_\text{s}^2\, d\Omega^2, \tag{3.123}$$

where $d\Omega^2 = d\theta^2 + \sin^2\theta\, d\phi^2$. The value of the Schwarzschild radial coordinate is $r_\text{s} = R(t, r_\text{c})$ at the boundary $r = r_\text{c}$. Also, $m(r_\text{c}) = M$, where M is the total Schwarzschild mass enclosed within the dust ball of coordinate radius $r = r_\text{c}$.

The apparent horizon, which indicates the boundary of all trapped surfaces that define the region that allows no light to escape, forming in the spacetime during the gravitational collapse in the interior dust ball, lies at $R = F(r)$. The corresponding time $t = t_\text{ah}(r)$ for the apparent horizon is given by

$$t = t_\text{ah}(r) = \frac{r^{3/2} G(-p)}{\sqrt{F}} - FG(-f). \tag{3.124}$$

The emissions from the shell-focusing singularity $R(t_0, r_0) = 0$ for all $r_0 > 0$ would lie in the region above $t = t_\text{ah}$, that is, $t_0 > t_\text{ah}$ for all $r_0 > 0$, with t_0 being the time when the singularity at $r = r_0$ occurs. Hence, all radiations would be future trapped from the shell-focusing singularities at $r > 0$. At $r = 0$ however, $t(0) = t_\text{ah}(0)$ and the singularity could be at least locally visible. Any light ray terminating at this singularity in the past goes to the future infinity if it reaches the surface of the cloud $r = r_\text{c}$ earlier than the apparent horizon at $r = r_\text{c}$. In such a case the singularity would be globally naked. The nature of this central spacetime singularity has to be examined in terms of its visibility or otherwise.

As seen from the Einstein equations above in the dust case, once the mass function is specified at the initial epoch, the energy function $f(r)$ fully specifies the velocity distribution of the in-falling shells. Also, the energy condition then implies that $F' \geq 0$ and the collapsing matter cloud condition implies that $\dot{R} < 0$. For dust clouds it follows from the equations of motion that once $\dot{R} < 0$ at the initial epoch from where the collapse commences, then, at all epochs, the same condition holds, and thus there is a continual collapse without any reversal until the shell-focusing singularity at $R = 0$ is reached. In other words, there is no bounce possible in the dust collapse once the collapse has initiated with $\dot{R} < 0$.

The scaling independence of the comoving coordinate r can again be used to give

$$R(t, r) = rv(t, r), \tag{3.125}$$

where

$$v(t_i, r) = 1, \qquad v(t_s(r), r) = 0, \qquad \dot{v} < 0. \tag{3.126}$$

This means the coordinate r has been scaled in such a way that, at the initial epoch, $R = r$, and at the singularity, $R = 0$. Therefore, $R = 0$ both at the regular center $r = 0$ of the cloud, and at the spacetime singularity where all matter shells collapse to a zero physical radius. The regular center is then distinguished from the singularity by suitable behavior of the mass function $F(r)$, so that the density remains finite and regular there at all times until the singular epoch. The introduction of the parameter v then allows the spacetime singularity to be distinguished from the regular center, with $v = 1$ all through the initial epoch, including the center $r = 0$, and which then monotonically decreases with time as the collapse progresses to the value $v = 0$ at the singularity $R = 0$.

From the equations of motion, it is evident that to have a regular solution over all space at the initial epoch, the two free functions $F(r)$ and $f(r)$ must have the following forms

$$F(r) = r^3 \mathcal{M}(r), \quad f(r) = r^2 b(r), \tag{3.127}$$

where $\mathcal{M}(r)$ and $b(r)$ are at least C^1-functions of r for $r = 0$, and at least C^2-functions for $r > 0$. This is dictated by the condition that the density and energy distributions must be regular at the initial epoch and should not blow up. This is because if the mass function F did not go as a power of at least r^3 closer to the origin, then, as implied from the equations of motion, the density will be singular at the origin $r = 0$ as it will diverge there. This cannot be accepted as regular initial data for the collapse. Similarly, (3.107) implies that $f(r)$ is determined once the velocity profile is specified, and vice versa, for a given initial density distribution. Since the center of the cloud is taken to be at rest in any spherically symmetric distribution, the leading term in the energy profile must be at least r or higher. Then again, (3.107) implies the behavior for $f(r)$ as above.

3.6.2 Collapse endstates

The continual collapse of a dust cloud to a final shell-focusing singularity at $R = 0$, where all matter shells collapse to a zero physical radius, is now considered. In particular, the nature of the central singularity at $R = 0, r = 0$ is specifically analyzed in detail to determine when it will be covered by the event horizon, and when it is visible and causally connected to outside observers. If there are future directed families of non-spacelike curves coming out from the singularity and reaching faraway observers, then the singularity will be naked. The absence of such families will give a covered case when the result is a blackhole.

In the following, the collapse as discussed above is considered and it will be shown how the initial density and energy distributions, prescribed at the initial epoch from which the collapse commences, completely determine if there will be families of non-spacelike trajectories emerging from the singularity. It will be shown that for a generic situation, given an initial density distribution for the collapse to develop, it is always possible to choose an energy profile so that the collapse of this density profile ends in a naked singularity. Alternatively, another class of energy distribution could also be chosen so that the same density distribution will end up collapsing in order to create a blackhole. That is, given an initial density profile, the outcome in terms of either a blackhole or a naked singularity as endstates really depends on the class of the energy distributions chosen. The converse will also be seen to hold true, that is, given the initial energy function, classes of density profiles, subject to the weak energy condition can be chosen so as to give rise to either the blackhole or naked singularity endstates, depending on the choice made. These results generalize naturally to any higher-dimensional dust collapse models in general relativity.

With the regular initial conditions as above, (3.52) can be written as

$$v^{1/2}\dot{v} = -\sqrt{\mathcal{M}(r) + vb(r)}. \tag{3.128}$$

Here, the negative sign implies that $\dot{v} < 0$, that is, the matter cloud is collapsing. Integrating the above equation with respect to v,

$$t(v,r) = \int_v^1 \frac{v^{1/2}dv}{\sqrt{\mathcal{M}(r) + vb(r)}}. \tag{3.129}$$

Note that the coordinate r is to be treated as a constant in the above equation. Expanding $t(v, r)$ around the center,

$$t(v,r) = t(v,0) + r\mathcal{X}(v) + r^2\frac{\mathcal{X}_2(v)}{2} + r^3\frac{\mathcal{X}_3(v)}{6} + \cdots, \tag{3.130}$$

where the function $\mathcal{X}(v)$ is given by

$$\mathcal{X}(v) = -\frac{1}{2}\int_v^1 \frac{v^{1/2}(\mathcal{M}_1 + vb_1)dv}{(\mathcal{M}_0 + vb_0)^{3/2}}, \tag{3.131}$$

where

$$b_0 = b(0), \quad \mathcal{M}_0 = \mathcal{M}(0), \quad b_1 = b'(0), \quad \mathcal{M}_1 = \mathcal{M}'(0). \tag{3.132}$$

Therefore, the time taken for the central shell to reach the singularity is given by

$$t_{s_0} = \int_0^1 \frac{v^{1/2}dv}{\sqrt{\mathcal{M}_0 + vb_0}}. \tag{3.133}$$

From the above equation it is clear that for t_{s_0} to be defined,

$$\mathcal{M}_0 + vb_0 > 0. \tag{3.134}$$

In other words, the continual collapse condition implies the positivity of the above term. Hence, the time taken for other shells to reach the singularity can be given by the expansion

$$t_s(r) = t_{s_0} + r\mathcal{X}(0) + \mathcal{O}(r^2). \tag{3.135}$$

Also, from the equation for $\dot{v}$ above and (3.131), for small values of r along constant v surfaces,

$$v^{1/2}v' = \sqrt{(\mathcal{M}_0 + vb_0)}\left(\mathcal{X}(v) + r\mathcal{X}_2(v) + \cdots\right). \tag{3.136}$$

Now, it can easily be seen that the value of $\mathcal{X}(0)$ depends on, and is completely characterized by, the functions $M(r)$ and $b(r)$, which in turn specify fully the initial mass and energy distributions for the collapsing matter. Specifying these functions is equivalent to specifying the regular initial data for the collapse on the initial surface $t = 0$. In other words, a given set of density and energy distributions completely determines the slope to the singularity curve at the origin, which is the central singularity. Also, it is evident that, given any one of these two profiles, the other one can always be chosen in such a manner that the quantity $\mathcal{X}(0)$ will be either positive or negative.

In order to determine the visibility, or otherwise, of the central singularity, the behavior of the non-spacelike curves in the vicinity of the singularity and the causal structure of the trapped surfaces now need to be analyzed.

The boundary of the trapped surface region of the spacetime is given by the apparent horizon within the collapsing cloud, which is given by the equation

$$\frac{F}{R} = 1. \tag{3.137}$$

One now needs to determine when there will be families of non-spacelike paths coming out of the singularity, reaching faraway observers, and when there will be none. The visibility, or otherwise, of the singularity is decided accordingly. Broadly, it can be stated that if the neighborhood of the center gets trapped earlier than the singularity, then it is covered, otherwise it is naked with families of non-spacelike future directed trajectories escaping away from it. By determining the nature of the singularity curve and its relation to the initial data, it is possible to deduce whether the trapped surface formation in the collapse takes place before or after the singularity. It is this causal structure that determines the possible emergence, or otherwise, of non-spacelike paths from the singularity, and settles the final outcome in terms of either a blackhole or a naked singularity.

To consider the possibility of the existence of such families, and to examine the nature of the singularity occurring at $R = 0$, $r = 0$ for the scenario under consideration, consider the outgoing radial null geodesics equation

$$\frac{dt}{dr} = e^{\psi}. \tag{3.138}$$

The singularity occurs at a point $v(t_{\rm s}(r), r) = 0$, which corresponds to $R(t_{\rm s}(r), r) = 0$. Therefore, if there are any future directed null geodesics terminating in the past at the singularity, then $R \to 0$ as $t \to t_{\rm s}$. Now, writing the null geodesics equation in terms of the variables $(u = r^{\alpha}, R)$ where $\alpha > 1$,

$$\frac{dR}{du} = \frac{1}{\alpha} r^{-(\alpha-1)} R' \left[1 + \frac{\dot{R}}{R'} e^{\psi}\right]. \tag{3.139}$$

Choosing $\alpha = 5/3$, and using (3.129) together with the collapse condition $\dot{R} < 0$,

$$\frac{dR}{du} = \frac{3}{5}\left(\frac{R}{u} + \frac{v' v^{1/2}}{\left(\frac{R}{u}\right)^{1/2}}\right)\left(\frac{1 - \frac{F}{R}}{e^{-\psi} R' \left(e^{-\psi} R' + |\dot{R}|\right)}\right). \tag{3.140}$$

If the null geodesics terminate at the singularity in the past with a definite tangent, then at the singularity $dR/du > 0$ in the (u, R) plane, and this must have a finite value.

In the case under consideration, all singularities for $r > 0$ are covered since $F/R \to \infty$ in the limit of the approach to the singularity, and hence $dR/du \to -\infty$. Therefore only the singularity at the central shell could be naked.

In order to see the possible emergence of null geodesics from the central singularity, (3.141) needs to be analyzed. The limits of the concerned functions in (3.141) are calculated at the central singularity. In the TBL case, in the limit of $t \to t_s, r \to 0$, $e^{-\psi}R' \to 1$. Also, from the equation for $\dot{v}$ above and (3.137), in this limit $\dot{R} \to 0$. It then follows in general, from the Einstein equations discussed above, that the term F/R goes to zero in this limit.

It would be interesting to find out when there will be future directed null geodesics emerging from the central singularity with a well-defined and definite positive tangent in the (t, r) or (R, u) plane, thus making the singularity visible. The tangent to the null geodesic at the singularity can be defined as

$$x_0 = \lim_{t \to t_s} \lim_{r \to 0} \frac{R}{u} = \left. \frac{dR}{du} \right|_{t \to t_s; r \to 0} . \tag{3.141}$$

Using (3.141) and (3.137), together with the required limits as above,

$$x_0^{3/2} = \frac{3}{2} \sqrt{\mathcal{M}_0} \mathcal{X}(0). \tag{3.142}$$

The necessary and sufficient conditions for a naked singularity to exist, that is, for the null geodesics with a well-defined tangent to come out from the central singularity can be deduced. Suppose $\mathcal{X}(0) > 0$, then, from (3.142), $x_0 > 0$ always; then in the (R, u) plane the equation for the null geodesic that comes out from the singularity is given by

$$R = x_0 u. \tag{3.143}$$

In other words, (3.143) is a solution of the null geodesic equation in the limit of the central singularity. Therefore, given $\mathcal{X}(0) > 0$, a solution of radially outgoing null geodesics emerging from the singularity can be constructed. This makes the central singularity visible. In the (t, r) plane, the above null geodesics near the singularity will be given as

$$t - t_s(0) = x_0 r^{5/3}. \tag{3.144}$$

It follows that $\mathcal{X}(0) > 0$ implies $x_0 > 0$ and radially outgoing null geodesics emerge from the singularity, giving rise to the central naked singularity.

On the other hand, if $\mathcal{X}(0) < 0$, then the singularity curve is a decreasing function of r. Hence, the region around the center gets singular before the central shell, and after that it is no longer in the spacetime. Now, if there would have been any *outgoing* null geodesic from the central singularity, it must then go to a singular region or outside the spacetime, which is impossible. Hence, when $\mathcal{X}(0) < 0$, there is always a blackhole solution.

If $\mathcal{X}(0) = 0$ then the next higher order non-zero term in the singularity curve equation will have to be taken into account, and a similar analysis by choosing a different value of α in (3.140) will have to be carried out.

It has thus been shown above that $\mathcal{X}(0) > 0$ is the necessary and sufficient condition for null geodesics to emerge from the central singularity with a definite positive tangent. It should be noted, however, that in general the dependence of R on r along the outgoing null geodesics from the singularity does not necessarily have to be of a power law form. However, in order to satisfy the regularity and physical relevance, examining the trajectories that come out with a regular and well-defined tangent is physically more appealing, which is the case examined here.

Shown below for completeness is that if null geodesics of any form come out at all, then those with a definite tangent also must emerge from the central singularity. Towards this end, consider the equation of the apparent horizon. From the equations $F = R$ and (3.130), this is given by

$$t_{\rm ah}(r) = t_{s_0} + r\mathcal{X}(0) + r^2\frac{\mathcal{X}_2(0)}{2} + \cdots - \mathcal{O}(r^3). \tag{3.145}$$

Since the apparent horizon is a well-behaved surface as one initiates close to $r = 0$, for a spherical dust collapse it can be said that the singularity curve for the collapse and the derivatives around the center are also well-defined, as the same coefficients are present in both the apparent horizon equation and (3.131). Also, this shows that whenever $\mathcal{X}(0)$ is negative, the region around the center gets trapped before the central singularity, giving a sufficient condition for a blackhole to develop. It follows that if null geodesics are emerging, then at least one of the coefficients $\mathcal{X}$ must be non-vanishing and positive. Then, as already shown, null geodesics with a definite tangent will emerge from the central singularity.

From the above it follows that in the absence of a null geodesic with a definite tangent there cannot be any null geodesics emerging from the singularity.

It is also clear now from (3.131) that whether $\mathcal{X}(0) > 0$ or otherwise, it is fully determined by the regular initial data for the collapse, in terms of the given initial density and energy distributions for the collapsing shells. It thus follows that the initial data here completely determines the final fate of the collapse in terms of the blackhole and naked singularity endstates. This will be discussed further in the next section.

3.6.3 Structure of singularities

The naked singularity forming in the dust collapse has been analyzed extensively, and many of its properties are well-understood. It is known, for example, that while the models analyzed by Christodoulou (1984) and Newman (1986) had a naked singularity that was necessarily gravitationally weak, in general the TBL collapse admits many classes where the naked singularity is a powerfully strong curvature singularity that is not removable. It is known also that, once the singularity is visible, families of non-spacelike trajectories do come out of it. These are future directed, and in the past they terminate at the singularity. The global visibility of the singularity has also been analyzed. It is seen that while the local visibility of the singularity is decided by the behavior of the density and energy profiles of the cloud close to the center, the global visibility depends on the nature and behavior of the mass function $F(r)$ and the velocity function $f(r)$ away from the center of the cloud (Joshi and Dwivedi, 1993a).

Here, the dust collapse has been treated under fairly general differentiability conditions on the mass and velocity functions. Sometimes much stronger conditions are assumed on these functions, taking them to be necessarily smooth and analytic. While it may be quite convenient to deal with smooth and analytic density profiles, especially when it comes to numerical models, it should not be forgotten that, after all, this is only an extra assumption, and that the basic equations of general relativity do not demand any such constraints. Neither is it clear astrophysically that the interiors of the stars must necessarily have analytic density or energy distributions. In certain equilibrium cases, the field equations imply that these have to be smooth, but this need not be true in general, and, the dynamically developing collapse situations especially, could be quite different. The conclusions here also apply under such stronger conditions.

Consider the scenarios when some of the above assumptions break down. It is easily seen, using considerations such as those above, that when the first derivative of density ρ_1 is non-vanishing, then the collapse ends in a naked singularity in all dimensions, including $N=4$. Again, the considerations above immediately imply that whenever the spacetime is *not* marginally bound, the collapse always results in both the blackhole and naked singularity phases as collapse endstates, depending on the nature of the initial data, irrespective of whether ρ_1 is either zero or non-zero. That is, in a more generic non-marginally bound case, the condition $\rho_1=0$ does not save the cosmic censorship. However, if, on physical grounds, $b_1=\rho_1 = 0$ are chosen, then from the above discussion it can be seen that $\mathcal{X}(0)=0$ and cosmic censorship could be preserved.

Consider the situation when it must be believed somehow that, in the later stages of the collapse, the form of matter cannot be dust-like, and that non-dust forms of matter, and the effects of pressures must be suitably taken into account. In such a case, as pointed out above, it is seen that even if only homogeneous initial density profiles (with non-zero initial pressures) are considered, then the pressure by itself can also cause sufficient distortions in the formation of the apparent horizon so as to cause a naked singularity as the endstate of the collapse, rather than a blackhole. However, if the equation of state is homogeneous, together with the initial data having a homogeneous density profile, then no naked singularity will appear.

Note that the formalism given here brings out the role of the initial data in causing the blackhole and naked singularity endstates for the four-dimensional dust collapse in a clear and transparent manner. To be specific, it is seen, using (3.131), that given any initial density distribution for the cloud, a suitable energy profile can be chosen so that the evolution could end in either a blackhole or a naked singularity, depending on the choice made. In other words, there is a non-zero measure of energy distributions that will take the given density profile to a blackhole. The same holds for a naked singularity to evolve from the same initial density. The converse is also true, namely that given any initial energy distribution, the density profiles that give rise to either of these endstates can be chosen.

3.6.4 Self-similar collapse

Self-similar models have been discussed and analyzed extensively in general relativity (see for example, Carr and Coley, 1999, and references therein). Here, a self-similar example of a spherically symmetric inhomogeneous dust cloud collapse as described by the TBL spacetimes is considered in some detail. A class of these models (Joshi and Singh, 1995), to find that the endstate of the collapse is either a blackhole or a naked singularity, depending on the parameters of the initial density distribution, which are $\rho_{\rm c}$, the initial central density of the massive body, and $r_{\rm b}$, the initial boundary, will be discussed. The collapse ends in a blackhole if a certain dimensionless quantity constructed out of this initial data is greater than a critical value, and ends in a naked singularity if it is less than this number. It is possible to interpret this result in terms of the strength of the gravitational potential at the starting epoch of the collapse.

The method to determine either the blackhole or naked singularity final states in terms of the quantities calculated from the initial data parameters for the density or the velocity profiles of the dust cloud was discussed above. Essentially, this amounts to examining the behavior of the singularity curve that forms at the termination of the collapse, or equivalently one has to examine the existence of real positive roots of an equation involving these initial parameters as shown below. In order to be able to ascertain the astrophysical implications of such a result, it is necessary to translate the condition for the existence of positive real roots into the actual constraints on the initial density distribution in the cloud.

This issue is investigated here, and it would be expected that the degree of inhomogeneity in the matter distribution plays a role in determining the final fate of the collapse. This is seen by working out the explicit conditions for collapse to an endstate that is either a blackhole or a naked singularity, depending on the initial conditions chosen. It turns out that for the class under consideration here, these outcomes are characterized in terms of the existence of real positive roots of a quartic equation. This enables the blackhole or naked singularity configuration as the endstate of the gravitational collapse to be related in terms of the initial density distribution $\rho(r)$ and radius $r_{\rm b}$ of the massive body.

As used earlier, the quantity $R(t_1, r_1)$ denotes the physical radius of a shell of collapsing matter at a coordinate radius r_1 and on the time slice $t = t_1$. The quantities F and f are arbitrary functions of r. In further discussion, the class of solutions is restricted to $f(r) \equiv 0$, which comprise the marginally bound TBL models. Similar considerations can be developed for the models with $f > 0$ or $f < 0$. As the collapsing cloud is of interest here, $\dot{R}(t, r) < 0$. The epoch $R = 0$ denotes a physical singularity where the spherical shells of matter collapse to a zero radius, and where the density blows up to infinity. The time $t = t_0(r)$ corresponds to the value $R = 0$ where the area of the shell of matter at a constant value of coordinate r vanishes. This singularity curve $t = t_0(r)$ corresponds to the time when the matter shells meet the physical singularity. In the case of a finite cloud of dust, there will be a cut off at $r = r_{\rm b}$, where the metric is matched smoothly with a Schwarzschild exterior, as discussed earlier.

In TBL models, the freedom to relabel the dust shells arbitrarily exists. It is given by $r \to g(r)$ for any shell with $r =$ const. on any $t =$ const. epoch. Therefore, at any constant time surface, say at $t = t_0$, $R(r, t_0)$ can be chosen to be an arbitrary function of r. This arbitrariness reflects essentially the freedom in the choice of units. The choice of scaling at $t = 0$ as given by $R(r, 0) = r$ is made for the sake of convenience in the calculation. With this scaling, the $\dot{R}$ equation (with $f = 0$), can be integrated to obtain

$$R^{3/2}(r,t) = r^{3/2} - \frac{3}{2}\sqrt{F(r)}t, \tag{3.146}$$

and the energy equation is given by

$$\epsilon(r,t) = \frac{4/3}{\left(t - \dfrac{2}{3}\dfrac{G(r)}{H(r)}\right)\left(t - \dfrac{2}{3}\dfrac{G'(r)}{H'(r)}\right)}, \tag{3.147}$$

where $G(r) = r^{3/2}$, $G'(r) = (3/2)r^{1/2}$, and $H(r) = \sqrt{F(r)}$.

Now, write $F(r) \equiv r\lambda(r)$, and assume $\lambda(0) \equiv \lambda_0 \neq 0$ and is finite, which is the class of models considered by Dwivedi and Joshi (1992). This means that, near the origin, $F(r)$ goes as r in the present scaling, and the density at the center behaves with time as $\epsilon(0,t) = 4/3t^2$. This is a general class of models that includes all self-similar solutions, as well as a wide range of non-self-similar models, which are found to be quite adequate for the purpose of the present investigation. The central density becomes singular at $t = 0$, and the singularity is interpreted as having arisen from the evolution of the dust collapse, which had a finite density distribution in the past on an earlier non-singular initial epoch.

To check if the singularity could be naked, the possibility that future directed null geodesics would come out of the singularity at $t = 0$, $r = 0$ needs to be examined. The equations of the outgoing radial null geodesics in the spacetime, with k as the affine parameter, can be written as

$$\frac{dK^t}{dk} + \dot{R}' K^r K^t = 0, \tag{3.148}$$

$$\frac{dt}{dr} = \frac{K^t}{K^r} = R', \tag{3.149}$$

where $K^t = dt/dk$ and $K^r = dr/dk$ are tangents to the outgoing null geodesics. The partial derivatives R' and $\dot{R}'$ that occur in the above equations can be worked out from the solution given above, and these are most suitably written as

$$R' = \eta P - \left[\frac{1-\eta}{\sqrt{\lambda}} + \eta\frac{t}{r}\right]\dot{R}, \tag{3.150}$$

$$\dot{R}' = \frac{\lambda}{2rP^2}\left[\frac{1-\eta}{\sqrt{\lambda}} + \eta\frac{t}{r}\right], \tag{3.151}$$

where

$$R(r,t) = rP(r,t), \quad \eta = \eta(r) = \frac{rF'}{F}. \tag{3.152}$$

The functions $\eta(r)$ and $P(r)$ are introduced because they have a well-defined limit in the approach to the singularity.

If the outgoing null geodesics terminate in the past with a definite tangent at the singularity (in which case the singularity would be naked), then using (3.150) and l'Hospital's rule,

$$X_0 = \lim_{t\to 0,\, r\to 0} \frac{t}{r} = \lim_{t\to 0,\, r\to 0} \frac{dt}{dr} = \lim_{t=0,\, r=0} R', \tag{3.153}$$

where $X = t/r$ is a new variable. The positive function $P(r,t) = P(X,r)$ is then given by

$$X - \frac{2}{3\sqrt{\lambda}} = -\frac{2P^{3/2}}{3\sqrt{\lambda}}. \tag{3.154}$$

Now, Q is defined as $Q = Q(X) = P(X,0)$. If the future directed null geodesics come out of the singularity at $t = 0, r = 0$, meeting the singularity in the past with a definite tangent $X = X_0$, as given above, then it follows from (3.154) that such a value X_0 must satisfy

$$X_0 < \frac{2}{3\sqrt{\lambda_0}}. \tag{3.155}$$

Furthermore, it is noted, by using the definition $F = r\lambda(r)$, that as $r \to 0$, $\eta \to 1$. Also, for $f = 0$, $\lim \dot{R} = -\sqrt{\lambda_0/Q}$. Using these results in the expression for R' implies that the condition (3.153) is simplified to the equation

$$V(X_0) = 0, \tag{3.156}$$

where

$$V(X) \equiv Q + X\sqrt{\frac{\lambda_0}{Q}} - X. \tag{3.157}$$

In order to be the past end point of the outgoing null geodesics, at least one real positive value of X_0 must satisfy (3.156), subject to the constraint implied by (3.155), as stated above, and which is given by $X_0 < 2/3\sqrt{\lambda_0}$. In general, it was shown by Dwivedi and Joshi (1992) that if the equation $V(X_0) = 0$ has a real positive root, the singularity would be naked. Whenever this is not realized, the evolution will lead to a blackhole. Such a singularity could be either locally or globally naked depending on the global features of the function $\lambda(r)$.

The sense in which the terms naked singularity and blackhole are used needs to be clarified. When there are no positive real roots to (3.156), the central singularity is not naked, because there are no outgoing future directed null geodesics from the singularity in this case. Furthermore, as discussed earlier, the shell-focusing singularity $R = 0$ for $r > 0$ is always covered. Hence, in the absence of positive real roots, the collapse will always lead to a blackhole. On the other hand, if there are positive real roots, it follows that the singularity is at least locally naked. The global visibility aspect of such a locally naked singularity will be discussed in the next subsection. Such a locally naked singularity would also be globally naked when the outgoing trajectories could reach arbitrarily large values of r, that is, the signals reach faraway observers. Otherwise, the collapse outcome would still be a blackhole when these trajectories that emerge from the singularity fall back to the singularity again without emerging from the horizon. This is a violation of weak censorship only. The occurrence of either of these situations will really depend on the nature of the function $\lambda(r)$. The conditions under which this locally naked singularity could be globally naked have been discussed, for instance, by Joshi and Dwivedi (1993a). The occurrence of positive real roots implies the violation of strong cosmic censorship, though not necessarily of weak cosmic censorship. In other words, a blackhole and locally naked singularity are not mutually exclusive alternatives. It can also be shown that whenever there is a positive real root as above, a family of outgoing null geodesics always terminates at the singularity in the past.

The condition for the occurrence of a naked singularity is now examined in some detail. The condition $V(X_0) = 0$ can be written as

$$Y^3 \left(Y - \frac{2}{3} \right) - \alpha (Y - 2)^3 = 0, \tag{3.158}$$

where $Y = \sqrt{\lambda_0} X_0$, and $\alpha = \lambda_0^{3/2}/12$ have been set. Note that $F(r)$, and hence λ_0, are always necessarily positive. Using standard results, it can be shown that this quartic equation has positive real roots if and only if $\alpha > \alpha_1$ or $\alpha < \alpha_2$, where

$$\alpha_1 = \frac{26}{3} + 5\sqrt{3} \approx 17.3269$$

and

$$\alpha_2 = \frac{26}{3} - 5\sqrt{3} \approx 6.4126 \times 10^{-3}. \tag{3.159}$$

To derive this condition, first note that if it has a real root, it must be positive, as negative values of Y do not solve this equation. Writing the general quartic as $ax^4 + 4bx^3 + 6cx^2 + 4dx + e = 0$ one defines $H = ac - b^2$, $I = ae - 4bd + 3c^2$, $J = ace + 2bcd - ad^2 - eb^2 - c^3$, and $\Delta = I^3 - 27J^2$. If $\Delta < 0$, the quartic has two real and two imaginary roots. If $\Delta > 0$, all roots are imaginary unless $H < 0$ and $(a^2 I - 12H^2) < 0$, in which case they are all real. The application of such a procedure to the quartic in (3.158) leads to the conditions on α given above.

It follows, however, from earlier discussions, that along any such outgoing null geodesics $\sqrt{\lambda_0} X_0 = Y < 2/3$. Then, (3.155) implies that the larger range of α for the existence of roots, that is $\alpha > 17.3269$, is ruled out in the sense that no outgoing trajectories can meet the singularity with this larger value of the tangent X_0. This is seen by writing α as a function of Y, which shows that if $\alpha > \alpha_1$, then $Y > 2$. It thus follows that a naked singularity arises if and only if $\alpha < \alpha_1$, or equivalently, if and only if $\lambda_0 < 0.1809$. Whenever the limiting value λ_0 does not satisfy this constraint, the gravitational collapse of the dust cloud must end in a blackhole. The physical interpretation for the quantity λ_0 can be obtained from the equation for the time evolution of the energy density. If the collapse starts at a time $-t_0 < 0$, and ρ_c is the initial energy density at the center, then $\rho_c = 4/3t_0^2$. If ρ'_c is the initial density gradient at the center, then $\lambda_0 = 16\rho_c^3/3\rho_c'^2$. Putting in the units gives

$$\lambda_0 = \frac{8\pi G}{c^4}\frac{16\rho_c^3}{3\rho_c'^2}. \tag{3.160}$$

Defining $\beta = \lambda_0/16$, it is found that the blackhole arises whenever

$$0.0113 < \beta, \tag{3.161}$$

and the naked singularity results for the values given by $\beta < 0.0113$. The occurrence of one or the other outcome is governed by the conditions of a combination of the initial central density and the initial density gradient at the center. The cosmic censorship hypothesis of Penrose could, in the present context, be translated to the conjecture that values of β smaller than 0.0113 do not occur in a realistic collapse.

Now, the value of the parameters β or λ_0, given the initial density profile for the collapsing massive star, need to be calculated. Given the initial central density, the initial density gradient $d\rho/dr = \rho'_c$ can be evaluated at the center as follows. First, note that the expression for $d\rho/dr$ can be written, and it is seen that in the limit of $r \to 0$, this always goes to a finite quantity that is proportional to $1/\sqrt{\lambda_0}$. Now, given the initial data in the form of the density distribution $\rho(R)$ for the body at an initial non-singular epoch of time, in terms of the physical radius R, integration can give the form of the mass function $F(R)$. Then, the above provides a functional relationship $r(R)$ that can be inverted, in principle, to express R in terms of r. The mass profile can then be written explicitly as $F(r)$, and λ_0 is evaluated as the limit of $F(r)/r$ as $r \to 0$.

This model can be generalized in many ways, for instance by considering the most general class of functions $F(r)$ and $f(r)$. Also, it has been shown earlier that the pattern of a transition from the blackhole configuration to the naked singularity configuration persists in models with more general equations of state. It is desirable to cast results for these models in terms of constraints on the initial density distribution. For a discussion on non-self-similar collapse, see Lake (1991).

3.6.5 Strength and global visibility

Here, the aspects related to the structure of the naked singularity, namely its curvature strength and the visibility for faraway observers, is discussed in some detail. The question related to when a non-zero measure set of non-spacelike trajectories would emerge from the singularity, as opposed to isolated trajectories emerging, will be discussed in a general manner in the next chapter.

The curvature strength of the naked singularity is examined, and a wide class of TBL models that has a strong curvature singularity, an important indicator for the physical significance is shown. This is carried out in terms of the strong curvature condition, which ensures that all the volume forms must be crushed to a zero size in the limit of the approach to the singularity. Also, the divergence of the Kretschmann scalar $\mathcal{K} = R^{ijkl}R_{ijkl}$ in this limit is seen. It is shown that the general class discussed here contains subclasses of solutions that admit strong curvature naked singularities in either of the senses stated above. The conditions are also discussed for the ultra-dense regions to be globally visible. An implication for the fundamental issue of the final fate of the gravitational collapse is that naked singularities need not be considered as artifacts of geometric symmetries of spacetime such as self-similarity, but that they arise in a wide range of gravitational collapse scenarios once the inhomogeneities in the matter distribution are taken into account. It follows that these need not be gravitationally weak, as was widely conjectured towards formulating a censorship statement (see for example, Joshi, 1993, for more details and references). Discussed further in the next chapter will be the stability and genericity aspects for naked singularities in order to see that these arise from open sets of non-zero measures of matter and velocity initial data, from which the collapse develops.

Clearly, any possible rigorous formulation of the cosmic censorship conjecture has to be designed so as to avoid the features above. This is why gravitational collapse studies are crucial for the cosmic censorship hypothesis. Possibilities in this direction are discussed in the next chapter, while indicating that the analysis presented here should be useful for any such attempts. On the other hand, these developments also give important insights into the formation and structure of naked singularities in continual gravitational collapse scenarios.

A genuine strong curvature singularity forms as the collapse final state in the Oppenheimer–Snyder case of a completely homogeneous dust cloud collapse with zero pressure. The homogeneous case can, however, be viewed as a subcase of a set of zero measures in the general inhomogeneous class given by the TBL solutions. It thus becomes imperative to study TBL models in greater detail to understand the final fate of a collapsing massive body, when the effects of inhomogeneities are taken into account. Therefore, the nature and strength of the singularity for a TBL collapse is now examined.

The numerical simulations of Eardley and Smarr (1979) imply that naked singularities arise in the marginally bound TBL collapse, and subsequently a class of such models was studied analytically by Christodoulou (1984) to draw similar conclusions. However, these singularities were shown to be gravitationally weak, and hence possibly removable, by Newman (1986), who studied the curvature strengths of these naked singularities and conjectured that nature avoids strong curvature naked singularities. The occurrence of naked singularities may not be considered a problem from the point of view of cosmic censorship if these were always gravitationally weak in a suitable sense.

The existence of an incomplete non-spacelike geodesic, or an inextendible non-spacelike curve that has a finite length, as measured by a generalized affine parameter, implies the existence of a spacetime singularity. The generalized affine length for such a curve is defined as (Hawking and Ellis, 1973)

$$L(\lambda) = \int_0^a \left[\sum_{i=0}^{3} (X^i)^2 \right]^{1/2} ds, \tag{3.162}$$

which is a finite quantity. The X^i values are the components of the tangent vector to the curve in a parallel propagated tetrad frame along the curve. Each such incomplete curve defines a boundary point of the spacetime, which is a singularity. In order to know if this is a genuine physical singularity, one would typically like to associate it with unboundedly growing spacetime curvatures. If all curvature components and the scalar polynomials formed out of the metric and the Riemann curvature tensor remain finite and well-behaved in the limit of approach to the singularity along an incomplete non-spacelike curve, it may be possible to remove the singularity by extending the spacetime when the differentiability requirements are lowered (Clarke, 1986).

There are several ways to formalize such a requirement. For example, a parallel propagated curvature singularity is the end point of at least one non-spacelike curve on which the components of the Riemann curvature tensor are unbounded in a parallel propagated frame. On the other hand, a *scalar polynomial singularity* has a scalar polynomial in the metric and Riemann tensor taking an unbounded large value along at least one non-spacelike curve with this singular end point. This includes the cases such as the Schwarzschild singularity where the Kretschmann scalar $R^{ijkl}R_{ijkl}$ blows up in the limit as $r \to 0$. Will genuine curvature singularities occur in general relativity? The answer, for the case of parallely propagated curvature singularities, is provided by a theorem of Clarke (see for example, Clarke, 1993). This theorem shows that for a globally hyperbolic spacetime that is C^{0-}-inextendible, when the Riemann tensor is not very specialized in the sense of not being type D and electrovac at the singular end point, then the singularity must be a parallely propagated curvature singularity. Similarly, a class of physically relevant strong curvature singularities was analyzed by Tipler (1977), Tipler, Clarke, and Ellis (1980), and Clarke and Królak (1986).

The idea here is to define a physically all-embracing strong curvature singularity in such a way so that all the objects falling within the singularity are destroyed and crushed to zero volume by the infinite gravitational tidal forces. This notion is formulated as below. Let $\lambda(t)$ be a timelike or null geodesic that is incomplete at an affine parameter value $t = 0$. Let K^i denote the tangent vector to λ and $\mu(t) = \boldsymbol{Z}_1 \wedge \boldsymbol{Z}_2 \wedge \boldsymbol{Z}_3$ be a volume form defined along $\lambda(t)$ where $\boldsymbol{Z}_1,\ \boldsymbol{Z}_2,\ \boldsymbol{Z}_3$ are linearly independent Jacobi vectors defined along the curve λ orthogonal to K^i. (If λ is null then $\mu(t)$ is defined as a two-form.) A real-valued map from the space of all such three-forms can be defined by $\triangle(\boldsymbol{A} \wedge \boldsymbol{B} \wedge \boldsymbol{C}) = \det[A^i,\ B^i,\ C^i]$. Denote $\triangle(\mu(t))$ by $V(t)$, which defines a volume element along $\lambda(t)$ and is independent of the choice of basis. The singularity at $t = 0$ is then called a *strong curvature singularity* if $V(t) = 0$ in the limit as $t \to 0$ for all possible $\mu(t)$, that is, for all possible choices of linearly independent Jacobi fields.

This definition effectively captures the notion that all objects falling into a strong curvature singularity are crushed to zero volume. For a strong curvature singularity, at least one non-spacelike geodesic terminating at the singularity along which the above curvature condition is satisfied exists. However, a strong curvature singularity can be defined in a much stronger sense also (Tipler, 1977), by requiring that the strong curvature condition above must be satisfied along *all* non-spacelike geodesics terminating in the singularity. Necessary and sufficient conditions for the occurrence of strong curvature singularities are derived by Clarke and Królak (1986), and which are shown to involve the tetrad components of the Riemann, Ricci and Weyl tensors, and also the divergence of their integrals along non-spacelike geodesics running into the singularity. It follows from their analysis that an incomplete non-spacelike geodesic does not define a strong curvature singularity unless either the Weyl or Ricci tensor components diverge sufficiently fast along such a trajectory.

A sufficient condition for this to happen is, if t is the affine parameter,

$$R_{ij} V^i V^j \geq A/t^2, \tag{3.163}$$

for some fixed constant A along the trajectory in the limit of approach to the singularity as $t \to 0$. This provides a sufficient condition for all the two-forms $\mu(k)$, defined along the singular null geodesic, to vanish as the singularity is approached, and implies a very powerful curvature growth establishing a strong curvature singularity. For timelike geodesics this will imply that all the volume forms defined by the Jacobi fields along these trajectories must vanish in the limit of approach to the singularity, or they must vanish infinitely many times in this limit.

To fix these ideas, the case where $R_{ij}V^iV^j = K/t^2$ is discussed in some detail. It is convenient to define a length scale y associated with the volume $V(t)$ by defining $y^3 = V$. The propagation equation for $y(t)$ along $\lambda(t)$ is the Raychaudhuri equation, which is a second order differential equation, given in this case as

$$\frac{d^2y}{dt^2} + \frac{1}{3}(R_{ij}V^iV^j + 2\sigma^2)y = 0, \tag{3.164}$$

with $\sigma^2 = \sigma_{ij}\sigma^{ij}$ being the trace of the shear tensor σ_{ij}. (For the null case, a similar equation holds with 1/3 being replaced by 1/2.) Writing $F(t) = (R_{ij}V^iV^j + 2\sigma^2)/3$, and ignoring the effects of the shear tensor, which will enhance the focusing effect considered here, $F(t) = A/t^2$, where $A > 0$ is a fixed constant, is chosen. If a solution of the form $y = t^\alpha$ is tried, then the condition on A is obtained to be $A = \alpha - \alpha^2$. Since $V(t) \to 0$ in the limit of approach to the singularity $t = 0$, $\alpha > 0$. Again, $A > 0$ and the above expression for A implies that $0 < \alpha < 1$. Solving for α gives that, in order for α to be real, A satisfies $A \leq 1/4$. The solution for y is then given by

$$y = t^{[1\pm(1-4A)^{1/2}]/2}. \tag{3.165}$$

Therefore, depending on the value of A, it is seen that $y \sim t^\alpha$ with $1/2 \leq \alpha < 1$. The volume $V(t)$ then goes to zero near the singularity at least as fast as $t^{3/2}$.

Basically, the sufficient condition for a singularity to be strong in the sense of Tipler (1977), is that at least along one null geodesic with the affine parameter k, with $k = 0$ at the singularity, in the limit of approach to the singularity,

$$\lim_{k\to 0} k^2 R_{ab}K^aK^b > 0, \tag{3.166}$$

where K^a and K^b are tangents to the null geodesics. This is a sufficient condition for all two-forms $\mu(k)$ defined along the singular null geodesic to vanish as the singularity is approached, and implies a very powerful curvature growth, establishing a strong curvature singularity. For the timelike geodesics this implies that all the volume forms defined by the Jacobi fields along these trajectories must vanish in the limit of approach to the singularity, or they must vanish infinitely many times in this limit.

The important physical consequences of the existence of a singularity are related to its strength. If the singularity is gravitationally weak, it may be possible to extend the spacetime. On the other hand, when there is a strong curvature singularity forming as above, the gravitational tidal forces associated with it are so strong that any object trying to cross it gets destroyed. Therefore, as argued by Ori (1992), the extension of spacetime becomes meaningless for such a strong singularity that destroys to zero size all the objects terminating at the singularity. From this point of view, the strength of the singularity may be considered crucial to the issue of classically extending the spacetime and thus avoiding the singularity, because for a strong curvature singularity, no continuous extension of the spacetime may be possible.

For the general class of TBL models under consideration,

$$\Psi \equiv R_{ab}K^aK^b = \frac{F'(K^t)^2}{R^2R'} = \frac{F'(K^t)^2}{R^2R'}. \tag{3.167}$$

For radial null geodesics, using l'Hospital's rule and the TBL equations as given earlier, and the fact that in the limit of approach to the singularity $r \to 0, X \to X_0$,

$$\lim_{k\to 0} k^2\Psi = \frac{4\eta_0\Lambda_0}{H_0X_0^2\left(2\left(3\alpha - \eta_0\right)\sqrt{1+f_0} - N_0\right)^2}, \tag{3.168}$$

where η and Λ are as defined in (3.118) and (3.119). Hence, it is seen from the definition of Λ in (3.120) that

$$\lim_{k\to 0} k^2\Psi = 0 \quad \text{for} \quad \alpha < \eta_0, \tag{3.169}$$

$$\lim_{k\to 0} k^2\Psi \neq 0 \quad \text{for} \quad \alpha \geq \eta_0. \tag{3.170}$$

However, from earlier conclusions, a naked singularity occurs only when $\alpha \leq \eta_0$, therefore the strong curvature condition is satisfied along the singular geodesics for the classes where $\alpha = \eta_0$. As noted earlier, for the special class considered by Christodoulou (1984) and Newman (1986), $\alpha = 7/3$ and $\eta = 3$, and hence the naked singularity turns out to be gravitationally weak. On the other hand, it is clear from the above that, for a wide variety of TBL solutions satisfying the condition $\alpha = \eta_0$, the singularity will be a strong curvature singularity in the above sense. In general, it is also possible that non-radial null or timelike curves could terminate at the naked singularity. Then, a similar calculation along non-spacelike geodesics in general gives

$$\lim_{k\to 0} k^2\Psi \propto \left(r^{\eta_0-\alpha}\right)_{r=0}. \tag{3.171}$$

So, as discussed above, it can be concluded that the condition for strong curvature is satisfied along non-spacelike geodesics as well if $\alpha = \eta_0$, and if such families meet the naked singularity in the past.

The Kretschmann scalar $R^{ijkl}R_{ijkl}$ along the geodesics in the TBL spacetimes goes as

$$\mathcal{K} \propto r^{2(\eta_0 - 3\alpha)}. \tag{3.172}$$

Hence, the singularity is a scalar polynomial singularity as long as $\alpha > \eta_0/3$.

A self-similar dust collapse example was discussed in an earlier section. The self-similar TBL models are, in general, defined by the conditions $f(r) =$ const. and $\eta(r) = 1 = \eta(0) = \alpha$. It follows from the above that the naked singularity forming in this class will be a strong curvature singularity along all the families of radial null geodesics. A detailed discussion on the families of future directed non-spacelike geodesics, terminating at the naked singularity in the past, is given by Joshi and Dwivedi (1992, 1993b). Various different families of non-spacelike geodesics do terminate at the naked singularity in the past in this case, along which the strong curvature condition is also necessarily satisfied.

As discussed in the previous section, the existence of a real positive root of the equation $V(X) = 0$ shows the singularity to be at least locally naked. The singularity can be globally visible also, and to examine this, note that the apparent horizon lies at $R(t, r) = F(r)$ within the collapsing cloud. Therefore, if a geodesic gets inside the apparent horizon as it is emerging from the singularity, it becomes ingoing, with $R < F$ along the geodesics and dR/dr becoming negative. Eventually this trajectory must fall back to the singularity. Therefore, if a light ray is to reach future infinity in order for the singularity to be globally naked, it must cross $r = r_{\mathrm{c}}$, which is the boundary of the dust cloud before the time when the apparent horizon reaches this boundary as it evolves during collapse. Hence, all escaping non-spacelike geodesics that reach the boundary $r = r_{\mathrm{c}}$ with $R(r_{\mathrm{c}}) > F(r_{\mathrm{c}})$ would reach the future infinity. Since geodesics emerge from the singularity with the tangent value $\boldsymbol{X}_0$ and the apparent horizon has a tangent $\boldsymbol{\Lambda}_0$ at the singularity, it follows that $\boldsymbol{X}_0 > \boldsymbol{\Lambda}_0$.

As a result, because of the generality of the function $F(r)$, which is the free initial data subject only to regularity conditions, r_{c} and $F(r_{\mathrm{c}}) = 2M$, with M being the Schwarzschild mass of the cloud, can be chosen suitably such that the geodesics reach the boundary of the cloud $r = r_{\mathrm{c}}$ with $R(r_{\mathrm{c}}) > F(r_{\mathrm{c}})$, making the singularity globally visible. However, given a boundary $r = r_{\mathrm{c}}$ and $F(r_{\mathrm{c}}) = 2M$, the singular geodesics that actually reach future infinity depend on the global properties of the functions $F(r)$ and $f(r)$.

An explicit class of TBL models where the singularity is shown to be globally visible is now discussed. Owing to the complexity of the equations, exact geodesic solutions are few, even for simple forms of the functions $F(r)$ and $f(r)$, except for the Friedmann models of complete homogeneity. However, a self-similar marginally bound collapse ($f = 0$) can be considered, and its formalism is illustrated here. This example can be analyzed using a special null trajectory that is the Cauchy horizon given by $X =$ const., which is the first null geodesic emerging from the singularity. It is shown that actually the geodesic equations can be integrated completely for this self-similar case, in order to obtain radial null families. As pointed out above, the collapse will end in a naked singularity when $V(X) = 0$ has two real positive and two complex roots. Let x_1, $x_2\,(x_1 > x_2)$ be two such positive roots of this equation. The equation of geodesics, in the form $r = r\,(x)$, $X = x^2$ is given by (Joshi and Dwivedi, 1993a)

$$r = r(X) \equiv r(x) = D\frac{(x - x_2)^{n_2}}{(x - x_1)^{n_1}} f_1(x), \tag{3.173}$$

where

$$f_1(x) = \exp\left(-\int \frac{Ax + B}{x^2 + D_1 x + D_2} dx\right). \tag{3.174}$$

Here, n_1, n_2, A, B, D_1, and D_2 are constants given by

$$x^4 + \frac{\sqrt{\mathbf{\Lambda}_0}}{2} x^3 - x + \sqrt{\mathbf{\Lambda}_0} = (x - x_1)(x - x_2)(x^2 + D_1 x + D_2), \tag{3.175}$$

$$\frac{3x^3}{x^4 + \frac{\sqrt{\mathbf{\Lambda}_0}}{2} x^3 - x + \sqrt{\mathbf{\Lambda}_0}} = \frac{n_1}{x - x_1} - \frac{n_2}{x - x_2} + \frac{Ax + B}{x^2 + D_1 x + D_2}, \tag{3.176}$$

and D is the constant that labels the different geodesics. The constants n_1, and n_2 are positive. For the case $\mathbf{\Lambda}_0 = 7/17$,

$$x_1 = 0.658303,\ x_2 = 0.5,\ n_1 = 2.09356,\ n_2 = 1.08511,\ D_1 = 1.36419, \tag{3.177}$$

$$D_2 = 1.2509, \quad A = -1.99154, \quad B = -1.26354. \tag{3.178}$$

It is clear from (3.173) that geodesics reach $r = 0$ at $x = x_2$ and $r = \infty$ at $x = x_1$, making the singularity globally naked. Note that $\eta(r)\Lambda(r) = F_0 < x_2$ and therefore all the trajectories that are emitted in the region $x_1 > x > x_2$ reach the future infinity. In fact, $x = x_1$ and $x = x_2$ are also geodesics that cross the boundary of the cloud and escape to future infinity.

To discuss conditions that ensure the global visibility of the ultra-dense regions in general, suppose the functions η and β are at least C^0 in the interval $r_c \geq r > 0$. Since these involve the first derivatives of f and F as f'/f and F'/F, this implies that f and F have at least first order continuous derivatives. The C^2-differentiability of the metric ensures this. If $V(X) = 0$ has only one simple root $X = X_0$, a family of curves terminates at the singularity, that is, $h_0 > 1$, with this value of the tangent. Let $\eta(r)\Lambda(r) < \alpha X_0$ for $r_c \geq r > 0$. Then the singularity would be globally naked. To see this, consider the equation of geodesics (3.173), where the constant D labels different geodesics terminating at the singularity, as determined by the boundary conditions. For a singular geodesic that reaches the boundary of the dust cloud $u = u_c = r^\alpha = r_c^\alpha$ with $X = (R_c/r_c^\alpha) = X_c$,

$$X_c - X_0 = D u_c^{h_0 - 1} + u_c^{h_0 - 1} \int_{u_c} S u^{(-h_0+1)}\, du, \tag{3.179}$$

and hence the equation of such a geodesic can be written as

$$X - X_0 = (X_c - X_0)\left(\frac{u}{u_c}\right)^{h_0 - 1} + u^{h_0 - 1} \int_{u_c}^{u} S u^{(-h_0+1)}\, du. \tag{3.180}$$

The event horizon is represented by the geodesic for which $X_c = \Lambda(r_c)$. Since it is outgoing, $dR/d(r^\alpha)$ is positive at $r = 0$ and ejected into the region $R > F$ where dR/dr is positive. Therefore, all the geodesics that reach $r = r_c$, where the TBL metric is matched with the Schwarzschild exterior, with $X_c > \Lambda(r_c)$ would escape to infinity, while others would become ingoing. It follows that the geodesics that reach future infinity with their past end point at the singularity are given by (3.180) with $X_c > \Lambda_c = \Lambda_{r_c}$. Hence, in the case when a family of geodesics terminates at the singularity with a tangent $X = X_0$ and $\eta(r)\Lambda(r) < \alpha X_0$, for $r_c \geq r > 0$, the singularity would be globally visible, as there would always be some geodesics that escape to infinity.

Consider the case now when the equation $V(X) = 0$ has two positive roots X_1 and X_2 ($X_1 > X_2$). In such a case, families of curves would emerge from the singularity with the tangent either X_1 or X_2. Let $\eta(r)\Lambda(r) < \alpha X_2$ for $r_c \geq r > 0$, then it ensures that some geodesics would cross the boundary of the cloud with $X_c > \Lambda(r_c)$ making the singularity globally naked. The same holds even when more than two positive roots exist. Therefore, if the family of geodesics do terminate at the singularity with tangent X_0, then the condition $\eta(r)\Lambda(r) < \alpha X_0$ for $r_c \geq r > 0$ implies the global visibility of the singularity.

3.6.6 Cosmological constant

For the inhomogeneous dust collapse models, a singularity always develops necessarily in an initially collapsing configuration, and rebounce or halting the collapse are not possible. The singularity can be naked, or hidden behind an event horizon, depending on the nature of the initial density and velocity profiles for the collapsing cloud.

In the above, the Einstein equations were used with a vanishing cosmological term. It is possible, however, that this is not the case and there may be some observational support that the universe is dominated by an energy component with a negative pressure (Perlmutter *et al.*, 1998, 1999; Zehavi and Dekel, 1999). A plausible explanation for such a dark energy present in the universe could be in terms of the presence of a non-zero value of the cosmological term Λ. This will give rise to a vacuum energy density, corresponding to a positive sign of Λ. This would represent a spatially uniform time independent energy density distribution, and its positive value acts as a globally repulsive force field. Also, the cold dark matter models with a substantial component supplied by Λ can fit the current observational data (Ostriker and Steinhardt, 1995).

Therefore, it is useful and interesting to consider the Einstein equations with a non-zero Λ, to investigate the gravitational collapse, and also the implications in cosmology. With such a perspective, the structure of a singularity in the spherically symmetric dust models with a non-vanishing cosmological constant is discussed here. Dust models with a cosmological term are discussed in the literature (see for example, Krasinski, 1997). These models can be matched with the Schwarzschild–de Sitter spacetime at the boundary of the cloud (Omer, 1965; Lake and Roeder, 1979; Lake, 2000.) Here, the TBL collapse models are analyzed with a cosmological constant, and provide the general solution to the Einstein equations with dust as the source term.

The weak energy condition, $T_{ij}V^iV^j \geq 0$, for all non-spacelike vectors V^i, is assumed in the spacetime for the matter to be physically reasonable, especially in the case of the gravitational collapse. The Einstein equations are analyzed to check whether there are globally regular solutions, as have been shown to exist in the case of homogeneous density dust solutions with a positive cosmological constant (Markovic and Shapiro, 2000). The general solution to the Einstein equations in the case corresponding to the marginally bound models in the TBL spacetimes, which is when the energy function $f = 0$, is also derived, and the condition for avoiding shell-crossings is discussed. The structure of the singularity can be investigated by studying the outgoing radial null geodesics near it. Then, it can be examined how the given initial conditions, in the presence of the cosmological term, affect the final state results in terms of the formation of a blackhole or a naked singularity. It can also be seen why the naked singularity turns out to be stable to the introduction of the cosmological term.

Here, the gravitational collapse of a dust cloud in the presence of a cosmological term is discussed in some detail, and how it may affect the dynamical evolution of the collapse is examined. In particular, it is found that there is a non-trivial detailed dynamical structure, especially for the $f < 0$ case, and this provides a strong motivation for the present study, namely how the introduction of a cosmological term non-trivially changes the structure of the collapse. In most cases, there is no rebounce, and then the nature of the singularity forming as the endstate of the collapse is studied. At a more technical level, the change in the dynamical structure of the collapse, when there is a non-vanishing Λ term present, is reflected in the change in the nature and structure of the singularity, which is far from obvious. There are some important changes in the collapse dynamics with the introduction of the cosmological constant. In particular, the analysis of the roots equation referred to earlier shows that part of the naked singularity spectrum is covered when a positive cosmological term is introduced, but it is never removed altogether irrespective of its value. In this sense, the naked singularity is stable to the introduction of the cosmological term.

In fact, there has been hope through the last three decades or so that it may be possible to avoid singularities by the introduction of a cosmological term with a positive sign (see for example, Hawking and Ellis, 1973, pp. 139, 362), especially in the cosmological case. Hence, it is important to investigate the collapse dynamics, especially on a large or medium scale in the universe, in the presence of a cosmological term. In fact, as shown by Deshingkar *et al.* (2001), for a certain range of initial data, there is a rebounce possible due to the presence of the positive cosmological term. Such a possibility will be especially important in the case of the collapse of either the clusters or superclusters of galaxies, where the densities may be sufficiently low at the initial epoch, and such a rebounce can become reality.

Towards the effects of a non-vanishing cosmological term on the final fate of an inhomogeneous collapsing dust cloud, it is shown that, depending on the nature of the initial data from which the collapse evolves, and for a positive value of the cosmological constant, it is possible to have a globally regular evolution where a bounce develops within the cloud. The initial data causing such a bounce is characterized in terms of the initial density and velocity profiles for the collapsing cloud. In the other cases, the result of the collapse is either the formation of a blackhole, or a naked singularity, as the endstate of the collapse. It is also found that a positive cosmological term can cover a part of the singularity spectrum that is visible in the corresponding dust collapse models for the same initial data. The basic set of the Einstein equations and the regularity conditions for the collapse is discussed, and then the possibility when there is a non-vanishing cosmological term present, when an initially collapsing cloud rebounces at a later epoch so that a singularity does not form, is investigated. Therefore, there is an occurrence of three phases, the collapse, the reversal, and the subsequent dispersal. A general solution with a non-zero Λ term for the marginally bound case is also given.

The model for a self-gravitating, spherically symmetric, inhomogeneous dust cloud with cosmological constant is given by the metric

$$ds^2 = -dt^2 + \frac{R'^2}{1+f(r)}dr^2 + R^2(d\theta^2 + \sin^2\theta\, d\phi^2), \tag{3.181}$$

where (t, r, θ, ϕ) is a comoving coordinate system, with notations as earlier. The equation of state for matter in the interior of the cloud is that of dust, with the stress–energy tensor given by $T^i_j = \epsilon(r,t)\delta^i_t\delta^t_j$, with $\epsilon(r,t)$ being the energy density of matter. The weak energy condition is assumed for the matter, which implies that $\epsilon(r,t) \geq 0$. This is equivalent to the strong energy condition in this case, as the principal pressures are zero.

The Einstein equations in the presence of a cosmological constant can be written as

$$\dot{R}^2 = \frac{F(r)}{R} + f(r) + \frac{\Lambda}{3}R^2, \tag{3.182}$$

$$\epsilon(t,r) = \frac{F'}{R^2 R'}. \tag{3.183}$$

The collapse situation with $\dot{R} < 0$ is mainly considered. The two free functions that characterize the dust cloud are those representing the total mass $F(r)$, and the energy $f(r)$, inside the shell labeled by the comoving coordinate r. The cosmological term Λ can, in principle, be of either sign, however the recent observations indicated above seem to favor the positive sign. The above equation for $\dot{R}$ can, in principle, be integrated, and after integration, a constant of integration is obtained. The constant of integration can be fixed by using the scaling freedom. Here, this is fixed by setting $R = r$ on the initial hypersurface ($t = 0$). The two free functions $F(r)$ and $f(r)$ can be fixed by prescribing the initial density and velocity profiles through (3.183) and (3.182) respectively. Assume these free functions have the form

$$\begin{aligned} F(r) &= F_0 r^3 + F_n r^{3+n} + \text{higher order terms}, \\ f(r) &= f_0 r^2 + f_{n_1} r^{2+n_1} + \text{higher order terms}, \end{aligned} \tag{3.184}$$

where the choice of the first non-vanishing term is made in order to have regular initial data on the initial surface $t = 0$ from which the collapse evolves. From (3.183), it is clear that it is possible to have both shell-crossing ($R'(t_{\rm sc}(r), r) = 0$), and shell-focusing singularities ($R(t_{\rm sf}(r), r) = 0$) in these spacetimes, depending on the dynamics of the shells given by (3.182). Here, only cases where there are no shell-crossings are considered, as these are generally not considered to be genuine singularities, and the main interest here is in studying the nature of the physical singularity corresponding to $R = 0$ where the matter shells shrink to zero radius. This puts some restrictions on the initial data.

The metric exterior to the collapsing cloud has the Schwarzschild–de Sitter form,

$$ds^2 = -g dt^2 + g^{-1} dr^2 + r^2(d\theta^2 + \sin^2\theta\, d\phi^2), \tag{3.185}$$

where $g = 1 - (2M/r) - (\Lambda r^2/3)$. Matching the solutions (3.181) and (3.185) at the boundary $r_{\rm b}$ of the collapsing dust cloud, $2M = F(r_{\rm b})$ is obtained. The boundary can be made to bounce from an initially collapsing phase by choosing appropriate initial mass and cosmological terms, as seen below. This behavior can also be understood by analyzing the curve of the allowed motion (3.182).

First, the possibility of having globally regular solutions due to the presence of a cosmological constant is looked for. Equation (3.182), governing the dynamics of collapsing shells, can be conveniently written as

$$\dot{R}^2 = \frac{V(R,r)}{3R}. \tag{3.186}$$

Here $V(R,r)$, defined as

$$V(R,r) = 3F(r) + 3f(r)R + \Lambda R^3, \tag{3.187}$$

is an analog of the Newtonian effective potential governing motion of the shells. The allowed region of motion corresponds to $V \geq 0$, as $\dot{R}^2$ has to be greater than zero.

If an initially collapsing state is the starting point, a rebounce will occur if $\dot{R} = 0$ for a given shell before that shell becomes singular. This can happen only if the equation

$$3R\dot{R}^2 = V(R,r) = 0 \tag{3.188}$$

has two real positive roots. In the following, it is convenient to define a quantity $\zeta(t,r) = R/r$. Equation (3.187) can then be rewritten as

$$V = 3F + 3f\zeta r + \Lambda\zeta^3 r^3, \tag{3.189}$$

which is a cubic equation with three roots in general. From the theory of cubic equations, if all the three roots of the equation above are real, then at least one of them has to be positive and at least one negative. Note that $V(R=0, r>0) = 3F > 0$. Hence, any regular region between $R = 0$ and the first zero of (3.189), that is, $\zeta_1 > 0$, always becomes singular during collapse and so two real positive roots for (3.189) are required for the possibility of a rebounce. The region between the two real positive roots is a forbidden region. So, starting on the right side, the collapsing shells bounce back and then there is continuous expansion. Since one of the real roots has to be negative and the region between two real positive roots is forbidden, it is not possible to have oscillating solutions in these spacetimes.

In the case of $\Lambda = 0$, it is well-known that a rebounce is not possible, and the collapse necessarily results in a singularity. The cubic then reduces to a linear equation and the solution is given as $R = R_{\max}(r) = -F/f$. So, only in the case where $f < 0$ can $\dot{R} = 0$ for positive R. This corresponds to the maximum possible physical radius for a given shell, that is, even if an initial expansion is the starting point, a given shell will reach the maximum radius $R_{\rm max}(r)$, and then it will recollapse.

In the case when $\Lambda < 0$, one, and only one, root is always positive. The other two roots are negative if $9F^2 < -(4f^3/\Lambda)$, or else they are complex conjugates. Therefore, any initial configuration becomes singular in this case. The real positive root in this case gives an upper bound on the radius $R = R_{\rm max}(r)$ of a shell labeled r. This upper bound occurs as the negative (attractive) contribution from the Λ term keeps on increasing with an increasing R, while the contribution from gravitational attraction keeps on decreasing, and so at some point for any value of f, the attraction due to Λ starts to dominate. Hence, even if there is an initially expanding configuration, finally there must always be a collapse in this case.

For $\Lambda > 0$ and $f(r) \geq 0$, it can easily be seen that it is never possible to have $\dot{\zeta}^2 = \dot{R}^2/r^2 = 0$, so a singularity always forms if an initial collapse is always started from.

For the case when $\Lambda > 0$ and $f(r) < 0$, one root of the cubic above is always negative. If

$$F^2 > -\frac{4f^3}{9\Lambda}, \tag{3.190}$$

then the other two roots are complex conjugates. So, the singularity always forms in such a case in an initially collapsing configuration. On the other hand, if the initial data is such that

$$F^2 < -\frac{4f^3}{9\Lambda}, \tag{3.191}$$

then the other two roots are real and positive. Denote these roots by ζ_1 and ζ_2 with $\zeta_1 < \zeta_2$, and the region between these two roots is forbidden. The entire space of allowed dynamics is given by the two disjoint regions $[0, \zeta_1]$ and $[\zeta_2, \infty]$. If the initial scale factor ζ_0 lies in the first section, i.e. if $\zeta_0 < \zeta_1$, then the singularity is always the end point of the collapse. Here, $r\zeta_1$ represents the upper bound for the physical radius of a shell in this region. If ζ_0 lies in the second section, that is, $\zeta_0 > \zeta_2$, then there will be a rebounce from the initial collapsing configuration. After the rebounce, the physical radius of the shell keeps on increasing forever. There is no upper limit for the maximum value of ζ in this region and $r\zeta_2$ gives the lower bound for the physical radius of a shell, that is, the shell rebounces at $R = r\zeta_2$.

From the above discussion, it can be seen that a rebounce is possible only in the case when $\Lambda > 0$ and $f < 0$, and when the following two conditions are satisfied:

$$F^2 < -\frac{4f^3}{9\Lambda} \tag{3.192}$$

and $\zeta_0 > \zeta_2$. With the scaling used here, the condition (3.192) can be written as

$$1 > 2\left(-\frac{f}{r^2\Lambda}\right)\cos\left[\frac{1}{3}\arccos\sqrt{-\frac{9F^2\Lambda}{4f^3}}\right]. \tag{3.193}$$

The physical quantities such as central density, $\rho_c = F_0/\xi^3(t)$, and curvature scalars,

$$R^{ijkl}R_{ijkl} = 12\left(\frac{\dot{\zeta}^4+\zeta^2\ddot{\zeta}^2}{\zeta^4}\right),\ R^{ij}R_{ij} = 12\left(\frac{\dot{\zeta}^4+\zeta\dot{\zeta}^2\ddot{\zeta}+\zeta^2\ddot{\zeta}^2}{\zeta^4}\right), \tag{3.194}$$

and

$$g^{ij}r_{ij} = 6\left(\frac{\dot{\zeta}^2+\xi\ddot{\zeta}}{\zeta^2}\right) \tag{3.195}$$

stay finite, as $\zeta(t,r) > 0$ for the regular models. For further details see Deshingkar *et al.* (2001) and Arun Madhav *et al.* (2005).

There are several features here that are worth noting, and which have an interesting physical significance as far as the dynamics of the collapse is concerned. These also illustrate the effects that a non-vanishing cosmological term may have towards determining the final fate of a collapsing cloud of matter. First, with a negative value of the cosmological term, all the solutions become closed and a singularity always forms in the future even if one starts with initial expansion. This is to be expected because such a value will only contribute in a positive manner to the overall gravitational attraction of matter, and it just acts as a constant positive energy field helping the collapse. Next, there is a range of initial data where the collapse necessarily ends in a singularity however large the positive value of the cosmological term is. This is contrary to the belief sometimes expressed that a positive cosmological constant can always cause a rebounce, provided it is sufficiently large. Finally, it is clear from the above that there can be a rebounce only if the initial density is sufficiently low for a given positive value of Λ. This is so because the cosmological term becomes dominant with increasing distances, and gravity dominates at higher densities. Therefore, as the cloud is more disperse, with a lesser density but a larger size, the effect of the cosmological term is greater.

It can therefore be seen that while a bounce and a regular solution occur for a specific range of the initial data, for the majority of the regular initial data space, the collapse results in a spacetime singularity where the densities and curvatures blow up. While for $\Lambda = 0$ the structure of this singularity and when it will be naked or covered is known in detail, the above discussion gives some understanding of the effects of a non-zero Λ towards the structure of the singularity forming in the collapse. To see this more clearly for a dust collapse, the case $f(r) = 0$ is now analyzed explicitly in some detail. While $f = 0$ has been chosen for the simplicity and clarity of the considerations, similar behavior would also be expected in other cases. Equation (3.182) can now be written as

$$t - t_c(r) = \pm \int \left(\frac{F(r)}{R} + \frac{\Lambda}{3} R^2 \right)^{-1/2} dR, \tag{3.196}$$

where $t_c(r)$ is an integration function that represents the time at which a given shell, at a value r, becomes singular, i.e. $R(t_{sf}(r), r) = 0$. The positive or negative sign respectively corresponds to the expanding and collapsing branches of the solution. Here, only the negative sign is considered because clouds that are collapsing with $\dot{R} < 0$ initially are being considered. The integral (3.196) can be written as an infinite series in R near the center as

$$t - t_c(r) = -\frac{2}{3} \frac{R^{3/2}}{\sqrt{F(r)}} \left[1 + \sum_{m=1}^{\infty} \frac{(-1)^m (2m-1)!!}{2^m (2m+1) m!} \left(\frac{\Lambda R^3}{3F(r)} \right)^m \right]. \tag{3.197}$$

Using the scaling freedom in this solution, $R(t = 0, r) = r$ can be set. This determines $t_c(r)$ as

$$t_c(r) = t_{sf}(r) = \frac{2}{3} \frac{r^{3/2}}{\sqrt{F(r)}} \left[1 + \sum_{m=1}^{\infty} \frac{(-1)^m (2m-1)!!}{2^m (2m+1) m!} \left(\frac{\Lambda r^3}{3F(r)} \right)^m \right]. \tag{3.198}$$

From the above expressions,

$$R' = \frac{F'}{3F} R + \left(1 - \frac{rF'}{3F} \right) \left(\frac{3F + \Lambda R^{3/2}}{3F + \Lambda r^{3/2}} \right)^{1/2} \left(\frac{r}{R} \right)^{1/2}. \tag{3.199}$$

Therefore, the condition $R' > 0$, implying no shell-crossings, can be satisfied if $F' > 0$ and $1 - (rF'/3F) > 0$. This means that the total mass inside a shell and the matter density are increasing and decreasing functions of r respectively. Incidently, the above expression also gives the conditions for no shell-cross singularities to occur in the dust collapse when the cosmological term also vanishes. It would appear that, in any physically reasonable collapse picture, these conditions will be satisfied and shell-crossings will not occur. The weak energy condition, implying positivity of energy density, guarantees that mass is an increasing function of r. Also, for any realistic density distributions, it would be physically reasonable that the density is higher at the center, decreasing away from the center. Therefore, decreasing density profiles will be worked with here. No shell-crossings thus occur in the spacetime before the occurrence of the central shell-focusing singularity for the choice of initial data.

As seen below, the behavior of collapsing shells near the center would depend only on the first non-vanishing derivatives of the density and velocity profiles. Therefore, the local visibility conditions are unaffected by the boundary conditions such as the initial choices of the mass function and the actual value of the Λ term. On the other hand, the global behavior of the trajectories emerging from the singularity can change due to the addition of a non-zero cosmological term.

To analyze the nature of the central singularity, the method developed by Joshi and Dwivedi (1993a), which gives a necessary and sufficient condition for the local visibility of the singularity is followed. The main idea here is to see if outgoing radial null geodesics in the spacetime, meeting the singularity in their past with a well-defined real positive tangent vector in a suitable plane are possible.

The equation for these null geodesics in the spacetime (3.181), for $f(r) = 0$ is given by

$$\frac{dt}{dr} = R'. \tag{3.200}$$

For convenience, this can be written in the (u, R) plane as

$$\frac{dR}{du} = \frac{R'}{\alpha r^{\alpha-1}}\left(1 - \sqrt{\frac{F}{R} + \frac{\Lambda}{3}R^2}\right), \tag{3.201}$$

where $u = r^\alpha$, and $\alpha \geq 1$ is a constant to be determined later. From (3.201), it is clear that there are no such outgoing paths existing from the non-central part of the singularity curve, because the first term under the square root sign goes to $-\infty$ as t approaches $t_{\rm sf}(r)$. Hence, it is only the central singularity that can be possibly visible.

Define $X = R/u$. The purpose here is to check if a well-defined tangent for (3.201), at $r = 0$, $R = 0$ in the limit of t approaching the singular epoch $t_{\rm s}(0)$ is possible. Using l'Hospital's rule,

$$X_0 = \lim_{u\to 0,\, R\to 0} \frac{R}{u} = \lim_{u\to 0,\, R\to 0} \frac{dR}{du} = \lim_{u\to 0,\, R\to 0} \frac{R'}{\alpha r^{\alpha-1}}$$
$$\left(1 - \sqrt{\frac{F}{R} + \frac{\Lambda}{3}R^2}\right) \equiv U(X_0, 0), \tag{3.202}$$

where the subscript 0 denotes the value of the quantities at $u = 0$. The constant α can be uniquely fixed by demanding that $R'/r^{\alpha-1}$ is non-zero and finite (Joshi and Dwivedi, 1993a). In this case, $R'/r^{\alpha-1}$ remains finite if

$\alpha = 1 + 2n/3$ is chosen, and (3.202) can be written as

$$\frac{1}{\alpha}\left[X - \frac{nF_n}{3F_0}\frac{1}{\sqrt{X}}\right]\left[1 + \sqrt{\frac{\lambda_0}{X}}\right] = 0, \tag{3.203}$$

where $\lambda = F/u$ and λ_0 is the limit as $r \to 0$. If the above has a real positive root X_0 then there will be at least one null geodesic emerging from the singularity at $R = 0, u = 0$ with the root $X = X_0$ as a tangent in the (u, R) plane.

When $n < 3$, $\alpha < 3$ and therefore $\lambda_0 = 0$, and the above equation reduces to

$$X_0^{3/2} = -\frac{F_n}{2F_0\sqrt{1 + \Lambda F_0/3}}, \tag{3.204}$$

which always has a real positive root. Apart from an additional Λ term, this equation is analogous to the corresponding equation obtained for the TBL models, and reduces to the same case for $\Lambda = 0$. Therefore, when either the first or the second derivative of density is non-zero, the singularity is always at least locally visible. The addition of the cosmological term changes only the value of the tangent (X_0) to the outgoing radial null geodesics from the singularity, but not the visibility property itself. Therefore, the corresponding dust naked singularity spectrum is stable to the addition of a positive cosmological constant.

In this case, as in the earlier studies on TBL models, the smaller root will be along the apparent horizon direction and a family of geodesics will come out along this direction. In this case, set $\alpha = 3$, and the first term in (3.201) blows up, and the second term goes to zero such that the product is F_0. For the value $n = 3$, which corresponds to the critical case in the TBL models as discussed above, $\lambda_0 = F_0$ is obtained. Introducing $X = F_0 x^2$ and $\xi_\Lambda = F_3/F_0^{5/2}\sqrt{1 + \Lambda F_0/3}$, the root equation above becomes

$$2x^4 + x^3 - \xi_\Lambda x + \xi_\Lambda = 0. \tag{3.205}$$

This equation is similar to the corresponding case in the TBL models, with a modification in the definition of ξ. From the theory of quartic equations, this equation admits a real positive root for $\xi_\Lambda < \xi_{\text{crit}} = -(26 + 15/\sqrt{3})/2$. Therefore, for a given central density F_0, and the inhomogeneity parameter as given by F_3, the naked dust singularity can be partly covered by a positive Λ, because there is an additional positive term in the denominator in ξ_Λ. But it is interesting to note that, however large, a finite Λ term cannot completely cover the corresponding visible part of dust models.

In a similar manner, it is easy to see that a negative Λ will open up some covered part in the dust collapse final state spectrum. As such, the discussion above does not use the positivity of Λ, so the discussion goes through even for negative Λ. For all values of $n > 3$, $\lambda_0 = \infty$, and there cannot be a real positive root in these cases. The final singularity is then hidden behind the event horizon.

It can therefore be seen that the gravitational collapse of a dust cloud with a non-zero cosmological term can develop into a blackhole, a naked singularity, and even a globally regular solution, as the final outcome of a collapse. Each of these outcomes is determined by the choice of initial parameters, given in terms of the density and velocity profiles of the cloud. Although for simplicity, this was restricted to the $f(r) = 0$ case while analyzing the structure of the singularity, the results can be extended to the general case. The main aim here has been to examine how the presence of a non-zero cosmological term affects the dynamical evolution of the collapse. It has been found that there is a non-trivial detailed dynamical structure, especially for the $f < 0$ case, and this provides a strong motivation for the present study, namely how the introduction of a cosmological term non-trivially changes the collapse outcomes. In most cases, there is no rebounce, and then the nature of the singularity forming as the endstate of the collapse is studied. At a more technical level, this is reflected in the change in the structure of the roots that characterize the outgoing null geodesics, as discussed above.

In particular, this analysis shows that it is possible to cover a part of the naked singularity spectrum in the corresponding critical branch of solutions in the TBL models, when a positive cosmological term is introduced. However, the naked singularity of dust collapse is never removed, irrespective of the value of Λ. In this sense, the naked singularity remains stable to the introduction of a cosmological term.

It can be seen that there are certain important changes in the collapse dynamics with the introduction of a cosmological term, and the study here brings this out, allowing the implications of a non-zero Λ towards the final outcome of gravitational collapse to be understood. These results are of interest in view of the recent observational claims about a non-vanishing cosmological constant.

3.7 Equation of state

Here, type I matter fields, which is a rather general form of matter, have been worked with, and no specific equation of state has been imposed so far. However, it is important to note that suitable care must be taken in interpreting these results. Whilst it has been shown that the initial data and dynamical evolutions chosen do determine the blackhole and naked singularity endstates for collapse, not all these dynamical variables are explicitly determined by the initial data given at the initial epoch in terms of the matter and metric variables. Hence, these functions are fully determined only as a result of the time development of the system from the initial data, provided the relation between the density and pressures, that is, a given 'equation of state' is known.

In principle, it is possible to choose these functions freely. For example, one could specify the matter and velocity profiles at the initial epoch and also the dynamical evolutions, such as $F(v,r)$ and $\nu(v,r)$, only subject to an energy condition and regularity. Note that v here plays the role of a time coordinate and this then fully determines the collapse evolution. One can then calculate the energy density, and the radial and tangential pressures for the matter. However, in such a case, the resultant 'equation of state' could be quite strange in general. If any equation of state of the form $p_r = f(\rho)$ and $p_\theta = g(\rho)$ is given, then it is clear from (3.28) and (3.29) that there would be a constraint on the otherwise arbitrary functions $\mathcal{M}$ and A, which specify the required class, if the solutions of the constraint equations exist. As for the physics of very high density matter, presently, there is very little idea on what kind of an equation of state the matter would follow, especially at very high densities. Closer to the collapse endstates ultra-high energy densities and pressures are certainly important. Hence, if the possibility that the property of the matter fields or the equation of state could freely be chosen as above is allowed for, then the analysis is certainly valid and gives several useful conclusions on possible collapse endstates. In such a case, it is also possible that the chosen equation of state will be in general such that the pressures may explicitly depend not only on the energy density, but also on the time coordinate.

From such a perspective, it is now shown and pointed out below that the analysis, as given above, does in fact include several well-known equations of state and useful classes of collapse models as special cases. It is also seen that the energy conditions are satisfied throughout the collapse.

Discussed above was the idealized class of dust collapse models where the pressures were taken to be vanishing. This has been studied extensively and has yielded many important insights into collapse evolutions. In this special case, the Einstein equations can be solved completely and some remarks are made below on the N-dimensional generalization of the usual TBL dust collapse models. The metric in this case is given as

$$ds^2 = dt^2 - \frac{R'^2}{1 + r^2 b_0(r)} dr^2 - R^2(t,r) d\Omega^2_{N-2}. \tag{3.206}$$

The equations of motion are then written as

$$\frac{(N-2)F'}{2R^{(N-2)}R'} = \rho \tag{3.207}$$

and

$$\dot{R}^2 = \frac{F(r)}{R^{(N-3)}} + f(r). \tag{3.208}$$

In the case of dust, the mass function must be $F = F(r)$, and hence the regularity condition implies that

$$F(r) = r^{(N-1)}\mathcal{M}(r). \tag{3.209}$$

The energy condition here gives, in the range $0 < r < r_{\mathrm{b}}$, $\mathcal{M}(r) \geq 0$ and $3\mathcal{M} + r\mathcal{M}_r \geq 0$. In this case now, the function $\mathcal{X}(v)$ that was discussed earlier is given as

$$\mathcal{X}(v) = -\frac{1}{2}\int_v^1 \frac{v^{(N-3)/2}(\mathcal{M}_1 + v^{(N-3)}b_1)dv}{(\mathcal{M}_0 + v^{(N-3)}b_{00})^{3/2}}, \tag{3.210}$$

and the time taken for the central shell to reach the singularity is

$$t_{s_0} = \int_0^1 \frac{v^{(N-3)/2}dv}{\sqrt{\mathcal{M}_0 + v^{(N-3)}b_{00}}}. \tag{3.211}$$

It is now seen clearly that any given sets of density and velocity profiles at the initial epoch completely determine the tangent to the singularity curve at the central singularity. Also, (3.80) becomes

$$x_0^{(N-1)/2} = \frac{N-1}{2}\sqrt{\mathcal{M}_0}\mathcal{X}(0). \tag{3.212}$$

It therefore follows that, given any specific density profile of the collapsing dust cloud, a velocity profile can be chosen so that the endstate of the collapse would be either a naked singularity or a blackhole, depending on the choice made, such that the energy conditions are also satisfied throughout the collapse. The converse also holds, namely a given velocity profile can be chosen for the cloud at the initial epoch, and then there are density profiles that will lead the collapse to either the naked singularity or blackhole final states. As seen, these conclusions hold irrespective of the number of dimensions of the spacetime. In this sense, this treatment unifies and generalizes the earlier results of the dust collapse. The interesting point that comes out here is, given an initial density profile for the collapsing cloud, the space of velocity profile functions is divided into the regions that lead the collapse either to a blackhole or a naked singularity evolution, depending on the choice made, and the converse holds similarly.

While the dust equation of state discussed above is fairly standard and extensively used, it is widely believed that pressures would play an important role in gravitational collapse considerations (Jhingan and Magli, 2000;

Goncalves and Jhingan, 2001). Discussed below is a class of collapse models with non-zero pressures, which is, however, idealized in the sense that while the tangential pressure is allowed to be arbitrary, the radial pressure is taken to be static. A special case of this situation would be the zero radial pressure case. These are the purely tangential pressure collapse models that have been analyzed extensively to investigate the gravitational collapse final states in terms of blackhole and naked singularity formation.

As pointed out above, if the classes of collapse in which the radial pressure remains static is considered, the constraint equation for $\mathcal{M}$ has the solution

$$\mathcal{M}(r,v) = m(r) - p_r(r)v^3. \tag{3.213}$$

In addition to this, if there is an equation of state of the form $p_\theta = f(\rho)$, then this gives the constraint equation for the function $A(r,v)$ as

$$2f(\rho) = p_r + p_r'(R/R') + A(r,v)_{,v}[\rho(r,v) + p_r]. \tag{3.214}$$

For this class of models, the energy conditions are given by

$$3[m - p_r v^3] + r[m_{,r} - p_{r,r}v^3 - 3p_r v^2 v'] \geq 0, \tag{3.215}$$

$$3[m - p_r v^3] + r[m_{,r} - p_{r,r}v^3 - 3p_r v^2 v'] + p_r R^2 R' \geq 0, \tag{3.216}$$

$$\frac{\rho}{2} + \frac{1}{2}\left[(\rho + p_r)(A_{,v} + 1) + p_r'\frac{R}{R'}\right] \geq 0. \tag{3.217}$$

As shown in Goswami and Joshi (2002), classes of functions m, p_r, and A exist such that the naked singularity is the endstate for the collapse, and also the above three energy conditions are satisfied. It follows that, given the initial matter profiles, classes of collapse evolutions satisfying the energy conditions as seen above exist such that either the blackhole or naked singularity endstates can result, subject to the above equation of state.

Finally, also discussed here is the perfect fluid with a linear equation of state. This has been widely used in astrophysical considerations and a linear equation of state is well-studied. In general, there have been many studies that discuss the gravitational collapse of fluids (see for example, Cahill and Taub, 1971; Ori and Piran, 1987, 1990; Foglizzo and Henriksen, 1993; Harada, 1998; Rocha and Wang, 2000; Harada and Maeda, 2001; Goswami and Joshi, 2002; Ghosh and Deshkar, 2003; Giambo *et al.*, 2004). Discussed below is how the formalism outlined above applies to this case in order to find the blackhole and naked singularity configurations as perfect fluid collapse endstates.

For an isentropic perfect fluid, whose pressure is a linear function of the density only, the equation of state of the collapsing matter is given by

$$p_r(t,r) = p_\theta(t,r) = k\rho(t,r), \tag{3.218}$$

where $k \in [-1,1]$ is a constant. The case $k = 0$ gives the dust case, and $k = 1$ is the stiff fluid case. Presently only the case of positive pressures is considered. In this case $k > 0$, and the energy conditions give

$$M_{,v} < 0. \tag{3.219}$$

From the above equation of state and the Einstein equations it can immediately be seen that the function $\mathcal{M}$ is now the solution of the equation

$$(N-1)k\mathcal{M} + kr\mathcal{M}_{,r} + Q(r,v)\mathcal{M}_{,v} = 0, \tag{3.220}$$

where

$$Q(r,v) = (k+1)rv' + v. \tag{3.221}$$

Now, (3.220) has a general solution of the form

$$\mathcal{F}(X,Y) = 0, \tag{3.222}$$

where $X(r,v,\mathcal{M})$ and $Y(r,v,\mathcal{M})$ are the solutions of the system of equations

$$-\frac{d\mathcal{M}}{3k\mathcal{M}} = \frac{dr}{kr} = \frac{dv}{Q}. \tag{3.223}$$

Therefore, it can easily be seen that (3.220) admits classes of solutions when $v' > 0$. Also, solving the equation for the central shell $r = 0$, with boundary conditions $\rho \to \infty$ as $v \to 0$,

$$\mathcal{M}(0,v) = \frac{m_0}{v^3 k}, \tag{3.224}$$

where m_0 is a constant. By choosing $m_0 > 0$, the central shell can be made to satisfy the energy condition $\rho(t,0) > 0$ for all epochs. Then, by the continuity of the density function, it can be said that an ϵ-ball exists around the central shell, for which $v'(t,r) > 0$ and also $\rho(t,r) > 0$. But at the central singularity, $\sqrt{v}v' \approx \mathcal{X}(0)$. This implies that classes of solutions exist that satisfy the energy conditions and also admit a naked singularity as the collapse endstate. For a further discussion on perfect fluid collapse, and the details of blackhole and naked singularity formation, see Goswami and Joshi (2006) and references therein.

It is seen from the above discussion that several well-known classes of collapse models and equation of states form subcases of the considerations given here. Along with these models, the above analysis would work for any other models with other equations of state, if solutions to the constraint equations on $\mathcal{M}$ and A are permitted. Hence, it follows that the above provides an interesting framework for the study of dynamical collapse, which is one of the most important problems in gravitation physics today.

4
Cosmic censorship

As indicated here, the cosmic censorship hypothesis is fundamental to the basic theory and applications in blackhole physics. It follows from the previous considerations that the cosmic censorship, if it does hold, is not valid in any obvious and plain manner in general relativity, but it has to be carefully designed and formulated. This is important for any clear basis and foundation for blackhole physics. As yet, such a mathematical formulation is not available, and much work is needed to achieve one. In such a case, the existence and formation of blackholes as the final state of a gravitational collapse, whenever a massive star collapses in the universe on exhausting its nuclear fuel, cannot be taken as automatic.

Basically, cosmic censorship is a statement about the causal structure of spacetime, as related to the dynamical collapse scenarios that are fundamental processes in astrophysics and cosmology. Therefore, in this chapter, which aims to discuss various aspects related to the censorship conjecture and its possible formulations, are considered in Section 4.1 some basic aspects of the causal structure of spacetime. The physics related to gravity becomes much more important in situations such as gravitational collapse and those of the early universe phases in cosmology. The general theory of relativity predicts the occurrence of spacetime singularity in such situations, which are the extreme gravity regions where densities, spacetime curvatures, and other physical quantities take extreme values. The quantum gravity effects would become much more prominent in such regions. The occurrence of spacetime singularities in collapses and cosmology is discussed here in Section 4.2.

The problem of cosmic censorship is then to ensure that the spacetime singularities developing in a gravitational collapse scenario are necessarily hidden within the event horizons of gravity, thus ensuring that the final product is a blackhole only. Based on this assumption, some aspects of blackhole physics are discussed in Section 4.3. The validity, or otherwise, of censorship in higher-dimensional spacetimes is discussed in Section 4.4. The major difficulties in formulating a censorship hypothesis and any possible proof are highlighted, and various possible avenues towards developing a mathematical statement are considered in Section 4.5.

The basic approach here has been that such a formulation, as well as any possible proof for the censorship, cannot be achieved without an extensive study and investigation of the gravitational collapse scenarios, such as those conducted in the previous chapter. Even if naked singularities do form in the collapse of massive matter clouds in violation of censorship conjecture, their genericity and stability still remain important issues. There are, however, no definite criteria in general relativity to test stability and genericity, which are rather wide spectrum notions in gravitation theory today, and a range of possibilities are yet to be discussed and explored. These issues are considered in Section 4.6.

The conclusion that emerges from these considerations on cosmic censorship here is that one may not hope to formulate this conjecture in any definitive manner mathematically without an extensive study of gravitational collapse models in general relativity. On the other hand, a study of the properties and structure of naked singularities found in the models studied so far, and especially that of the physical processes in the vicinity of these extreme gravity regions that are visible, could lead to interesting physical implications. This latter possibility is discussed in the next chapter.

4.1 Causal structure

The causal relationships between events in a spacetime are determined by the Lorentzian metric, which decides the past and future light cones at every event of the manifold. While, locally, the geometry is always that of special relativity, the global behavior of light cones now depends on the choice of the metric tensor, which is a function of space and time coordinates. Therefore, even though locally the speed of light is not to be exceeded, globally the phenomena, such as the occurrence of closed timelike curves and causality violations, may be allowed in principle. The possibility of causality violations really depends on the overall global topology of the spacetime. Various causality conditions are imposed on the model under consideration in order to ensure a regular causal structure of the spacetime, as will be specified here.

While the Einstein equations govern the local dynamics of the matter and spacetime curvature, they say nothing about the global topology or causal structure of the spacetime universe. These are free to be chosen as per the appropriate model under consideration and the cosmological observations of the universe as a whole. As has been realized in the last few decades, these elements, namely the topology and causal structure, can have significant physical implications for cosmology and the universe as observed on larger scales.

The spacetime model worked with must satisfy several physical reasonability conditions in order to represent the physical universe as observed. One of these is that the spacetime should have a regular causal structure. The physical condition that no material particle signals can travel faster than the velocity of light fixes the causal structure for the Minkowski spacetime. Also, in general relativity, which uses the framework of a general spacetime manifold, it was noted that locally the causality relations were the same as in the Minkowski spacetime, which is the background manifold for special relativity. However, globally, there could be important differences in the causality structure and other physical properties, due to a different spacetime topology, as compared with that of the Euclidean space R^4 of the Minkowski spacetime (Geroch, 1967).

The strong gravitational fields that arise either in cosmology or gravitational collapse, as signaled by the occurrence of spacetime singularities in general relativity, also may have significant causality and topology implications for the structure of the spacetime in the vicinity of these ultra-strong gravity regions. The causal structure of the spacetime has been studied in detail, especially from the perspective of the occurrence of spacetime singularities in collapse and cosmology (see for example, Geroch, 1970b; Penrose, 1972; Hawking and Ellis, 1973). Here, in this section, this causal structure and its relationship with the spacetime topology are discussed. The known results and definitions required for later chapters are also reviewed and some results that are of intrinsic interest for the global aspects of spacetimes, and which emphasize the close interplay between the causal structure and topology, are given. Basic causal structure ideas and definitions are given, and topological properties of several spacetime sets and various causality conditions are stated in order to arrive at a unified causality statement for a reasonable model of spacetime. The results in the present section are valid for manifolds of arbitrary dimensions greater than or equal to two.

An event p *chronologically precedes* event q, denoted by $p \ll q$, if there is a smooth, future directed timelike curve from p to q. If such a curve is non-spacelike, namely timelike or null, p *causally precedes* q, or $p < q$. The *chronological future* $I^+(p)$ of p is the set of all points q such that $p \ll q$. The *chronological past* of p is defined similarly. Therefore,

$$I^+(p) = \{q \in M \mid p \ll q\}, \tag{4.1}$$

$$I^-(p) = \{q \in M \mid q \ll p\}. \tag{4.2}$$

The *causal future* (*past*) for p is defined similarly:

$$J^+(p) = \{q \in M \mid p < q\}, \tag{4.3}$$

$$J^-(p) = \{q \in M \mid q < p\}. \tag{4.4}$$

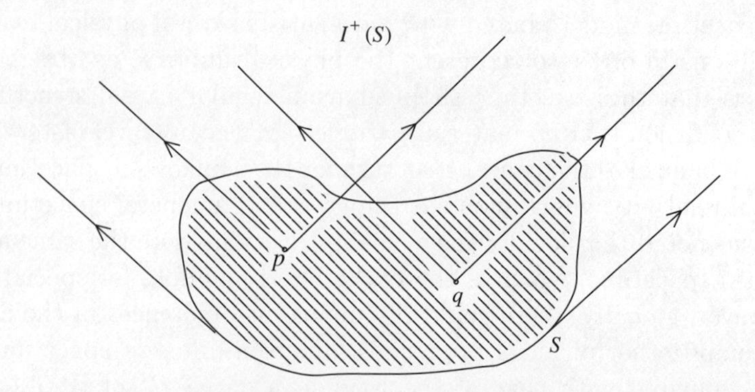

Fig. 4.1 The future of a set S is the union of the futures of all the events in this set.

The relations $\ll$ and $<$ are transitive, and for events p, q, and r, $p \ll q$ and $q < r$ or $p < q$ and $q \ll r$ implies $p < r$ (Penrose, 1972). It is seen from this that

$$\overline{I^+(p)} = \overline{J^+(p)}, \tag{4.5}$$

and also

$$\dot{I}^+(p) = \dot{J}^+(p), \tag{4.6}$$

where, for a set A, $\bar{A}$ is the closure of A and $\dot{A}$ denotes the topological boundary. The chronological (causal) future of any set $S \subset M$ is defined similarly (see Fig. 4.1),

$$I^+(S) = \bigcup_{p \in S} I^+(p), \tag{4.7}$$

$$J^+(S) = \bigcup_{p \in S} J^+(p), \tag{4.8}$$

and the chronological (causal) pasts of the subsets of the spacetimes are defined similarly. Such dual definitions or results will often be taken as granted.

Suppose there is a future directed timelike curve from p to q. There is a local region containing q that the timelike curve must enter in which special relativity is valid. Therefore, there is a neighborhood N of q such that any point of N can be reached by a future directed timelike curve from p. It is thus seen that for any event $p \in M$, the sets $I^+(p)$ and $I^-(p)$ are open in M. The above also implies that the sets $I^{\pm}(S)$ are open, as they are the union of open sets in M. However, the sets $J^{\pm}(p)$ are neither open nor closed in general (see for example, Hawking and Ellis, 1973, or Joshi, 1993). Therefore, all the points in the boundary of $J^+(S)$ are not necessarily connected to a point in S by a null geodesic generator.

In Minkowski spacetime, the set $I^+(p)$ is the set of those points that are reached by future directed timelike geodesics from p, and the boundary of this set is generated by null geodesics from p. As seen above, this is not true for an arbitrary spacetime in general, but locally this property is still valid as shown by the following result from Hawking and Ellis (1973). If (M, g) is a spacetime manifold and N is a convex normal neighborhood of $p \in M$, then, for any $q \in N \cap I^+(p)$, there is a timelike geodesic from p to q in N, and the boundary of $I^+(p)$ in N is generated by future directed null geodesics from p in N.

To derive properties of more general boundaries, a future set can be defined. A set F of M is called a *future set* if $F = I^+(S)$ for some subset S of M. An equivalent criterion is $I^+(F) \subset F$. Past sets are defined similarly. Clearly, future sets are open in M as they are the union of $I^+(p)$ for all $p \in S$.

To see the properties of future sets, note that if F is a future set, $\overline{F}$ is the set of all points x such that $I^+(x) \subset F$. To see this, suppose x is such that $I^+(x) \subset F$. Then, a sequence in $I^+(x)$ and hence in F can be constructed with the limit point x. Therefore $x \in \overline{F}$. Conversely, if $x \in \overline{F}$, then take $y \in I^+(x)$. This gives $x \in I^-(y)$, which contains an open neighborhood of x. Then, this neighborhood contains points of F, that is, $y \in I^+(F)$, which implies $I^+(x) \subset F$.

Again, the boundary of a future set F is made of all events x such that $I^+(x) \subset F$ but $x \notin F$. If $x \in \dot{F}$, then clearly $x \notin F$ as F is an open set. But $x \in \overline{F}$, and as seen above, $I^+(x) \subset F$. Conversely, let $x \notin F$ but $I^+(x) \subset F$. Then a sequence in $I^+(x)$ converging to x can clearly be constructed, which implies $x \in \overline{F}$. Therefore, x must be in the boundary of F.

A set S is called *achronal* if no two points of S are timelike related, that is, $I^+(S) \cap S = \emptyset$. If F is a future set, the following result from Hawking and Ellis (1973) shows that the boundary of F is a well-behaved achronal manifold. Let F be a future set, then the boundary of F is a closed, achronal C^0-manifold that is a three-dimensional embedded hypersurface.

The following result from Penrose (1972) then shows that the achronal boundary of a future set F is always generated by null geodesics that are either past endless or always have a past end point on $\overline{F}$. Let $S \subset M$ and $p \in \dot{I}^+(S) - \overline{S}$. Then, a null geodesic contained in the boundary of $I^+(S)$ exists with a future end point p, and which is either past endless or has a past end-point on $\overline{S}$.

From the above, it follows now that if $q \in J^+(p) - I^+(p)$, then any non-spacelike curve joining p and q must be a null geodesic, and that the boundary of $I^+(p)$ or $J^+(p)$ is generated by null geodesics that have either a past end point at p or are past endless.

Causal relations in a spacetime are defined by the existence of smooth non-spacelike curves between pairs of events. It is, however, useful to extend this to define causality by means of continuous curves. This is carried out by requiring that pairs of points on a curve are locally joined by a smooth timelike or causal curve. To be precise, a *continuous* curve λ is called a *future directed timelike (or non-spacelike)* curve if each $x \in \lambda$ is contained in a convex normal neighborhood N, such that if $\lambda(t_1), \lambda(t_2) \in N$ with $t_1 < t_2$, then there is a smooth future directed timelike (non-spacelike) curve in N from $\lambda(t_1)$ to $\lambda(t_2)$. Such curves are regarded as equivalent under a one–one continuous reparametrization.

It is useful to introduce the notion of future and past inextendible non-spacelike curves, which are effectively the trajectories that have no future or past end points. Let λ be a non-spacelike curve. Then $p \in M$ is called a *future end point* of λ if, for every neighborhood N of p, a value of the curve parameter t' exists such that for all $t > t'$, $\lambda(t) \in N$. The past end point is defined similarly. It is clear that if λ has an end point, it must be unique because M is Hausdorff. The curve λ is called *future* or *past inextendible* if it has no future or past end point respectively in M.

An inextendible causal curve might be running off to infinity, or it might end up in a spacetime singularity, or it could enter a compact set in which it could be trapped to go round and round for ever.

Note that the study of casual relations in a spacetime (M, g) is equivalent to that of the conformal geometry of M. Let (M, g) be the physical spacetime, and consider the set of all conformal metrics $\bar{g}$, where $\bar{g} = \Omega^2 g$, with Ω being a non-zero C^r-function. Then if $p \ll q$ or $p < q$ in (M, g), the same relation is preserved in $(M, \bar{g})$ for any conformal metric $\bar{g}$. Therefore, casual relationships are invariant under a conformal transformation of the metric. However, non-spacelike geodesics in (M, g) will no longer be geodesics in $(M, \bar{g})$ unless they are null. The null geodesics are conformally invariant up to a reparametrization of the affine parameter along the curve. Therefore, specifying causal relations in M fixes the spacetime metric up to a conformal factor. Let p be an event in M, and N be a convex normal neighborhood of p. Minkowskian coordinates $\{x^i\}$ can be introduced in N, in which case it follows from the above that the set of events in N that are causally connected to p satisfy

$$-(x^0)^2 + (x^1)^2 + (x^2)^2 + (x^3)^2 \leq 0. \tag{4.9}$$

The boundary of these points define the null cone in the tangent space T_p. Suppose now that $\boldsymbol{W}$ and $\boldsymbol{Z}$ are any two non-null vectors in T_p, then,

$$g(\boldsymbol{W}, \boldsymbol{Z}) = \tfrac{1}{2}\left[g(\boldsymbol{W} + \boldsymbol{Z}, \boldsymbol{W} + \boldsymbol{Z}) - g(\boldsymbol{W}, \boldsymbol{W}) - g(\boldsymbol{Z}, \boldsymbol{Z})\right]. \tag{4.10}$$

Now, if $\boldsymbol{X}, \boldsymbol{Y} \in T_p$ are a timelike and a spacelike vector respectively, then the equation

$$g(\boldsymbol{X}+\lambda\boldsymbol{Y}, \boldsymbol{X}+\lambda\boldsymbol{Y}) = g(\boldsymbol{X},\boldsymbol{X}) + 2\lambda g(\boldsymbol{X},\boldsymbol{Y}) + \lambda^2 g(\boldsymbol{Y},\boldsymbol{Y}) = 0 \quad (4.11)$$

has two distinct roots λ_1 and λ_2, as $g(\boldsymbol{X},\boldsymbol{X}) < 0$ and $g(\boldsymbol{Y},\boldsymbol{Y}) > 0$. The knowledge of the null cone then implies that λ_1 and λ_2 can be determined in principle. But $\lambda_1\lambda_2 = g(\boldsymbol{X},\boldsymbol{X})/g(\boldsymbol{Y},\boldsymbol{Y})$. Hence, the null cone gives the ratio of the magnitudes of a timelike and a spacelike vector. Therefore, each term in the equation above is determined up to a factor, and so $g(\boldsymbol{W},\boldsymbol{Z})$ is determined up to a factor.

The local causality principle implies that, over small regions of space and time, the causal structure is the same as in the special theory of relativity. However, on a larger scale global pathological features, such as the violation of time orientation, possible non-Hausdorff nature or non-paracompactness, disconnected components of spacetime, and such others, may show up. Such pathologies are to be ruled out by means of reasonable topological assumptions, and one would like to ensure that the spacetime is causally well-behaved. This is carried out by means of introducing various causality conditions such as the non-occurrence of closed timelike or non-spacelike curves (causality), and the stability of this condition under small perturbations in the metric (stable causality). In fact, Carter (1971) pointed out that there is an infinite hierarchy of such causality conditions for a spacetime.

It would appear reasonable to demand that physically realistic spacetimes do not allow either closed timelike or closed non-spacelike curves, as this would give rise to the phenomenon of entering one's own past. However, general relativity and Einstein's equations as such do not rule out such a possibility on their own. For example, the Gödel universe (Gödel, 1949) has closed timelike curves through each point of the spacetime. Again, the global topology of M can cause closed timelike curves. For example, for the cylinder $M = S^1 \times R$, obtained from the Minkowski spacetime by identifying $t = 0$ and $t = 1$ hypersurfaces with the metric given by $ds^2 = -dt^2 + dx^2$, the circles $x =$ const., are closed timelike curves. In fact, for all $p \in M$ here, $I^+(p) = I^-(p) = M$. Such examples could be discarded as mathematical pathologies in the spacetime topology, and the Gödel universe may be termed unrealistic because it is a rotating model that does not correspond to the observed universe.

More difficult to rule out are the Kerr solutions of a spinning gravitational source (Kerr, 1963), which contain closed timelike curves if the rotation is sufficiently fast in comparison with the value of the mass parameter. These could possibly represent the final fate of a massive collapsing star that is rotating. If a star failed to get rid of enough spin during the process of collapse, it would give rise to a time machine in the spacetime. Wormholes in a spacetime representing the multiply connected nature of the topology of space could also give rise to closed timelike curves, as indicated earlier. The physical significance and acceptability of such causality violations have been examined by Morris, Thorne, and Yurtsever (1988), and by Friedman *et al.* (1990). It follows from these considerations that such wormhole spacetimes do admit unique solutions to at least simple field equations, such as a single non-interacting scalar field. This shows that even though the causality violation is perceived as contradictory to the predictability requirements in the spacetime, in general this need not be so, as there are wormhole spacetimes where this is not the case.

Spacetimes with closed non-spacelike curves are avoided by requiring M to satisfy the *causality condition*; that is, (M, g) does not admit any closed timelike or null curves. A spacetime M is said to be *chronological* when it admits no closed timelike curves, that is, $p \notin I^+(p)$ for all $p \in M$. When M admits no closed non-spacelike curves, it is said to be causal. If $M = S^1 \times R$ and the metric is chosen to be $ds^2 = dt\,dx$, then M is chronological but not causal. The circles $x =$ const. are null geodesics here.

The causally well-behaved nature of M turns out to be closely related to the topological structure of M in that if M is chronological, M cannot be compact. To see this, suppose M is compact. The sets $\{I^+(p) \mid p \in M\}$ cover M. Now, compactness implies that a finite set of points $p_1, \ldots, p_n$ exists, such that the set $I^+(p_1) \cup \cdots \cup I^+(p_n)$ covers M. Now, $p_1, \ldots, p_n$ must be in the cover implies that $p_1 \gg p_{i_1}$ for some i_1 in $1, \ldots, n$. Therefore, $p_1 \gg p_{i_1} \gg \ldots \gg p_{i_{n-1}}$, where all n-points have been exhausted. Hence, $p_{i_k} \gg p_{i_k}$ for some i_k that violates chronology.

Even though a spacetime may be causal, it could be on the verge of violating causality. Consider (M, g) given by $ds^2 = dt\,dx + t^2 dx^2$. To see the behavior of light rays, consider the null geodesics of this spacetime that are given by $dt/dx = -t^2$. At $t = 0$, $dt/dx = 0$, and so one arm of the light cone will lie along the x-axis. In this situation, there is a spacetime that is causal, but a non-spacelike curve from p can enter arbitrary small neighborhoods of p. For distinct events p and q along the x-axis, $I^+(p) = I^+(q)$. To avoid such a causal pathology, the *distinguishing condition* can be imposed, namely that for all p and q in M, $I^+(p) = I^+(q)$ implies $p = q$ and $I^-(p) = I^-(q)$ implies $p = q$. A similar but stronger condition is *strong causality* (Penrose, 1972), which states that for all events $p \in M$, every neighborhood of p contains a neighborhood of p that no non-spacelike curve in M intersects more than once. If strong causality is violated at p, then there are non-spacelike curves from neighborhoods of p that come arbitrary close to intersecting themselves.

The relevance of strong causality is that it determines the topology of spacetime. First, note that the *Alexandrov topology* or the interval topology on M is defined by taking the collection $\{I^+(p) \cap I^-(q) \mid p, q \in M\}$ as the basis for the topology on M. The following from Penrose (1972) can then be shown. Let (M, g) be a strongly causal spacetime. Then, the manifold topology on M is the same as the Alexandrov topology.

Therefore, for a strongly causal spacetime, the causal relations determine the topological structure of M. It can also be shown that the Alexandrov topology will be Hausdorff for a strongly causal spacetime. Next, strong causality imposes a regularity on M in the sense that if γ is any future inextendible non-spacelike curve, then γ cannot be either totally or partially future imprisoned in any compact set S in the sense that γ will enter and remain within S, or it will continually re-enter S. Therefore, for a strongly causal M, the future endless curve γ must go either to infinity or terminate in a spacetime singularity, that is, it must go to the boundary of the spacetime in either case.

Even though (M, g) is strongly causal, it is possible to create examples where a small perturbation in the metric tensor components g_{ij} will give rise to closed timelike curves. Therefore, a strongly causal spacetime could still be on the verge of causality violation. Now, general relativity is supposed to be a classical approximation of some, as yet unknown, quantum theory of gravity in which a precise measurement of the metric components at a single event would not be possible. Therefore, one would like the causality of spacetime to be preserved under small perturbations in the metric. This is achieved by requiring the spacetime to be *stably causal*. Let (M, g) be a spacetime, then another metric $g' > g$ exists, such that there is no closed non-spacelike curve in g' if there was none in g. Here, $g' > g$ means that every non-spacelike vector in g is always timelike in g'. It is of course clear, by definition, that if $g' < g$ then g' will have no closed non-spacelike curves if g had none.

Of the infinite hierarchy of causality requirements available for a spacetime, stable causality may be considered to be the most relevant physically and the one that provides a unified criterion for basic causal regularity. Such a spacetime is causal and its stability is ensured in a suitable manner. A stably causal M is strongly causal and hence causality determines the topology of the spacetime. Finally, a stably causal spacetime admits the existence of a *global time function* on M, that is, a smooth scalar field f, for which the gradient is always timelike (Hawking and Ellis, 1973). Such a function assigns a value of 'time' to each point in M so that this time is strictly increasing along all timelike curves. This is precisely the property that is desired for a global assignment of a time coordinate in M. Of course, such a time function is not unique.

A spacetime (M, g) is called *globally hyperbolic* if the sets $J^+(x) \cap J^-(y)$ are compact for all x and y in M, and M is causal. For a globally hyperbolic spacetime, the light cones are necessarily closed. To see this, suppose $J^+(x)$ is not closed, and let $y \in \overline{J^+(x)}$, but $y \notin J^+(x)$. Let $z \in I^+(y)$. Then, $y \in \overline{J^+(x) \cap J^-(z)}$, but $y \notin J^+(x) \cap J^-(z)$. This is a contradiction as

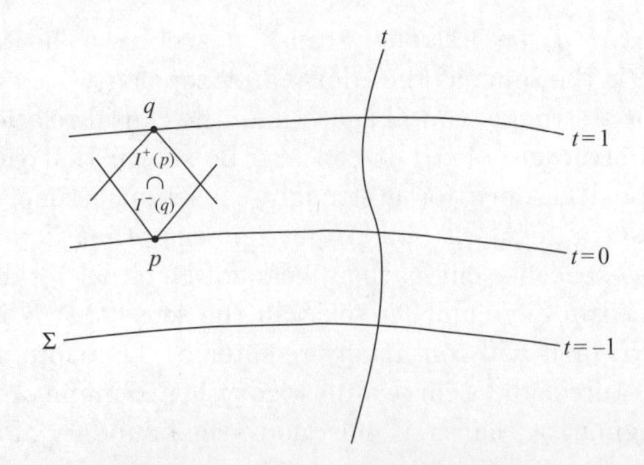

Fig. 4.2 A globally hyperbolic spacetime has a fixed topology $M = \Sigma \times R$, and is foliated entirely by spacelike surfaces Σ. For any events p and q, the intersection $I^+(p) \cap I^-(q)$ is compact. Such a spacetime obeys strong cosmic censorship and allows no naked singularities.

$J^+(x) \cap J^-(z)$ is closed as it is compact. It can also be shown that if S is a compact set, $J^+(S)$ will be closed in M, and if S_1 and S_2 are any two compact subsets, $J^+(S_1) \cap J^-(S_2)$ must be compact.

From earlier discussions, it can be shown that if M is globally hyperbolic, then it is strongly and stably causal. In fact, global hyperbolicity is a rather strong condition on M that uniquely fixes the overall topology of the spacetime (see Fig. 4.2). The spacetime then has a very regular global behavior, and all the various pathological features stated in the above are removed. Physically interesting spacetimes such as the Schwarzschild solution, the Friedmann–Robertson–Walker cosmological solutions, and the steady state models are all globally hyperbolic.

Another strong motivation for global hyperbolicity is the cosmic censorship hypothesis, which rules out timelike or naked singularities. It turns out that if even the locally naked singularities are ruled out, the resulting spacetime structure must be globally hyperbolic. In the absence of cosmic censorship, the deterministic structure or global hyperbolicity breaks down as a Cauchy horizon develops.

The global hyperbolicity of M is closely related to the future or past development of initial data from a given spacelike hypersurface in the spacetime. Let S be a closed achronal set. The *edge* of a set S, $E(S)$, is defined as a set of points $x \in S$ such that every neighborhood of x contains $y \in I^+(x)$ and $z \in I^-(x)$ with a timelike curve from z to y that does not meet S. When S is a closed achronal set without an edge, the result is that S is a three-dimensional, embedded, C^0-submanifold of M. A *partial Cauchy surface* S is defined as an acausal set without an edge. Therefore, no non-spacelike curve intersects S more than once and S is a spacelike hypersurface. The *future domain of dependence* of S, denoted by $D^+(S)$, is defined as the set of all points $x \in M$ such that every past inextendible non-spacelike curve from x intersects S. It is clear that $S \subset D^+(S) \subset J^+(S)$ and as S is achronal, $D^+(S) \cap I^-(S) = \emptyset$. The *past domain of dependence* $D^-(S)$ for a partial Cauchy surface S is defined similarly. The full *domain of dependence* for S is defined as

$$D(S) = D^+(S) \cup D^-(S). \tag{4.12}$$

A partial Cauchy surface is called a *Cauchy surface* or a *global Cauchy surface* if $D(S) = M$. Clearly, for a Cauchy surface S, edge$(S) = \emptyset$. Every non-spacelike curve in M must meet S once (and exactly once) if S is a Cauchy surface. The relationship between the global hyperbolicity of M and the notion of a Cauchy surface is that M is globally hyperbolic if and only if it admits a spacelike hypersurface S that is a Cauchy surface for M.

Note that even when M is globally hyperbolic, all spacelike surfaces in M need not be Cauchy surfaces. For example, for the Minkowski spacetime, the spacelike surfaces $t =$ const. are global Cauchy surfaces, but the hyperboloids

$$t^2 - x^2 - y^2 - z^2 = \text{const.} \tag{4.13}$$

are not, as the past or future null cones of the origin are boundaries of the domain of dependence for these spacelike surfaces. Again, if a point from the Minkowski spacetime is deleted, the resulting M admits no Cauchy surface and M is not globally hyperbolic.

It was shown by Geroch (1970a) that a globally hyperbolic M has a unique topological structure, and it admits no topology change in that M is then homeomorphic to $R \times S$, where S is a three-dimensional submanifold and for each $t \in R$, $\{t\} \times S$ is a Cauchy surface for M.

The basic idea of the proof for the above involves introducing a finite measure μ on M so that $\mu(M) = 1$. Then, a function $h : M \to R$ is introduced by

$$h^+(p) = \frac{\mu(J^+(p))}{\mu(J^-(p))}. \tag{4.14}$$

The function h^- is defined similarly. The sets $h^\pm =$ const. are seen to be Cauchy surfaces for M. Such functions are called *causal functions* on a spacetime.

It can be asked if the converse is true in some sense, that is, whether the direct product spacetimes (M, g) with $M = S \times T$, where each $S \times \{t\}$ is spacelike and each $\{x\} \times T$ is timelike, are always globally hyperbolic. The answer is no; refer to Clarke and Joshi (1988), who discuss a spacetime that is a direct product as above, but not globally hyperbolic.

Next, let S be a partial Cauchy surface. Then $N = D^+(S) \cup D^-(S) \neq M$ and N must be a proper subset of M. The boundary of N in M can be divided into two regions $H^+(S)$ and $H^-(S)$ that are respectively called the *future* and *past Cauchy horizons* of S,

$$H^+(S) = \{x \mid x \in D^+(S), I^+(x) \cap D^+(S) = \emptyset\}. \tag{4.15}$$

The past horizon $H^-(S)$ is defined in a similar manner. Even though M may not be globally hyperbolic and S is not a Cauchy surface, the region $\mathrm{Int}(D^+(S))$ or $\mathrm{Int}(D^-(S))$ is globally hyperbolic in its own right and the surface S serves as a Cauchy surface for the manifold $\mathrm{Int}(N)$. Therefore, $H^+(S)$ or $H^-(S)$ represent the failure of S to be a global Cauchy surface for M.

The set $H^+(S)$ is achronal and closed. To see this, suppose $x, y \in H^+(S)$ with $x \ll y$. Then, $x \in I^-(y)$ and a neighborhood $N_x \subset I^-(y)$ exists. Let $p \in N_x \cap I^+(x)$. Then, $p \ll y$ and every past directed timelike curve from p can be extended to be a past directed curve from y, which must meet S. Therefore, $p \in D^+(S)$, which is a contradiction to $D^+(S) \cap I^+(x) = \emptyset$. Hence, no two points of $H^+(S)$ are timelike related. Next, let x be a limit point of a sequence of points $\{x_n\}$ in $H^+(S)$. Suppose γ is a past directed timelike curve not meeting S, with a future end point at x. Then, for any $y \in \gamma$, $I^+(y)$ contains all points $\{x_n\}$ for some $n \geq k$. Then, coming from x_n to y and following γ, gives a non-spacelike curve from x_n not meeting S, which is contradictory to $x_n \in D^+(S)$.

As in the case of the boundary of a future set, $H^+(S)$ is generated by null geodesics that are either past inextendible in $H^+(S)$ or have a past end point on the edge of S (Geroch, 1970a). The basic physical relevance of global hyperbolicity on M is that it implies a deterministic structure for the spacetime in the sense characterized by the above results, given initial data defined on a Cauchy surface S. The notion of global hyperbolicity was introduced by Leray (1952), and it is seen from considerations on the Cauchy problem in general relativity (see for example, Wald, 1984, for a review), that if N is a globally hyperbolic subset of M, the wave equation for a δ-function source at any $p \in N$ has a unique solution that vanishes outside $N - J^+(p)$. The definition of global hyperbolicity given by Leray (1952) involves the space $C(p,q)$, which is the set of all C^0 non-spacelike curves from p to q that are the same up to a reparametrization. Defining a C^0-topology on this space of curves, by stating that the two curves are nearby if their points in M are close enough, allows the global hyperbolicity to be characterized in term of $C(p,q)$ as seen by Seifert (1967). Let M be a strongly causal spacetime. Then, M is globally hyperbolic if and only if $C(p,q)$ is compact for all $p, q \in M$.

An important property of globally hyperbolic spacetimes, or any globally hyperbolic subset in M, which is relevant for the singularity theorems is the existence of maximum length non-spacelike geodesics between pairs of causally related events. In a complete Riemannian manifold with a positive definite metric, any two points can be joined by a geodesic of minimum length, and, in fact, such a geodesic need not be unique. The analog of this result for Lorentzian metrics was given by Avez (1963) and Seifert (1967). If (M, g) is globally hyperbolic and $p, q \in M$ such that $p < q$, then there is a non-spacelike geodesic from p to q whose length is greater than or equal to that of any other future directed non-spacelike curve from p to q.

Therefore, in globally hyperbolic spacetimes, there is a finite upper bound on the proper time lengths of non-spacelike curves between two chronologically related events. It is clear that there is no lower limit of lengths for such curves except zero, because the chronologically related events can always be joined using broken null curves that could give an arbitrary small length curve between them. Similarly, if S is a Cauchy surface in a globally hyperbolic spacetime, then for any point p in the future of S, there is a past directed timelike geodesic from p orthogonal to S that maximizes the lengths of all non-spacelike curves from p to S.

The ideal points boundary of the spacetime (that is, the points at infinity and singularities) are now discussed. Although the points at infinity and singularities in a spacetime are not regular points for M, they can be attached to M as an additional boundary. Such a boundary construction including both the points at infinity and singularities was provided by Geroch, Kronheimer, and Penrose (1972), based only on the causal structure of spacetime. Here, this attached causal boundary or ideal points will be classified using the properties of causal functions.

A non-empty subset P of M is said to be a *past set* if there is some $A \subset M$ such that $I^-(A) = P$. If P cannot be expressed as the union of two proper past subsets, then it is called an *indecomposable past set* (IP). If P is an IP and if there is some $x \in M$ with $I^-(x) = P$, then P is known as a *proper* IP, or PIP. If an IP set is not a PIP, then it is termed a *terminal* IP, or TIP. The definitions of future sets, indecomposable future sets (IFs), PIFs, and TIFs are similar. For many of the statements and propositions the dual, or similar definitions, are taken for granted. Geroch, Kronheimer, and Penrose (1972) proved that a set $P \subset M$ is an IP if and only if there is a future directed timelike curve γ such that $I^-(\gamma) = P$. Now, let M^+ be the union of $\hat{M}$ and $\check{M}$, which are unions of all IPs and IFs in M respectively. Then, avoiding duplication in M^+, M^* can be defined as the quotient space M^+/R_{h}, where R_{h} is the intersection of all equivalence relations $R \subset M^+ \times M^+$ for which M^+/R is Hausdorff. In this case, M^* can be viewed as a spacetime with boundary $M \subset M^*$, and the topology of M can be looked upon as the induced topology of M^*. Throughout the discussion here, M has been assumed to be distinguishing.

A point $x \in M$ is said to be a *regular point* if it is represented by a PIP or PIF. All other points in M^* are represented by TIPs or TIFs and are called the *ideal* or *boundary points* of M. A curve γ in M is taken as a continuous map of a general interval into M. The result of Geroch, Kronheimer, and Penrose (1972) is useful here. Let $P = I^-(\gamma)$ be an IP where γ is a future directed timelike curve. Then P is a PIP if and only if γ has a future end point, and P is a TIP if and only if γ is future inextendible without a future end point.

Now, the PIPs (and PIFs) can be characterized using the causal functions introduced earlier. Let γ be a future directed timelike curve. Then, $I^-(\gamma)$ is a PIP if and only if h^- attains its maximum value along γ. To see this, if $I^-(\gamma)$ is a PIP, then γ has a future end point p as seen above. It follows that $I^-(\gamma) = I^-(p)$. Now, as M is distinguishing, h^- is strictly increasing along γ, and hence has a maximum value at the future end point p. Conversely, let h^- attain its maximum $h^-(p)$ for some $p \in \gamma$. Now, if p is not the future end point of γ, then some $q \gg p$ exists with $q \in \gamma$, and $h^-(q) > h^-(p)$ by the strict increasing nature of h^-, which contradicts the hypothesis.

The above characterization can be stated in the following manner also. For a future directed timelike curve γ, $I^-(\gamma)$ is a PIP if and only if h^+ attains its minimum value along γ. It now follows that for such a curve, $I^-(\gamma)$ is a TIP if and only if either $h^+ \to 0$ along γ, or $h^+ \to k$ with $0 \ldots k \ldots 1$ along γ, with $h^+(p) \neq k$ for any $p \in \gamma$. Along future endless curves in globally hyperbolic spacetimes, $h^+ \to 0$, whereas the other situation will be realized when points have been amputated from the spacetime, and the timelike curve converges to such a cut without having a future end point. Therefore, the causal boundary points in M^* are defined by precisely those timelike curves along which $h^\pm$ do not realize their extremum values.

Note that the ideal points in $M^* - M$ include both the points at infinity and singularities characterized by the TIPs and TIFs of the spacetime. A TIP is *non-singular* if a timelike curve generating it that has an infinite length exists. All other ideal points are singularities of the spacetime. Now, let x be a future (past) ideal point in M^* defined by a TIP, say $I^-(\gamma)$ (for a TIF, say $I^+(\lambda)$). Then x will be called a *future (past) 0-ideal point* if h^+ (or h^-) converges to zero along the curve, and x will be called a *future (past) k-ideal point* if h^+ (or h^-) converges to k with $0 \ldots k \ldots 1$ along γ (or λ).

Therefore, causal functions classify all the ideal points for M into the following eight categories:

(a) future (past) 0-ideal points that are singularities;
(b) future (past) k-ideal points that are singularities;
(c) future (past) 0-ideal points at infinity;
(d) future (past) k-ideal points at infinity.

As mentioned earlier, endless timelike curves in a globally hyperbolic spacetime define ideal points of type (c). Timelike curves falling into a Schwarzschild singularity provide an example for ideal points of type (a).

Finally, it is shown here that the naked singularities as defined by Penrose (1974a) are all the boundary points of type (b). It needs to be decided whether the singularities are always hidden behind an event horizon. The question is whether naked or timelike singularities, which are defined as follows, could also occur. Let $x \in M^*$ be a TIP given by $I^-(\gamma)$. Then, x is called a *future naked singularity* if the future endless timelike curve γ has a future singular end point and a point $p \in M$ exists such that $I^-(\gamma) \subset I^-(p)$. Now let x be a future singular point defined by a future endless timelike curve γ. As shown earlier, as γ is a TIP, either $h^+ \to 0$ along γ or $h^+ \to k$ with $0 \ldots k \ldots 1$. Now, consider any sequence $\{p_n\}$ on γ with $p_n \ll p_{n+1}$ for all n. Then $h^+(p_n)$ is strictly decreasing along this trajectory. Since $I^-(\gamma) \subset I^-(p)$, it follows that $p_n \ll p$ for all n. Therefore, $h^+(p_n)$ is bounded below by a non-zero fixed number $h^+(p)$, implying that $h^+ \not\to 0$. This shows that x is an ideal point of type (b).

4.2 Spacetime singularities

Now the nature and existence of spacetime singularities in a general spacetime are considered. After Einstein proposed the general theory describing the gravitational force in terms of spacetime curvature, and proposed the field equations relating the geometry and matter content of the spacetime manifold, the earliest solutions found for the field equations were the Schwarzschild metric representing the gravitational field around an isolated body such as a spherically symmetric star, and the Friedmann cosmological models. Each of these solutions contained a spacetime singularity where the curvatures and densities were infinite and the physical description would break down. In the Schwarzschild solution such a singularity was present at $r = 0$, whereas in the Friedmann models it was found at the epoch $t = 0$, which is the beginning of the universe and the origin of time where the scale factor $S(t)$ also vanishes and all objects are crushed to zero volume due to infinite gravitational tidal forces.

Even though the physical problem posed by the existence of such a strong curvature singularity in these solutions was realized, initially this phenomena was not taken seriously. It was generally thought that the existence of such a singularity must be a consequence of the very high degree of symmetry imposed on the spacetime while deriving these solutions, which led to many interesting physical applications. Subsequently, the distinction between a genuine singularity and a mere coordinate singularity became clear, and it was realized that the singularity at $r = 2m$ in the Schwarzschild spacetime was a coordinate singularity that could be removed by a suitable coordinate transformation. It was clear, however, that the genuine curvature singularity at $r = 0$ could not be removed by any such coordinate transformation.

The hope was then that, when more general solutions were considered with a lesser degree of symmetry requirements, such singularities would be avoided.

This issue was sorted out when a detailed study of the structure of a general spacetime and the associated problem of a spacetime singularity was taken up by Hawking, Penrose, and Geroch (see for example, Penrose, 1968; Geroch, 1971; Hawking and Ellis, 1973). It was shown by this work that a spacetime will admit singularities within a very general framework, provided that it satisfies certain reasonable assumptions, such as the positivity of energy, a suitable causality assumption, and a condition such as the existence of trapped surfaces. It thus follows that the spacetime singularities form a general feature of the relativity theory. In fact, these considerations also ensure the existence of singularities in other theories of gravity that are based on a spacetime manifold framework and satisfy the general conditions stated above.

4.2.1 The definition of a singularity

First, the meaning of a singular spacetime is discussed in some detail, and the notion of a singularity is specified. It turns out that it is the notion of geodesic incompleteness that characterizes the notion of a singularity in an effective manner for a spacetime, and enables its existence to be proved by means of general theorems. A variety of ways in which a spacetime exhibits singular behavior, and the related notions of singular TIPs and TIFs are then discussed.

The gravitational focusing caused by spacetime curvature in congruences of timelike and null geodesics turns out to be the main cause of the existence of a singularity in the form of non-spacelike incomplete geodesics in a spacetime. There are many theorems that establish the existence of the non-spacelike geodesic incompleteness for the spacetime, either in the past or future. However, these provide no information on the nature of these singularities or their properties. In particular, these singularities could be covered inside an event horizon of gravity, or they could be visible to external observers if the trapped surfaces are delayed while the singularity forms during dynamical processes in the spacetime.

The issue of the physical nature of a singularity is thus of much interest. There are many types of singular behaviors possible for a spacetime, and some of these could be regarded only as mathematical pathologies in the spacetime, rather than having any physical significance. This will be especially so if the spacetime curvatures and other similar physical quantities remained finite along an incomplete non-spacelike geodesic in the limit of approach to the singularity. The criterion of Tipler, Clarke, and Ellis (1980) as to when a singularity should be considered to be physically important in terms of the curvature growth along singular geodesics will also be specified here.

When should it be said that a spacetime manifold (M, g) is singular, or that it contains a spacetime singularity? As pointed out, several examples of singular behavior in the spacetime models of general relativity are known. Important exact solutions of the Einstein equations, such as the Friedmann–Robertson–Walker cosmological models and the Schwarzschild spacetime, contain a spacetime singularity where the energy density or the spacetime curvatures diverge strongly, and the usual description of the spacetime breaks down.

In the Schwarzschild spacetime, there is an essential curvature singularity at $r = 0$ in the sense that, along any non-spacelike trajectory falling into the singularity, as $r \to 0$, the Kretschmann scalar $\alpha = R^{ijkl}R_{ijkl} \to \infty$. Also, all future directed non-spacelike geodesics that enter the horizon at $r = 2m$ must fall into this curvature singularity within a finite value of the proper time (finite value of the affine parameter in the case of null geodesics). Therefore, all such curves are future geodesically incomplete.

In the Friedmann–Robertson–Walker models, the Einstein equations imply that if $\rho + 3p > 0$ at all times, where ρ is the total energy density and p is the pressure, there is a singularity at $t = 0$ that could be identified as the origin of the universe. If $\rho + p > 0$ at all times, then it is seen that along all the past directed trajectories meeting this singularity $\rho \to \infty$, and also that the curvature scalar $R = R_{ij}R^{ij} \to \infty$. Again, all the past directed non-spacelike geodesics are incomplete in the above sense. Therefore, there is an essential curvature singularity at $t = 0$ that cannot be transformed away by any coordinate transformation. In fact, similar behavior has been generalized to the class of spatially homogeneous cosmological models as shown by Ellis and King (1974), which satisfy the positivity of energy conditions $\rho \geq 0, \rho \geq 3p \geq 0$ and $1 \geq 3dp/d\rho \geq 0$.

The existence of such singularities, where the curvature scalars and densities diverge, implies a genuine spacetime pathology where the usual laws of physics must break down. The existence of the geodesic incompleteness in these cases implies that, for example, a timelike observer suddenly disappears from the spacetime after a finite amount of proper time.

Of course, singular behavior can also occur without bad behavior of the curvatures. A simple example is the Minkowski spacetime with a point deleted. With such a hole in the spacetime, there will be, for example, timelike geodesics running into the hole, and hence they will be future incomplete. This is clearly an artificial situation which one would like to rule out in general by requiring that the spacetime is *inextendible*, that is, it cannot be isometrically embedded into another larger spacetime manifold as a proper subset.

It is, however, possible to give a non-trivial example of the singular behavior of the above type, where a conical singularity exists in the spacetime as shown by Ellis and Schmidt (1977). Here, the spacetime is inextendible and the curvature components do not diverge in the limit of approach to the singularity. This is a behavior similar to that occurring in a Weyl type of solution. The metric is given by

$$ds^2 = -dt^2 + dr^2 + r^2(d\theta^2 + \sin^2\theta\, d\phi^2), \tag{4.16}$$

with the range of coordinates given by $-\infty < t < \infty, 0 < r < \infty, 0 < \theta < \pi$, but with $0 < \phi < a$, with $\phi = 0$ and $\phi = a$ identified and $a \neq 2\pi$. There is a conical singularity at $r = 0$ through which the spacetime cannot be extended and the singular boundary is related to the timelike two-plane $r = 0$ of the Minkowski spacetime.

The important question that arises is whether such singularities develop even when a spacetime of generality is considered and if so, under what conditions. In order to consider this question, it is first necessary to characterize more precisely what is meant by a spacetime singularity.

While trying to characterize a spacetime singularity, the first point to note is that by very definition, the metric tensor has to be well-defined at all the regular points of the spacetime. Since this is no longer true at a spacetime singularity such as those discussed above, a singularity cannot be regarded as a regular point of the spacetime, but must be treated as a boundary point attached to it. This situation causes difficulty when one attempts to characterize a singularity by the criterion that the curvatures must blow up near the singularity. The trouble is, since the singularity is not a part of the spacetime, it is not possible to define its neighborhood in the usual sense in order to discuss the behavior of curvature quantities in that region.

Characterizing the singularity could be attempted in terms of the divergence of the components of the Riemann curvature tensor along non-spacelike trajectories of the spacetime. The trouble with this is that the behavior of such components will, in general, change with the change of frames used, and this approach is not really of much help. The curvature scalars or the scalar polynomials in the metric and the Riemann tensor could be used, and they could be required to achieve unboundedly large values. This is the behavior encountered in the Schwarzschild and the Friedmann models. However, it is possible that such a divergence of curvature scalars occurs only at infinity for a given non-spacelike curve. In general, it looks reasonable to demand that some sort of curvature divergence must take place along the non-spacelike curves that encounter a spacetime singularity. However, a general characterization of the singularity in terms of the curvature divergence runs into various difficulties. For example, for the plane wave vacuum solutions, the polynomials in the curvature scalars vanish, but the curvature tensor is still allowed to be singular (Penrose, 1965). Another example is the Taub–NUT type of solutions given by Misner (1963, 1967). Here, the spacetime curvatures are bounded and the manifold is inextendible, but it is both null and timelike geodesically incomplete.

Considering these as well as similar situations, the occurrence of non-spacelike geodesic incompleteness has been generally agreed upon as the criterion for the existence of a singularity for a spacetime. This criterion does not cover all possible types of singular behavior. For example, Geroch (1968a) has given an example of a spacetime that is geodesically complete, but which contains a future inextendible timelike curve with a bounded acceleration and with a finite proper length. This could correspond to a rocket ship with enough fuel to disappear suddenly from the universe after a finite proper time. Also, one would not like to term all the geodesically incomplete models as containing a physically genuine singularity, especially if the curvatures are finite everywhere throughout the spacetime. This includes the Taub–NUT case mentioned above. In order to call a singularity physically genuine, one would like to demand some sort of curvature divergence along the incomplete non-spacelike geodesic. On the other hand, if there is a powerful curvature divergence along an incomplete non-spacelike geodesic, one would certainly like to call such a singularity physically significant.

It is clear, however, that if a spacetime manifold contains incomplete non-spacelike geodesics, there is a definite singular behavior present in the spacetime. In such a case, a timelike observer or a photon suddenly disappears from the spacetime after a finite amount of proper time, or after a finite value of the affine parameter. The singularity theorems that result from an analysis of gravitational focusing and global properties of a spacetime prove this incompleteness property for a wide class of spacetimes under a set of rather general conditions.

4.2.2 Gravitational focusing

Gravitational focusing in a spacetime plays an important and key role in the formation of trapped surfaces as the gravitational collapse develops. It is discussed how the matter fields with positive energy density affect the causality relations in a spacetime and how they cause focusing in the families of timelike and null trajectories. The essential phenomenon that occurs here is that matter focuses the non-spacelike geodesics of the spacetime into pairs of focal points or the conjugate points. The basic property of conjugate points is that if $p < q$ are two conjugate points along a non-spacelike geodesic, then $p \ll q$. Now, there are null hypersurfaces, such as the boundary of the future $I^+(p)$ for a point p, such that no two points of the hypersurface could be joined by a timelike curve. Therefore, the null geodesic generators of such surfaces cannot contain any conjugate points, and these must leave the hypersurface before encountering a conjugate point. This puts strong constraints on the nature of such surfaces, and the singularity theorems result from an analysis of these limits.

Consider a congruence of timelike geodesics in the spacetime. This is a family of curves such that precisely one timelike geodesic trajectory passes through each point p. Choosing the curves to be smooth, this defines a smooth timelike vector field on the spacetime. On the other hand, a given smooth vector field on the spacetime specifies a congruence of curves in the manifold.

Let V^i denote the timelike tangent vector to the congruence. Choosing the parameter to be the proper time along such timelike trajectories, this can be normalized to be a unit tangent vector,

$$V^i V_i = -1. \tag{4.17}$$

The *spatial part*, h_{ij}, of the metric tensor can be defined as

$$h_{ij} = g_{ij} + V_i V_j. \tag{4.18}$$

Then, $h^i{}_j = \delta^i{}_j + V^i V_j = g^{ik} h_{kj}$, and

$$h_{ij} V^i = h_{ij} V^j = h^i{}_j V_i = h^i{}_j V^j = 0. \tag{4.19}$$

Therefore, $h^i{}_j$ can be called the *projection operator* onto the subspace of T_p, orthogonal to the vector V^i. The indices of h are now raised and lowered, just as in the case of the metric tensor,

$$h_{ij} h^j{}_k = (g_{ij} + V_i V_j)(g^j{}_k + V^j V_k) = g_{ik} + V_i V_k = h_{ik}, \tag{4.20}$$

and also

$$h^{ij} h_{ij} = h^i{}_i = \delta^i{}_i + V^i V_i = 3. \tag{4.21}$$

For the given congruence of timelike geodesics, the *expansion*, *shear*, and *rotation* tensors are respectively defined as

$$\theta_{ij} = V_{(k;l)} h^k{}_i h^l{}_j, \tag{4.22}$$

$$\sigma_{ij} = \theta_{ij} - \tfrac{1}{3} h_{ij} \theta, \tag{4.23}$$

$$\omega_{ij} = h^k{}_i h^l{}_j V_{[k;l]}. \tag{4.24}$$

Here, the *volume expansion* θ is defined as

$$\theta = \theta_{ij} h^{ij} = V_{(k;l)} h^{lk} = \nabla_k V^k = V^k{}_{;k}. \tag{4.25}$$

Furthermore, note that σ_{ij} and ω_{ij} are purely spatial quantities in the sense that

$$\sigma_{ij}V^i = \omega_{ij}V^i = 0. \tag{4.26}$$

Also, note that

$$\sigma^i{}_i = h^{ij}\sigma_{ij} = \theta - \tfrac{1}{3}h_{ij}h^{ij} = 0. \tag{4.27}$$

The covariant derivative of V is then expressed as

$$\nabla_j V_i = V_{i;j} = \tfrac{1}{3}\theta h_{ij} + \sigma_{ij} + \omega_{ij}. \tag{4.28}$$

This is verified by direct substitution from (4.22), (4.23), and (4.24).

Now, the geodesic equations imply that

$$V^k \nabla_k \nabla_j V_i = V^k \nabla_j \nabla_k V_i + R_{ilkj} V^l V^k. \tag{4.29}$$

Using the fact that V^k is a tangent to the geodesics, that is, $\nabla_j(V^k \nabla_k V_i) = 0$, the above equation can be written as

$$V^k \nabla_k \nabla_j V_i = -(\nabla_j V^k)(\nabla_k V_i) + R_{ilkj} V^l V^k. \tag{4.30}$$

Taking trace in the above,

$$\frac{d\theta}{d\tau} = V^k \nabla_k V^i{}_{;i} = -(V^k{}_{;i} V^i{}_{;k}) - R_{lk} V^l V^k, \tag{4.31}$$

where τ is the affine parameter along the geodesic.

Using (4.28) in the above and the anti-symmetry properties of the tensor ω_{ij}, after some simplification,

$$\frac{d\theta}{d\tau} = -R_{lk} V^l V^k - \tfrac{1}{3}\theta^2 - \sigma_{ij}\sigma^{ij} + \omega_{ij}\omega^{ij} \tag{4.32}$$

can be obtained, which can be written as

$$\frac{d\theta}{d\tau} = -R_{lk} V^l V^k - \tfrac{1}{3}\theta^2 - 2\sigma^2 + 2\omega^2. \tag{4.33}$$

Equation (4.33) above is called the *Raychaudhuri equation* (Raychaudhuri, 1955), which describes the rate of change of the volume expansion as the timelike geodesic curves in the congruence are moved along.

The second and third terms on the right-hand side involving θ and σ are always positive. Consider now the term $R_{ij}V^iV^j$, which by the Einstein equations can be written as

$$R_{ij}V^iV^j = 8\pi[T_{ij}V^iV^j + \tfrac{1}{2}T]. \tag{4.34}$$

The term $T_{ij}V^iV^j$ above represents the energy density measured by a timelike observer with the unit tangent V^i, which is the four-velocity of the observer. For all reasonable classical physical fields this energy density is generally taken as non-negative, and it is assumed that for all timelike vectors V^i,

$$T_{ij}V^iV^j \geq 0, \tag{4.35}$$

is satisfied. Such an assumption is called the *weak energy condition.* On the other hand, it is also considered reasonable to believe that the matter stresses will not be so large as to make the right-hand side of (4.34) negative. This will be satisfied when

$$T_{ij}V^iV^j \geq -\tfrac{1}{2}T \tag{4.36}$$

is satisfied. Such an assumption is called the *strong energy condition*, and it implies that for all timelike vectors V^i,

$$R_{ij}V^iV^j \geq 0. \tag{4.37}$$

By continuity, it can be argued that the same will then also hold for all null vectors.

Both the strong and weak energy conditions will be valid for well-known forms of matter, such as the perfect fluid, provided that the energy density ρ is non-negative and that there are no large negative pressures that are bigger or comparable to ρ, when converted into physical units.

An additional energy condition often required by the singularity theorems is the *dominant energy condition*, which states that in addition to the weak energy condition, the pressure of the medium must not exceed the energy density. This can be equivalently stated as, for all timelike vectors V^i, $T_{ij}V^iV^j \geq 0$ and the vector $T^{ij}V_i$ is a non-spacelike vector. Such a condition would be satisfied provided that the local speed of sound does not exceed the local speed of light.

With the strong energy condition being satisfied, the Raychaudhuri equation implies that the effect of matter on the spacetime curvature causes a focusing effect in the congruence of the timelike geodesics due to gravitational attraction. This, in general, causes the neighboring geodesics in the congruence to cross each other, which gives rise to caustics or conjugate points. This separation between nearby timelike geodesics is governed by the geodesic deviation equation,

$$D^2Z^j = -R^j_{kil}V^kZ^iV^l, \tag{4.38}$$

where Z^i is the separation vector between nearby geodesics of the congruence. Solutions of the above equation are called the *Jacobi fields* along a given timelike geodesic.

Suppose now γ is a timelike geodesic. Then, two points p and q along γ are called *conjugate points* if a Jacobi field along γ exists that is not identically zero, but vanishes at p and q. From the derivation of the Raychaudhuri equation given above, it is clear that the occurrence of conjugate points along a timelike geodesic is closely related to the behavior of the expansion parameter θ of the congruence. In fact, it can be shown that the necessary and sufficient condition for a point q to be conjugate to p is that for the congruence of timelike geodesics emerging from p, $\theta \to -\infty$ at q (see for example, Hawking and Ellis, 1973). The conjugate points along the null geodesics are also similarly defined. Consider, for example, a congruence of null geodesics emanating from a point p. If infinitesimally nearby null geodesics of the congruence meet again at some other point q in the future, then p and q are said to be *conjugate* to each other.

Similarly, let S be a smooth spacelike hypersurface, that is, it is an embedded three-dimensional submanifold. Consider a congruence of timelike geodesics orthogonal to S. Then, a point p along a timelike geodesic γ of the congruence is said to be *conjugate to* S along γ if a Jacobi vector field along γ exists that is non-zero at S, but vanishes at p. This means that there are two infinitesimally nearby geodesics orthogonal to S that intersect at p. Again, the equivalent condition for this to happen, in terms of the parameter θ, is that the expansion θ for the congruence orthogonal to S tends to $-\infty$ at p. If V^i denotes the unit timelike tangent vector field of the congruence of timelike geodesics, where V^i denotes the normal to S, then the *extrinsic curvature* χ_{ij} of S is defined as

$$\chi_{ij} = \nabla_i V_j, \tag{4.39}$$

which is evaluated at S. Clearly, $\chi_{ij}V^i = \chi_{ij}V^j = 0$. Also, the hypersurface orthogonality of the congruence implies that $\omega_{ij} = 0$. As a result, $\chi_{ij} = \chi_{ji}$, that is, this is a symmetric tensor. The trace of the extrinsic curvature, denoted by χ is given by

$$\chi = \chi^i{}_i = h^{ij}\chi_{ij} = \theta. \tag{4.40}$$

Therefore, $\chi = \theta$ at S, where θ is the expansion of the congruence orthogonal to S.

The behavior of the expansion parameter θ is governed by the Raychaudhuri equation as pointed out above. For example, consider the situation when the spacetime satisfies the strong energy condition and the congruence of timelike geodesics is hypersurface orthogonal. In such a case, $\omega_{ij} = 0$ and the corresponding term, ω^2, vanishes in (4.33). Then, the expression for the covariant derivative of ω_{ij} implies that it must also vanish for all future times. It follows from the above discussion that

$$\frac{d\theta}{d\tau} \leq -\frac{\theta^2}{3}, \tag{4.41}$$

which means that the volume expansion parameter must necessarily be decreasing along the timelike geodesics. If θ_0 denotes the initial value of the expansion, then the above can be integrated as $\theta^{-1} \geq \theta_0^{-1} + \tau/3$. It is clear from this that, if the congruence is initially converging and θ_0 is negative, then $\theta \to -\infty$ within a proper time distance $\tau \leq 3/\mid \theta_0 \mid$.

The following can then be seen from the above discussion. Let M be a spacetime satisfying the strong energy condition, and let S be a spacelike hypersurface with $\theta < 0$ at $p \in S$. If γ is the timelike geodesic of the congruence orthogonal to S passing through p, then a point q conjugate to S along γ exists within a proper time distance $\tau \leq 3/\mid \theta \mid$, provided γ can be extended to that value of the proper time.

Suppose now that the trace of the extrinsic curvature χ_{ij} (which is also sometimes called the *second fundamental form* of the surface S) is negative everywhere on S, that is, $\theta = \chi < 0$ on S, and it is bounded above by a negative value θ_{max}. In this case, it is clear from the above that all the timelike geodesics of the congruence orthogonal to S will contain a point conjugate to S within a proper time distance $\tau \leq 3/\mid \theta_{\text{max}} \mid$. Therefore, let M be a spacetime satisfying the above conditions, and let S be a spacelike surface in M. Let the trace of the extrinsic curvature $\chi = \theta < 0$ on S and be bounded above by a negative value θ_{max}. Then, all the timelike geodesics orthogonal to S have a point p conjugate to S within a proper time distance $\tau \leq 3/\mid \theta_{\text{max}} \mid$, provided the geodesics can be extended to that value of the proper time.

Consider now the congruence of timelike geodesics passing through a point p. As shown by Lemma 4.5.2 of Hawking and Ellis (1973), for any convex normal neighborhood of p, the trajectories of this congruence are orthogonal to the spacelike surfaces of the proper time $\tau =$ const. along the geodesics. Therefore, the congruence is hypersurface orthogonal, $\omega_{ij} = 0$, and it will also be zero for all future times. Then, the above discussion again implies that if the strong energy condition holds for all timelike vectors V^i, and if $p \in \gamma$ with $\theta = \theta_0 < 0$ at some point q in the future of p along this timelike geodesic, then γ contains a point r conjugate to p within a proper time distance $\tau \leq 3/\mid \theta_0 \mid$ from q, provided it can be extended to that value of the proper time.

The basic implication of the above results is that, once a convergence occurs in a congruence of the timelike geodesics, the conjugate points or the caustics must develop in the spacetime. These can be interpreted as the singularities of the congruence. Such singularities could occur even in Minkowski spacetimes and similar other perfectly regular spacetimes. However, when combined with certain causal structure properties of the spacetime, this implies the existence of spacetime singularities in the form of geodesic incompleteness. The gravitational focusing for congruences of null geodesics, or for null geodesics orthogonal to a spacelike two-surface, can be discussed similarly.

In general, note that even if $R_{ij}V^iV^j = 0$ throughout the spacetime, if $\sigma^2 > 0$ then a net focusing effect again results. This will be so if $R_{ijkl}V^jV^l \neq 0$ at least at one point in the spacetime. It is then possible from Hawking and Ellis (1973) to show that if $\lambda(t)$ is a non-spacelike geodesic, complete both in the future and the past with a range of the affine parameter t over $(-\infty, +\infty)$, with $R_{ij}V^iV^j + 2\sigma^2$ being continuous and non-negative, and if the latter quantity is positive for at least one value of t, then $\lambda(t)$ must contain a pair of conjugate points in the interval $(-\infty, +\infty)$.

The condition required above amounts to a statement that the non-spacelike trajectory must pass through some matter or radiation at least once throughout its history. This will happen if $R_{ij}V^iV^j \neq 0$ at some point on the trajectory, or the Weyl tensor is non-zero at this point in a suitable manner. A precise condition to ensure this is that every non-spacelike geodesic in M must contain a point at which $K_{[i}R_{j]el[m}K_{n]}K^eK^l \neq 0$, where $\boldsymbol{K}$ is the tangent to the non-spacelike geodesic. This is called the *generic condition*. Therefore, every timelike and null geodesic that is both future and past complete must contain a pair of conjugate points if the spacetime satisfies the generic condition.

Globally hyperbolic spacetimes have been discussed, and an important property of these is that if N is a globally hyperbolic subset, and if $p \ll q$ for $p, q \in N$, then a timelike geodesic from p to q exists that maximizes the lengths of all the non-spacelike curves from p to q. Such a maximal curve is related to the existence of conjugate points in that, if N is a globally hyperbolic subset of and $p, q \in N$, with $\gamma(t)$ being a timelike geodesic maximizing the lengths of all non-spacelike curves from p to q, then $\gamma(t)$ contains no points conjugate to p between p and q. The point is that, if there were a conjugate point r to p between p and q, then it can be shown using variational arguments, that a longer non-spacelike curve from p to q could be obtained by 'rounding off the corner' at r, generated due to the conjugate point r. This is contradictory to the maximality of the curve $\gamma(t)$.

4.2.3 Existence of singularities

There are several singularity theorems available that establish the non-spacelike geodesic incompleteness for a spacetime under different sets of conditions and that are applicable to different physical situations. However, the most general of these is the Hawking–Penrose theorem (Hawking and Penrose, 1970), which is applicable in both the collapse situation and the cosmological scenario. The main idea of the proof of such a theorem is now discussed. Using the causal structure analysis, it is shown that there must be maximal length timelike curves between certain pairs of events in the spacetime. As pointed out above, a causal geodesic that is both future and past complete must contain pairs of conjugate points if the spacetime satisfies the generic condition and an energy condition. This is then used to draw the necessary contradiction in order to show that M must be non-spacelike geodesically incomplete. This result can be stated as in the following.

A spacetime (M, g) cannot be timelike and null geodesically complete if the following are satisfied:

(1) $R_{ij}K^iK^j \geq 0$ for all non-spacelike vectors K^i;
(2) the generic condition is satisfied, that is, every non-spacelike geodesic contains a point at which $K_{[i}R_{j]el[m}K_{n]}K^eK^l \neq 0$, where $\boldsymbol{K}$ is the tangent to the non-spacelike geodesic;
(3) the chronology condition holds;
(4) in M, either a compact achronal set without edge or a closed trapped surface exists, or a point p exists such that for all past directed null geodesics from p, θ must eventually be negative.

The main idea of the proof is that it can be shown that the following three cannot hold simultaneously: (a) every inextendible non-spacelike geodesic contains pairs of conjugate points, (b) the chronology condition holds, (c) an achronal set $\mathcal{S}$ exists in the spacetime such that $E^+(\mathcal{S})$ or $E^-(\mathcal{S})$ is compact, where $E^+(\mathcal{S})$ and $E^-(\mathcal{S})$ denote the future and past of the edge of set $\mathcal{S}$.

If this is shown, then the theorem is proved, as (3) is the same as (b), (4) implies (c), and (1) and (2) imply (a). First, note that (a) and (b) imply the strong causality of the spacetime (see Prop. 6.4.6 of Hawking and Ellis, 1973). Next, it can be shown that if $\mathcal{S}$ is a future trapped set and if strong causality holds on $\overline{I^+(\mathcal{S})}$ then a future endless trip γ exists such that $\gamma \subset \operatorname{Int} D^+(E^+(\gamma))$. Now, $T = \overline{J^-(\gamma)} \cap E^+(\mathcal{S})$ is defined, T turns out to be past trapped, and hence λ, a past endless causal geodesic in $\operatorname{Int}(D^-(E^-(T))$, exists. Then, a sequence $\{a_i\}$ receding into the past on λ, and a sequence $\{c_i\}$ on γ to the future is chosen. The sets $J^-(c_i) \cap J^+(a_i)$ are compact and globally hyperbolic, so a maximal geodesic μ_i from a_i to c_i exists for each i. The intersections of μ_i with the compact set T have a limit point p and a limiting causal direction. The causal geodesic μ with this direction at p must have a pair of conjugate points. This is then shown to be contradictory to the maximality property of the geodesics stated above.

There is a more general way in which the singular points in a spacetime can be defined using the terminal indecomposable pasts (TIPs) and terminal indecomposable futures (TIFs), as discussed earlier (Penrose, 1974a, 1979). The spacetime is assumed to be strongly causal. Here, a curve means a map γ from an interval $[0, a)$ of the real line into M, where a could possibly be infinity. Therefore, the curve starts at an initial point $\gamma(0)$ with a definite

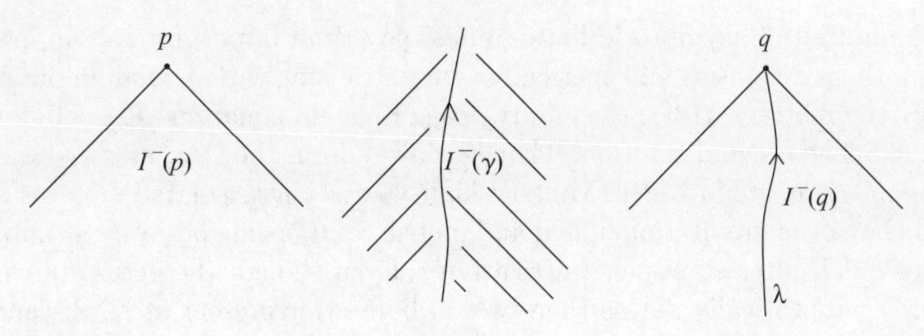

Fig. 4.3 The event p is a regular point in M and the set $I^-(p)$ defines a PIP. For a non-spacelike curve γ going to infinity, $I^-(\gamma)$ gives an ∞-TIP. If λ is a finite length curve going into a spacetime singularity, then $I^-(\lambda)$ is a singular TIP.

tangent but has no end point, as the interval is open at a. Such a curve will be called *extendible* if it is possible to extend the map γ to an end point $\gamma(a)$ in M, and otherwise it is called *inextendible.* Of particular interest here are the inextendible non-spacelike curves. The TIPs and TIFs are generated by future directed and past directed timelike curves respectively, and they give all the boundary points of spacetime that include both the singularities and points at infinity. Such a boundary point is called ∞-TIP, which is a point at infinity if it is generated by some timelike curve of infinite proper time length in the future. A *singular* TIP is one that is not generated by any such timelike curve of infinite length (see Fig. 4.3). Similarly, ∞-TIFs and singular TIFs can be defined. The existence of a singular TIP defines a singularity of spacetime, giving a class of future directed inextendible timelike curves that have a finite proper time length, but no future end point.

As pointed out by Clarke (1986), the basic requirement for the ideal end point of a timelike curve to be called a singularity, rather then a regular boundary point, is that there should be no extension of the spacetime possible in which the curve in question could be continued. If such an extension existed, then the singularity would be similar in some sense to the coordinate singularity in the Schwarzschild geometry at $r = 2m$. Therefore, the question of singularity depends on what type of extension is allowed for the spacetime. Therefore, a boundary point is called a C^k*-singularity* of the spacetime if there is no C^k-extension of M that removes it. Clarke then defines the index k as a measure of the strength of the singularity in the sense that the smaller the k, the stronger the singularity.

4.3 Blackholes

Assuming cosmic censorship in the form of an asymptotic predictability in the spacetime, much of the theoretical basic blackhole physics is carried out in an asymptotically flat and predictable model.

The notion of asymptotic flatness has played an important role in gravitation theory. Consider a spherically symmetric star with a vacuum outside, where the metric satisfies the empty space Einstein equations $R_{ij} = 0$, given by the Schwarzschild solution. Then, far away from the star, as $r \to \infty$, the components g_{ij} tend to the Minkowskian values. Such isolated systems and the behavior of gravitational field and metric components far away at infinity are of much interest in general relativity. Even though the actual universe is not asymptotically flat as there would be matter present at all distances, such an approximation is useful to model the geometry of an individual star and to study its gravitational field.

It was pointed out by the studies of Bondi, van der Burg, and Metzner (1962), and Sachs (1962) that the characteristic or null surfaces play an important role in understanding the asymptotic properties of gravitational fields for such isolated systems. They used characteristic surfaces to study the metric components and curvature tensor properties in the asymptotic limit. A coordinate free construction of null infinity and the notion of asymptotic flatness for a general spacetime were introduced by Penrose (1965, 1968) by means of conformal compactification of the spacetime. Here, the main idea is to attach a boundary to the spacetime in such a manner that its properties coincide with the geometric properties of the boundary $\mathcal{I}^+$ or $\mathcal{I}^-$ for the Minkowski spacetime discussed earlier. A general spacetime is then called asymptotically flat if it admits such a boundary attachment. Just as in the Minkowski spacetime, a conformal transformation Ω on the original spacetime M can be introduced so that $\Omega \to 0$ near infinity, and the new unphysical spacetime $(\overline{M}, \Omega^2 g_{ij})$ is compactified. In $(\overline{M}, \Omega^2 g_{ij})$, the boundary surface of $\overline{M}$ corresponds to the infinity of the spacetime M.

Specifically, a spacetime (M, g_{ij}) is called *asymptotically flat* if a new, unphysical spacetime $(\overline{M}, \bar{g})$ with a boundary $\mathcal{I}$ exists such that $\overline{M} - \mathcal{I}$ is diffeomorphic to M with $\Omega > 0$ and $\bar{g}_{ij} = \Omega^2 g_{ij}$, with the following conditions:

(1) the new unphysical manifold $\overline{M}$ is smooth everywhere including the boundary;
(2) the conformal factor Ω is smooth everywhere and $\Omega = 0$ on $\mathcal{I}$;
(3) all the maximally extended null geodesics in $\overline{M}$ have a future and a past end point on $\mathcal{I}$;
(4) there is a neighborhood of $\mathcal{I}$ in M where g_{ij} satisfies the vacuum Einstein equations $R_{ij} = 0$.

Such a construction of conformal compactification for a spacetime turns out to be particularly useful in order to study isolated sources in otherwise empty spacetimes. The surface $\mathcal{I}$ can be thought of as infinity in the sense that the affine parameter along every null geodesic in M grows unboundedly large near $\mathcal{I}$. The null geodesics of these two conformally related spacetimes are completely identical as point sets, as mentioned earlier. However, the affine parameters along the geodesics in M and $\overline{M}$ are related as $d\bar{v} = \Omega^2 dv$. Therefore, the affine parameter along the null geodesics in M must blow up near $\mathcal{I}^+$ in the future, and similar behavior holds near the past infinity.

The definition above is quite stringent in that it assumes that every null geodesic has two end points: in the future and in the past at $\mathcal{I}$. Although this is satisfied in the Minkowski spacetime, this does not hold in spacetimes such as the Schwarzschild and Reissner–Nordström cases, which contain event horizons and a blackhole region. Here, the future directed null geodesics that enter the blackhole must end in the singularity at $r = 0$ and cannot have an end point at $\mathcal{I}^+$. One would like to include these spacetimes in the general asymptotically flat class, and so a spacetime M is defined to be *weakly asymptotically simple and empty* (or a WASE spacetime) if an asymptotically flat spacetime $\overline{M}$ in the above sense exists such that there is a neighborhood of $\mathcal{I}$ in $\overline{M}$ that is isometric to an open set in M. This definition covers the Schwarzschild and the Reissner–Nordström cases and also the Kerr solutions.

Some important general properties of blackholes are now discussed. For a detailed treatment on blackhole physics and results such as the blackhole uniqueness theorems and details, see Hawking and Ellis (1973) and Wald (1984). As stated earlier, the fundamental motivation for the concept of a blackhole comes from a spherically symmetric homogeneous dust collapse that has two important features. First, for a star undergoing a complete gravitational collapse, a region of trapped surfaces forms below $r = 2m$, from which no light rays escape to an observer at infinity. Therefore, a blackhole forms in the spacetime. Second, the ultimate fate of the star undergoing the collapse is an infinite curvature singularity at $r = 0$, which is completely hidden within the trapped surface region and the blackhole. Hence, no emissions or light rays from the singularity could go out to any observer at infinity, and the singularity is causally disconnected from the outside spacetime.

The question now is whether these conclusions can be generalized for a non-spherically symmetric collapse, and whether they are valid at least for small perturbations from exact spherical symmetry. It is known from Hawking and Ellis (1973), using the stability of a Cauchy development in general relativity, that the formation of trapped surfaces, and hence of a blackhole, is a stable property when departures from spherical symmetry are taken into account. Considering a spherically symmetric collapse evolution from given initial data on a partial Cauchy surface S, the formation of trapped surfaces T in the form of all the spheres with $r < 2m$ in the exterior Schwarzschild geometry is found. The stability of the Cauchy development then implies that, for all initial data sufficiently near to the original data in the compact region $J^+(S) \cap J^-(T)$, the trapped surfaces must still occur. Then, the curvature singularity of a spherical collapse also turns out to be a stable feature as implied by the singularity theorems, which show that the closed trapped surfaces always imply the existence of a spacetime singularity under reasonable general conditions.

There is no proof available, however, that the singularity will continue to be hidden within the blackhole and remain causally disconnected from outside observers, even when the collapse departs from the homogeneous dust cases, or is not exactly spherical. If the singularity became visible to external observers, the predictability in the spacetime will be undermined because new information could come from the singularity where the densities and curvatures could be arbitrarily large.

Hence, in order to generalize the notion of blackholes to gravitational collapse situations other than exactly spherically symmetric homogeneous dust cases, it becomes necessary to rule out such naked singularities by means of an explicit cosmic censorship assumption. This could be stated as follows: if S is a partial Cauchy surface, then there are no naked singularities to the future of S that can be seen from the future null infinity $\mathcal{I}^+$. This is true for the spherical homogeneous dust collapse, where the breakdown of physical theory at the spacetime singularity does not disturb the prediction in the future for the outside asymptotically flat region.

This assumption is made precise by considering the spacetimes (M, g) that admit a weakly asymptotically simple and empty conformal completion $(\overline{M}, \bar{g})$. Then, (M, g) is said to be *future asymptotically predictable* from a partial Cauchy surface S if

$$\mathcal{I}^+ \subset \bar{D}^+(S, \overline{M}), \tag{4.42}$$

that is, the future null infinity $\mathcal{I}^+$ is contained in the closure of $D^+(S)$ in the conformal manifold.

Future asymptotic predictability ensures the cosmic censorship condition in the form that there are no singularities in the future of S that are 'naked', that is, visible from the future null infinity $\mathcal{I}^+$. In the spherical homogeneous dust collapse, the resulting spacetime is future asymptotically predictable and the censorship holds. Whether this is respected in any other general situations is not known, either as a proof for the future asymptotic predictability for general spacetimes, or of any other suitable version of the cosmic censorship hypothesis.

A *blackhole region* in the spacetime, which can be denoted by $\mathcal{B}$, for a future asymptotically predictable spacetime M is defined as

$$\mathcal{B} = M - J^-(\mathcal{I}^+). \tag{4.43}$$

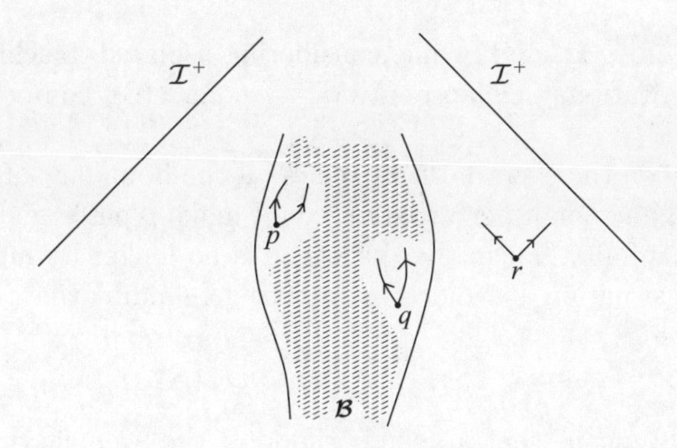

Fig. 4.4 No non-spacelike curves from the blackhole region $\mathcal{B}$ can reach the future infinity and the events faraway in the spacetime, so it is cut off from the outside universe. The events p and q are in a blackhole, whereas the event r that is outside the blackhole can send signals to infinity.

Therefore, this is a spacetime region from which no null or timelike curves can reach an observer at infinity (see Fig. 4.4). The boundary of $\mathcal{B}$ in M, given by

$$\mathcal{H} = \dot{J}^-(\mathcal{I}^+) \cap M, \tag{4.44}$$

is called the *event horizon.* As pointed out earlier, this horizon must be an achronal surface generated by null geodesics that could have past end points in M, but that have no future end points. For the Minkowski spacetime, $J^-(\mathcal{I}^+) = M$ and there is no blackhole. However, for the Schwarzschild case, $J^-(\mathcal{I}^+)$ is the region for the spacetime exterior to $r = 2m$, and the event horizon is given by the null hypersurface $r = 2m$, which is the boundary of the blackhole region $0 < r < 2m$.

By definition, no spacetime singularities are visible at null infinity in an asymptotically predictable spacetime. In fact, this is true also for trapped surfaces in this case, provided either the weak or strong energy condition is satisfied, which implies that all trapped surfaces must be fully contained within the blackhole region and not visible from $\mathcal{I}^+$ (Wald, 1984). Specifically, let (M, g_{ij}) be future asymptotically predictable from a partial Cauchy surface S, and $R_{ij}K^iK^j \geq 0$ for all null vectors K^i. Then, if T is a closed trapped surface in $D^+(S)$, $T \cap J^-(\mathcal{I}^+) = \emptyset$.

Let (M, g_{ij}) be an asymptotically flat spacetime with the associated unphysical conformal spacetime $(\overline{M}, \bar{g}_{ij})$. Suppose $p \in \mathcal{I}^+$ and $q \in M \cap J^-(p)$. Let γ be the future directed null geodesics generator of $\mathcal{I}^+$ through p, and let $r \in \gamma$ be any point. Then, $q \in I^-(r)$ in M. Therefore, $J^-(\mathcal{I}^+) = I^-(\mathcal{I}^+)$ in M, and hence $J^-(\mathcal{I}^+)$ is open in M. Therefore, the blackhole region $\mathcal{B} = M - J^-(\mathcal{I}^+)$ is closed in M. This implies that the event horizon is contained in $\mathcal{B}$. As such, the blackhole region $\mathcal{B}$ need not be connected in M, and while considering isolated blackholes forming out of a gravitational collapse in M, a connected component of $\mathcal{B}$ is worked with.

Any event p on the event horizon $\mathcal{H}$ lies on the boundary of the blackhole region, and so any small perturbation could make p enter $J^-(\mathcal{I}^+)$, causally connecting to infinity. Then, the spacetime is no longer asymptotically predictable. This situation is avoided by further demanding that, for the partial Cauchy surface S,

$$J^+(S) \cap \overline{J^-(\mathcal{I}^+)} \subset D^+(S). \tag{4.45}$$

This effectively means that a neighborhood of the event horizon could also be predicted from S, and is equivalent to the condition that the spacetime exterior to the blackhole region is globally hyperbolic.

In the case of collapsing dust, the event horizon is a null hypersurface generated by those null geodesics that just reach the surface of the star when it crosses the radius $r = 2m$, with m being the Schwarzschild mass of the star. The area of the horizon increases monotonically until the horizon reaches the surface of the star. Outside, this area is a constant given by $A = 16\pi m^2$. For the Kerr spacetime, the horizon is defined by $r = r_+$, the area being obtained by setting $t =$ const., $r = r_+$, which gives the metric on the surface. Then,

$$A = \int \sqrt{g}\, d\theta\, d\phi = 8\pi m\big[m + (m^2 - a^2)^{1/2}\big], \tag{4.46}$$

where m and a denote the mass and angular momentum parameters respectively. Therefore, the area of the horizon is a non-decreasing function.

In fact, for strongly asymptotically predictable spacetimes in general, the area of a blackhole horizon must either remain constant or must increase, provided $R_{ij}K^iK^j \geq 0$ for all null vectors K^i (Hawking, 1971). The basic argument leading to the proof of this result describes the evolution of the event horizon. First, note that the horizon $\mathcal{H}$ is generated by future inextendible null geodesics generators, as $\mathcal{H}$ is the boundary of the past of $\mathcal{I}^+$, and so in order to have a future endpoint, a generator must intersect $\mathcal{I}^+$, which is not possible. Next, for all the null geodesic generators of $\mathcal{H}$, the expansion θ must be non-negative everywhere, $\theta \geq 0$. This is because, if at some $p \in \mathcal{H}$, if $\theta < 0$, $\mathcal{H}$ can be deformed in the neighborhood of p so that $\theta < 0$ still in this neighborhood, and it enters the past of $J^-(\mathcal{I}^+)$. Now, choose a spacelike two-surface T in $\dot{J}^-(\mathcal{I}^+)$ in this neighborhood, then T intersects $J^-(\mathcal{I}^+)$. Then, generators of $\dot{J}^+(T)$, visible from $\mathcal{I}^+$ have past endpoints at T and are orthogonal to T. However, $\theta < 0$ implies that they must have a point conjugate to T within a finite affine distance and so cannot remain in the null boundary all the way to infinity, which is a contradiction.

Therefore, the area of a two-dimensional cross section of generators cannot decrease as $\mathcal{H}$ evolves to the future. From this, it is then possible to deduce the result that the area of the event horizon must be non-decreasing in the future.

In particular, if two or more blackholes merge to form a single blackhole, the area of its boundary must be greater than or equal to the sum of the original blackhole areas. An interesting implication of this result was pointed out by Hawking (1971), to get an upper limit on the energy that can be emitted in gravitational radiation when two blackholes coalesce, at most half the initial energy could be released in blackhole collisions.

According to the area theorem above, for all physically allowed processes, the total area of the blackholes cannot decrease, that is, $\delta A \geq 0$. This is very similar to the second law of thermodynamics, which says that for all physical processes, the total entropy of all the matter in the universe is non-decreasing, that is, $\delta S \geq 0$. Consider, for example, the Schwarzschild blackhole. The area for its event horizon is given by $A = 16\pi m^2$. The only way to reduce this area is to extract mass from the blackhole, which is impossible because no particles or photons can cross the event horizon to come out. On the other hand, the area could be increased by throwing particles in, which would increase the mass (see for example, Bekenstein, 1973, for details on the relationship between entropy and area of the event horizon).

Similar analogies have been developed between other laws of thermodynamics and the laws of blackhole physics. In fact, it has been shown by Bardeen, Carter, and Hawking (1973) that for any stationary axisymmetric blackhole in an asymptotically flat spacetime, it is possible to define a quantity called surface gravity that is constant on the horizon. Therefore, just as the temperature is constant throughout a body in thermal equilibrium (zeroth law of thermodynamics), so is the surface gravity constant over the horizon of a stationary blackhole. Again, for a Schwarzschild blackhole, the surface gravity turns out to be $1/4m$, and it is impossible to reduce it to zero by any physical process, just as it is impossible to achieve the zero temperature for a system by any physical process (third law of thermodynamics).

As such, the thermodynamic temperature of a blackhole in classical relativity will be absolute zero, since it is a perfect absorber that does not emit at all. So, it would appear that the surface gravity may not represent physical temperature. However, it was shown by Hawking (1975) that when quantum particle creation effects are taken into account, a blackhole actually radiates with a black body spectrum at a temperature proportional to the surface gravity. In this sense, the surface gravity represents the thermodynamic temperature of a blackhole.

These results on blackholes assume the spacetime to be asymptotically flat. In reality, however, the universe is not so, in view of the observed distribution of matter at the largest possible scales. Hence, it would be desirable to have the laws of blackholes given in a more general spacetime framework. Blackholes in a general globally hyperbolic spacetime were defined by Tipler, Clarke, and Ellis (1980), and Joshi and Narlikar (1982), who examined the laws of blackhole physics in such a framework. It is seen that for a closed globally hyperbolic spacetime with a compact Cauchy surface and a strong curvature crushing singularity in the future, the sum of the areas of the blackholes must decrease in the future. It is thus seen that in the situation of the spacetime not being asymptotically flat, the behavior of blackholes and the laws governing them may change.

Apart from general relativity, there is a motivation to consider blackholes in Newtonian theory also, in the sense that the gravitational field of a star could be so strong that it might stop even light (considered as particles) from escaping. For example, consider a spherically symmetric mass distribution of uniform density with radius R and total mass M. For a particle with a velocity v away from the center and at a distance r, its total energy, which is the sum of its kinetic and potential energies, is conserved. Suppose now the velocity v_0 of the particle is such that it is able to escape to infinity where it has a vanishing velocity. Since the total energy will be zero at $r = \infty$ with $v = 0$, which is conserved,

$$v_0^2 = \frac{2GM}{R}. \tag{4.47}$$

When the radial velocity of the particle is less than v_0, it must fall back to the body, otherwise it escapes to infinity. Therefore, if the mass distribution and radius of the body were such that $c^2 = 2GM/R$, where c is the velocity of light, then for any larger mass or smaller radius of the body, even light will not escape. This was realized by Laplace in 1798, who pointed out that, for a star with a density the same as the Sun but radius 250 times larger, no light could escape from its surface.

It is clear that a blackhole could not be observed directly, but gravitational effects exhibited by such an object must be looked for. Although there is no conclusive evidence available for the existence of blackholes at the moment, presently the best candidates seem to be the binary stars in which one of the partners is visible and the other is supposed to be a blackhole. Such a blackhole would suck matter from its visible component, in the process forming an accretion disk around the blackhole. Before the in-falling matter spirals down the blackhole, the inner, hot regions are believed to produce intense bursts of X-rays formed by synchrotron radiation. Therefore, the discovery of the X-ray source Cygnus XI in 1971, which shows rapid variations, indicates possible evidence for blackholes. Further to this, several other X-ray binaries have been proposed as possible candidates for blackholes.

4.4 Higher spacetime dimensions

Although there is no satisfactory proof or mathematical formulation of censorship available despite many efforts, as the discussion above shows, there are many classes of dynamical collapse models investigated so far that lead to a blackhole or a naked singularity as the collapse endstate (see for example, Clarke and Joshi, 1988; Newman and Joshi, 1988; Lake, 1992; Rendall, 1992; Szekeres and Iyer, 1993; Rein, Rendall, and Schaeffer, 1995; Wald, 1997; Penrose, 1998; Gundlach, 1999; Królak, 1999; Jhingan and Magli, 2000; Joshi, 2000; Celerier and Szekeres, 2002; Harada, Iguchi, and Nakao, 2002; Giambo *et al.*, 2003; Debnath, Chakraborty, and Barrow, 2004; Maeda, 2006).

One of the possibilities to recover cosmic censorship could be to consider the possibility that one may actually be living in a higher-dimensional universe. The recent developments in string theory and other field theories indicate that gravity is possibly a higher-dimensional interaction, which reduces to the general relativistic description at lower energies. Hence, it may be that censorship is restored in higher spacetime dimensions due to extra physical effects arising from the transition itself to a higher-dimensional spacetime continuum. The recent revival of interest in such a possibility is motivated partly by the Randall–Sundrum brane-world models (Randall and Sundrum, 1999).

It is shown below that certain naked singularities arising in a dust collapse from smooth initial data, which include those obtained by Eardley and Smarr (1979), Christodoulou (1984), and Newman (1986), are removed when the transition to higher-dimensional spacetimes is made. The cosmic censorship is then restored for dust collapses of these classes, which will always produce a blackhole as the collapse endstate for dimensions $N \geq 6$, under various conditions to be motivated physically, such as the smoothness of the initial data from which the collapse develops. This is of interest because the gravitational collapse of a dust cloud has served, over the last several decades, as a basic and fundamental paradigm in blackhole physics. It is seen explicitly that the above mentioned naked singularities of the dust collapse, for both marginally bound and non-marginally bound dust, are removed on going to higher (≥ 6) dimensions, if only smooth and analytic initial profiles are allowed for.

It is pointed out here that the increase in dimensions deforms the apparent horizon, which is the outer boundary of the trapped surfaces, in such a manner that the formation of the trapped surfaces is advanced. Hence, a critical dimension exists, beyond which the neighborhood of the center gets trapped before the central singularity and the final outcome of collapse is necessarily a blackhole (Goswami and Joshi, 2004b).

Consider a general spherically symmetric comoving metric in $N \geq 4$ dimensions that has the form

$$ds^2 = -e^{\nu(t,r)}dt^2 + e^{2\psi(t,r)}dr^2 + R^2(t,r)d\Omega^2_{N-2}, \tag{4.48}$$

where

$$d\Omega_{N-2}^2 = \sum_{i=1}^{N-2} \left[\prod_{j=1}^{i-1} \sin^2(\theta^j) \right] (d\theta^i)^2 \tag{4.49}$$

is the metric on the $(N-2)$ sphere. The energy–momentum tensor of the dust has the form

$$T^t{}_t = \rho(t,r), \quad T^r{}_r = 0, \quad T^{\theta^i}{}_{\theta^i} = 0. \tag{4.50}$$

Take the matter field to satisfy the weak energy condition, that is, the energy density measured by any local observer is non-negative, and so for any timelike vector V^i,

$$T_{ik}V^iV^k \geq 0, \quad \text{i.e. } \rho \geq 0. \tag{4.51}$$

In the case of a finite collapsing cloud, there is a finite boundary $0 < r < r_{\rm b}$, outside which the cloud is matched to a Schwarzschild exterior. The range of the coordinates for the metric is then $0 < r < r_{\rm b}$, and $-\infty < t < t_{\rm s}(r)$, where $t_{\rm s}(r)$ corresponds to the singular epoch $R = 0$. Solving the N-dimensional Einstein equations, the generalized Tolman–Bondi–Lemaitre (TBL) metric can be obtained as

$$ds^2 = -dt^2 + \frac{R'^2}{1+f(r)}dr^2 + R^2(t,r)d\Omega_{N-2}^2, \tag{4.52}$$

where $f(r)$ is an arbitrary function of the comoving radius r, and $f(r) > -1$. The equations of motion are then given by

$$\frac{(N-2)F'}{2R^{(N-2)}R'} = \rho, \quad \dot{R}^2 = \frac{F(r)}{R^{(N-3)}} + f(r). \tag{4.53}$$

Here, $F(r)$ is an arbitrary function of the comoving coordinate r and has the interpretation of mass function of the dust cloud, as discussed earlier, and $f(r)$ is the energy function, which specifies the velocity profiles for the collapsing shells for any given mass function. The energy condition then implies $F' \geq 0$. Consider the shell-focusing naked singularity at $R = 0$. The continual collapse condition is given as $\dot{R} < 0$.

Using the scaling independence of the comoving coordinate r,

$$R(t,r) = rv(t,r), \tag{4.54}$$

where

$$v(t_i, r) = 1, \quad v(t_{\rm s}(r), r) = 0, \quad \dot{v} < 0. \tag{4.55}$$

This means the coordinate r has been scaled in such a way that at the initial epoch, $R = r$, and at the singularity $R = 0$. As noted previously, $R = 0$ both at the regular center $r = 0$ of the cloud, and at the spacetime singularity, where all matter shells collapse to a zero physical radius. The regular center is then distinguished from the singularity by a suitable behavior of the mass function $F(r)$ so that the density remains finite and regular there at all times until the singular epoch. The introduction of the parameter v as above then allows the spacetime singularity to be distinguished from the regular center, with $v = 1$ at the initial epoch, including the center $r = 0$, and which then decreases monotonically with time as the collapse progresses to the value $v = 0$ at the singularity $R = 0$.

In order to ensure the regularity of the initial data, it is seen from the equations of motion that at the initial epoch the two free functions $F(r)$ and $f(r)$ must have the forms

$$F(r) = r^{(N-1)}\mathcal{M}(r), \quad f(r) = r^2 b(r), \tag{4.56}$$

where $F(r)$ and $f(r)$ are the mass function and the energy function respectively.

Following Goswami and Joshi (2006), now assume that the initial density and energy functions $\rho(r)$ and $f(r)$ are smooth and even, ensuring their analytic nature. Note that the Einstein equations do not impose any such restriction, which has to be physically motivated, and this implies a certain mathematical simplicity in the arguments to deal with a dynamical collapse situation. It follows that both $\mathcal{M}(r)$ and $b(r)$ are now smooth C^∞-functions, which means the Taylor expansions of these functions around the center must be of the form

$$\mathcal{M}(r) = \mathcal{M}_0 + r^2\mathcal{M}_2 + r^4\mathcal{M}_4 + \cdots, \tag{4.57}$$

and

$$b(r) = b_0 + r^2 b_2 + r^4 b_4 + \cdots, \tag{4.58}$$

that is, all odd terms in r vanish in these expansions, and the presence of even terms only would ensure smoothness.

To predict the final state of the collapse for a given initial mass and velocity distribution, the singularity curve resulting from the collapse of successive matter shells, and the apparent horizon developing in the spacetime are studied below (see also Goswami and Joshi, 2004a). A decreasing apparent horizon in the (t, r) plane then is a sufficient condition for a blackhole, as it shows the entrapment of the neighborhood of the center before the central singularity.

Consider the continual collapse of the dust cloud to a final shell-focusing singularity at $R = 0$, where all matter shells collapse to a zero physical radius. With the regular initial conditions as above, the Einstein equation for $\dot{R}$ can be written as

$$v^{(N-3)/2}\dot{v} = -\sqrt{\mathcal{M}(r) + v^{(N-3)}b(r)}. \tag{4.59}$$

Here, the negative sign implies that $\dot{v} < 0$, that is, the matter cloud is collapsing. Integrating the above equation with respect to v gives

$$t(v, r) = \int_v^1 \frac{v^{(N-3)/2}dv}{\sqrt{\mathcal{M}(r) + v^{(N-3)}b(r)}}. \tag{4.60}$$

Note that the coordinate r is to be treated as a constant in the above equation. Expanding $t(v, r)$ around the center,

$$t(v, r) = t(v, 0) + r\mathcal{X}(v) + r^2\frac{\mathcal{X}_2(v)}{2} + r^3\frac{\mathcal{X}_3(v)}{6} + \cdots. \tag{4.61}$$

Now, from the equation for the form of the mass function above and (4.58), $\mathcal{X}(v) = 0$. The function $\mathcal{X}_2(v)$ is then given as

$$\mathcal{X}_2(v) = -\int_v^1 \frac{v^{(N-3)/2}(\mathcal{M}_2 + v^{(N-3)}b_2)dv}{(\mathcal{M}_0 + v^{(N-3)}b_0)^{3/2}}, \tag{4.62}$$

where

$$b_0 = b(0), \quad \mathcal{M}_0 = \mathcal{M}(0), \quad b_2 = b''(0), \quad \mathcal{M}_2 = \mathcal{M}''(0). \tag{4.63}$$

Note that the value of $\mathcal{X}_2$ is determined fully by the initial values of the mass function $F(r)$ and the energy function $b(r)$. Therefore, the time taken for the central shell to reach the singularity is given by

$$t_{s_0} = \int_0^1 \frac{v^{(N-3)/2}dv}{\sqrt{\mathcal{M}_0 + v^{(N-3)}b_0}}. \tag{4.64}$$

From the above equation it is clear that for t_{s_0} to be defined,

$$\mathcal{M}_0 + v^{(N-3)}b_0 > 0, \tag{4.65}$$

that is, the continual collapse condition implies the positivity of the above term. Hence, the time taken for other shells close to the center to reach the singularity can now be given by

$$t_s(r) = t_{s_0} + r^2\frac{\mathcal{X}_2(0)}{2} + \cdots. \tag{4.66}$$

In order to determine the visibility, or otherwise, of the singularity at $R = 0$, the causal structure of the trapped surfaces, and the nature and behavior of the null geodesics in the vicinity need to be analyzed. If future directed null geodesics with a past end point at the singularity that go out to faraway observers in spacetime exist, then the singularity is naked. Otherwise, a blackhole results as the endstate of a continual collapse. The boundary of the trapped region of the spacetime is given by the apparent horizon within the collapsing cloud, which is given by the equation

$$\frac{F}{R^{N-3}} = 1. \tag{4.67}$$

Broadly, it can be stated that if the neighborhood of the center gets trapped earlier than the singularity, then it is covered, otherwise it is naked with families of non-spacelike future directed trajectories escaping away from it. For example, it follows from the above equation that along the singularity curve $t = t_{\mathrm{s}}(r)$ (which corresponds to $R = 0$), for any $r > 0$, $F(r)$ goes to a constant positive value, whereas $R \to 0$. Hence, it follows that trapping already occurs before the singularity develops at any $r > 0$ along the singularity curve $t_{\mathrm{s}}(r)$.

Now, there is a need to determine when there will be families of non-spacelike paths coming out of the central singularity at $r = 0$, $t = t_{\mathrm{s}}(0)$, reaching outside observers, and when there will be none. The visibility, or otherwise, of the singularity is decided accordingly. By determining the nature of the singularity curve and its relation to the initial data, it is possible to deduce whether the trapped surface formation in the collapse takes place before or after the central singularity. It is this causal structure that determines the possible emergence, or otherwise, of non-spacelike paths from the singularity, and settles the final outcome in terms of either a blackhole or a naked singularity.

From (4.67),

$$v_{\mathrm{ah}}(r) = [r^2 \mathcal{M}(r)]^{1/(N-3)}. \tag{4.68}$$

Using the above equation in (4.60),

$$t_{\mathrm{ah}}(r) = t_{\mathrm{s}}(r) - \int_0^{v_{\mathrm{ah}}(r)} \frac{v^{(N-3)/2} dv}{\sqrt{\mathcal{M}(r) + v^{(N-3)} b(r)}}. \tag{4.69}$$

As the behavior of the apparent horizon is under consideration close to the central singularity at $r = 0$, $R = 0$ (all other points $r > 0$ on the singularity curve are already necessarily covered), the upper limit of integration in the above equation is small, and hence it is possible to expand the integrand in a power series in v, and keep only the leading order term, which amounts to

$$t_{\mathrm{ah}}(r) = t_{\mathrm{s}_0} + r^2 \frac{\mathcal{X}_2(0)}{2} + \cdots - r^{(N-1)/(N-3)} \frac{2}{N-1} \mathcal{M}_0^{1/(N-3)}. \tag{4.70}$$

Since the apparent horizon is a well-behaved surface for a spherical dust collapse, it can be said that the singularity curve for the collapse and its derivatives around the center are also well-defined, as the same coefficients are present in both (4.70) and (4.61). It is now possible to analyze the effect of the number of dimensions on the nature and shape of the apparent horizon. As the number of dimensions is increased, and reaches dimensions higher than five, the negative term in (4.70) starts to dominate, thus advancing the trapped surface formation in time. This equation shows the behavior of the apparent horizon curve for the same initial data in different dimensions. As seen here, in the usual four-dimensional spacetime, the initial profile ensures an increasing apparent horizon and it can be explicitly shown that the end-state is a naked singularity. But, in six and more spacetime dimensions, the negative term starts to dominate, purely as a result of the increase in dimensions. This causes the trapped surfaces to form sufficiently early to cover the singularity. It follows that the number of spacetime dimensions causes an intriguing effect on the causal structure and nature of the apparent horizon, as can be clearly seen. Therefore, for smooth initial data and for dimensions higher than five, the apparent horizon becomes a decreasing function of r near the center. This implies that the neighborhood of the center gets trapped before the central singularity and the central singularity is then always covered, as opposed to the four-dimensional models of Eardley and Smarr (1979), Christodoulou (1984) and Newman (1986), in which case a positive $\mathcal{X}_2(0)$ makes the apparent horizon curve increase and hence ensures a naked singularity (see Fig. 4.5).

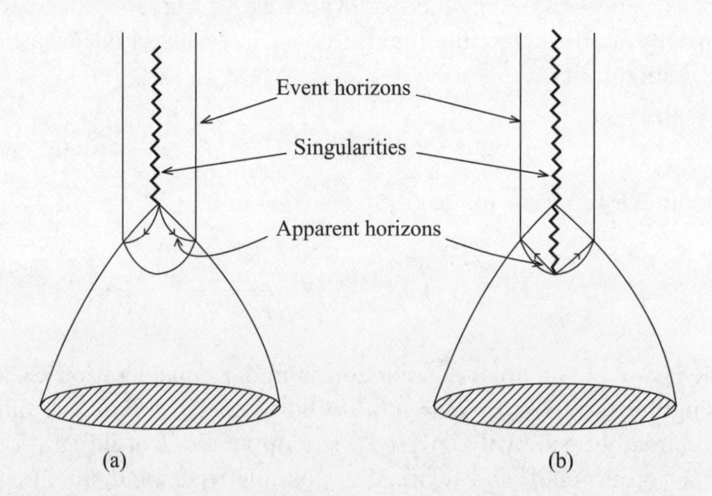

Fig. 4.5 Typical apparent horizon behavior in a blackhole (a) and a naked singularity (b) collapse. Whereas in the blackhole case the apparent horizon is decreasing away from the center, in the naked singularity situation it typically increases in time away from the center.

Specifically, suppose there is a future directed outgoing null geodesic energing from the central singularity at $R = 0$, $r = 0$. If (t_1, r_1) is an event along the null geodesic, then $t_1 > t_{s_0}$ and $r_1 > 0$. But for any such r_1, the trapped region already starts before $t = t_{s_0}$, hence the event (t_1, r_1) is already in the trapped region and the geodesic cannot be outgoing. Therefore, there are no outgoing paths from the central singularity, making it covered. It now follows in general, that is for both marginally bound and non-marginal cases, that for a dust collapse with smooth initial profiles, the final outcome is always a blackhole for any spacetime dimensions $N \geq 6$, and all the naked singularities occurring in lower dimensions are removed.

In five dimensions, there is an interesting scenario arising from Banerjee, Debnath, and Chakraborty (2003). As can be seen from (4.70), there is a critical value of $\mathcal{X}_2(0)$, below which the apparent horizon is decreasing, resulting in a blackhole endstate. Otherwise, a naked singularity can result. It is interesting to note also that the above results hold if the initial profiles, instead of being absolutely smooth C^∞-functions, are taken to be sufficiently smooth, namely at least C^2-functions.

Note that it may still be possible to have both blackhole and naked singularity outcomes as the collapse endstates, when the initial data and the metric are not assumed necessarily to be C^∞, or at least C^2, analytic functions. Hence, such smoothness conditions need to be investigated and carefully probed further to see if they could be strongly motivated from a physical perspective.

4.5 Formulating the censorship

It is clear from the discussions so far that the assumption of cosmic censorship is crucial and necessary to basic results in blackhole physics. In fact, when the gravitational collapse is considered in a generic situation, the very existence of blackholes requires this hypothesis (Penrose, 1969).

If the cosmic censorship is to be established by means of a rigorous proof, this of course requires a much more precise formulation of the hypothesis. The statement that the result of a complete gravitational collapse must always be a blackhole and not a naked singularity, or that all singularities of the collapse must be hidden in blackholes, causally disconnected from observers at infinity, is not rigorous enough. This is because, under completely general circumstances, the censorship or asymptotic predictability is false as it is possible to choose a spacetime with a naked singularity that would be a solution to Einstein's equations if

$$T_{ij} \equiv \frac{1}{8\pi} G_{ij}. \tag{4.71}$$

Therefore, at the minimum, certain conditions on the stress–energy tensor are required, for example, an energy condition. However, it turns out that to obtain an exact characterization of the restrictions that should be required on matter fields in order to prove a suitable version of the cosmic censorship hypothesis is an extremely difficult task and no such specific conditions are available presently. In other words, no plausible restrictions have been able to ensure any provable version of the censorship conjecture.

The requirements in blackhole physics and general predictability arguments have led to several different formulations of the cosmic censorship hypothesis. The version known as the *weak cosmic censorship* refers to the asymptotically flat spacetimes, the ones that are Minkowskian faraway from the source, and has reference to the null infinity, which is reached by null geodesics an infinite affine distance away. Weak censorship, or asymptotic predictability, effectively postulates that the singularities of gravitational collapse cannot influence events near the future null infinity $\mathcal{I}^+$. If S is the partial Cauchy surface on which the regular initial data for the collapse is defined, this is the requirement that $\mathcal{I}^+$ is contained in the closure of $D^+(S)$. Therefore, the data on S predict the entire future for faraway observers.

The other version, called the *strong cosmic censorship*, is a general predictability requirement on any spacetime, stating that all physically reasonable spacetimes must be globally hyperbolic (see for example, Geroch and Horowitz, 1979; Hawking and Israel, 1979a, 1979b; Penrose, 1979).

The weak cosmic censorship or the strong asymptotic predictability requirement means that the region of spacetime outside a blackhole must be globally hyperbolic (Tipler, Clarke, and Ellis, 1980). A precise formulation of this version of censorship will consist of specifying exact conditions under which the spacetime would be strongly asymptotically predictable. In its weak form, the censorship conjecture does not allow causal influences from the singularity to asymptotic regions in the spacetime to an observer at infinity, that is, the singularity cannot be globally naked. However, it could be locally naked in that an observer within the event horizon and in the interior of the blackhole could possibly receive particles or photons from the singularity.

Clearly, such a requirement must be formulated more precisely. For example, the metric on space should approach that of the Euclidean three-space at infinity, and matter fields should satisfy suitable fall off conditions at spatial infinity. Also, the null generators of $\mathcal{I}^+$ may need to be complete, and the exact meaning of 'physically reasonable' matter fields has to be specified. In fact, as far as the cosmic censorship hypothesis is concerned, it is a major problem in itself to find a satisfactory and mathematically rigorous formulation of what is physically desired to be achieved (see for example, Penrose, 1998). Developing a suitable formulation would probably be a major advance towards the solution of the main problem.

Since the interest here is mainly in the gravitational collapse scenario, it is required that the spacetime contains a regular initial spacelike hypersurface on which the matter fields, as represented by the stress–energy tensor T_{ij}, have a compact support and all physical quantities are well-behaved on this surface. Also, it is generally required that the matter satisfies a suitable energy condition, and that the Einstein equations are satisfied. Then, it can be said that the spacetime contains a naked singularity if there is a future directed non-spacelike curve which reaches a faraway observer or infinity in the future, and in the past it terminates at the singularity.

The main difficulty in proving the weak censorship appears to be that the event horizon is a feature depending on the whole future behavior of the solution over an infinite time period, whereas the present theory of quasi-linear hyperbolic equations guarantee the existence and regularity of the solutions over a finite time internal only (Israel, 1984). In this connection, the results of Christodoulou (1986) on generic spherically symmetric collapses of massless scalar fields are relevant, where it is shown using global existence theorems on partial differential equations that global singularity free solutions exist for weak enough initial data. In any case, even if it is true, the proof for a suitable version of the weak censorship conjecture would seem to require much more knowledge of the general global properties of Einstein's equations and solutions than is known presently. As such, various attempts to formulate and prove the weak censorship have not succeeded so far.

It would appear that sufficient data are not yet available on the various possibilities present for gravitationally collapsing configurations that would enable us to decide one way or the other on the issue of the censorship hypothesis. In this situation, a detailed investigation of the collapse scenarios and models within general relativity really appears necessary, and the possibilities arising in order to have insights into the issue of the final fate of the gravitational collapse need to be examined.

For example, shell-crossing naked singularities have been shown to occur in the spherical collapse of perfect fluids (Yodzis, Seifert, and Muler zum Hagen, 1973, 1974), where shells of matter implode in such a way that fast moving outer shells overtake the inner shells, producing a globally naked singularity outside the horizon. These are the singularities where shells of matter pile up to give two-dimensional caustics and the density and some curvature components blow up. The general point of view, however, is that such singularities need not be treated as serious counter-examples to the censorship hypothesis, as these are merely consequent to the intersection of matter flow lines. This gives a distributional singularity that is gravitationally weak in the sense that the curvatures and tidal forces remain finite near it (Tipler, Clarke, and Ellis, 1980).

On the other hand, there are shell-focusing naked singularities, as discussed earlier in detail, occurring at the center of the spherically symmetric collapsing configurations of dust, perfect fluids or radiation shells. These have to be taken more seriously, and they can be ruled out only by saying that the dust or perfect fluids are not really 'fundamental' forms of matter. However, if the cosmic censorship is to be established as a rigorous theorem, this objection has to be made much more precise in terms of a clear restriction on the stress–energy tensor, because these are forms of matter that otherwise satisfy reasonability conditions. These include the dominant energy condition (provided there are no large negative pressures) or a well-posed initial value formulation for the coupled Einstein matter field equations. Also, these forms of matter are widely used in discussing various astrophysical processes. For a review of various related developments and a detailed discussion, see Joshi (1993) and Krasinski (1997).

Discussed briefly now is the hypothesis of strong cosmic censorship (Penrose, 1979). Unlike the weak conjecture, the strong version demands that the singularities should not be visible, even to the observers within the black-hole. That is, they cannot be even locally naked, but are always spacelike and the spacetime must be globally hyperbolic. Therefore, unless they actually encounter them, the observer never sees the singular regions. The argument given in favor of such a strong principle is that if cosmic censorship is really a basic principle of nature, there should not be any special role given to the observer at infinity because physical laws operate at a local level. Again, this principle is to be carefully formulated because, by suitable cuts and identifications in the Minkowski spacetime, it is easy to generate an inextendible non-globally hyperbolic spacetime. In other words, any non-trivial topology for the spacetime would give rise to violations of global hyperbolicity, and therefore of the strong cosmic censorship. For any certain proposed formulations of strong censorship and difficulties encountered, see Penrose (1979). Again, no general proof is available for strong censorship, but it is sometimes argued that the Cauchy horizons, forming as a result of non-global hyperbolicity of the spacetime, turn out to be unstable in certain cases. For example, in the Reissner–Nordström case, the Cauchy horizon exhibits a so-called 'infinite blue-shift' instability (Chandrasekhar and Hartle, 1983). However, again the evidence is not uniform here. For example, Morris, Thorne, and Yurtsever (1988) have shown that for the wormhole spacetime that they constructed, the Cauchy horizon is immune to the Taub–NUT type of instability, and they conjecture that it is fully stable, thus providing a counter-example to strong censorship. An additional condition that is often required in both the weak and strong formulations of the censorship is that *stable* spacetimes do not admit naked singularities. This requires a suitable criteria for the stability of the spacetimes, which is again a major difficult problem in general relativity.

Various different formulations of the censorship principle have been tried out, based on different motivations. Therefore, a class of censorship conjectures suggested all naked singularities must be in some sense gravitationally weak (Tipler, Clarke, and Ellis, 1980; Israel, 1986a, 1986b; Newman, 1986; Newman and Joshi, 1988; Nolan, 1999). Some of these suggestions are discussed here. In order to avoid the difficulties associated with the question as to which forms of matter and equations of state should be considered reasonable, it is also suggested that one examines first a purely vacuum version of the censorship. That will show whether pure gravity allows naked singularities. In fact, Geroch and Horowitz (1979) have detailed several possible approaches to the censorship formulations, and pointed out difficulties in each case.

To summarize the situation, while the cosmic censorship hypothesis is a crucial assumption underlying all blackhole physics, gravitational collapse theory, and many important related areas in gravity physics, no proof and formulation is available today. The first major task here is actually to formulate rigorously a satisfactory version of the hypothesis. The proof of cosmic censorship would confirm the already widely accepted and applied theory of blackholes, while its overturn would throw the blackhole dynamics into serious doubt. Therefore, cosmic censorship turns out to be one of the most important issues for general relativity and gravitation theory today. Even if true, a proof for this conjecture does not seem possible unless some major theoretical advances of the mathematical techniques and the understanding of the global structure of the Einstein equations are made.

This situation leads to the conclusion that the first and foremost task here is to carry out a detailed and careful examination of gravitational collapse scenarios that possibly give rise to a naked singularity formation. Until this is done properly, trying out different formulations for censorship may not help, because without really knowing what is involved in the collapse processes, one may be in for a complete surprise as far as the final fate of a gravitational collapse is concerned. It is only such investigations of general collapse situations that could indicate which theoretical advances to expect for a proof, and which features to avoid while formulating the cosmic censorship.

Under this situation, while no mathematical formulation or proof of censorship is currently available, another alternative is to ask if it would be possible to impose suitable *physical constraints* on the gravitational collapse so as to ensure the validity of the censorship. In other words, *physically realistic* collapses should satisfy the censorship. Then again, the precise physical conditions under which the censorship is supposed to be holding are to be specified. The advantage then is that even if a certain set of physical conditions did not work towards proving the censorship, there is still the option of trying out another set of physical constraints to continue further efforts. Eventually, this may lead to an appropriate mathematical formulation of the censorship conjecture to be established. Many natural looking physical conditions can be proposed and tried out, with these being indicated as a remedy to rule out naked singularities.

Discussed below are several such physical constraints on a realistic gravitational collapse scenario, and the implications they have towards determining the final fate of the collapse. In particular, the motivation is to rule out a naked singularity as the final state by imposing such conditions. It turns out that a naked singularity cannot be ruled out with the help of such conditions considered so far. But the advantage of such an analysis is, first, that it clarifies the situation as to what the conditions can possibly achieve. Second, it serves as a pointer to something deeper that should be looked for if a censorship is to be established. This also could imply that, in fact, naked singularities do develop in wide classes of gravitational collapse scenarios under realistic physical conditions. An attempt is also made here to get an insight into why many of these conditions have not worked, or are unlikely to work in establishing a censorship, and why further more subtle alternatives must be explored. Finally, it is indicated that hope appears to lie in a detailed genericity and stability analysis only.

1. A suitable energy condition must be obeyed.

This is one of the basic conditions assumed in the classical gravity description, and it should be satisfied by the matter fields constituting the star, at least until the collapse has proceeded to such an advanced stage so as to enter a phase governed by quantum gravity. This is the stage when the classical description starts breaking down in one way or another.

As noted earlier, if completely arbitrary matter fields are allowed for, it is quite easy to produce naked singularities. For example, one possibility is to start with a geometry allowing families of future directed non-spacelike geodesics, which are future endless, but which terminate in the past at the singularity. Then, define the matter fields to be given by $T_{ij} \equiv (1/8\pi)G_{ij}$. It is thus obvious that in gravitational collapse consideration must be given to scenarios where matter fields do satisfy reasonable physical conditions. It would be hoped that a suitable energy condition would be one of these, as all observed classical fields do obey such a condition. A further motivation would be the energy conditions that have been used extensively in the singularity theorems in general relativity, and which predict the existence of singularities in gravitational collapse and cosmology.

It would be nice to see if the censorship is obeyed once the matter fields have been assumed to satisfy suitable energy conditions. It turns out, however, that there are several classes of collapse models where collapsing matter does satisfy a proper energy condition, but the collapse leads to an endstate that is a naked singularity.

Actually, there are classes of collapse models where satisfying the energy condition appears to be aiding the naked singularity formation as the final state of collapse, in turn making the naked singularity physically more interesting and serious. An example of this is the spherically symmetric self-similar collapse of a perfect fluid. The general form of the metric is

$$ds^2 = -e^{2\nu(r,t)} + e^{2\psi(r,t)} + r^2 S^2(r,t)(d\theta^2 + \sin^2\theta\, d\phi^2), \tag{4.72}$$

where the metric functions are taken to depend on $X = t/r$ due to self-similarity. The outgoing null geodesics (Joshi and Dwivedi, 1992, 1993a) can be worked out from the naked singularity, which turn out to be related to the density and pressure distributions in the spacetime via the Einstein equations. These are then given by

$$r = D(X - X_0)^{2/(H_0-2)}. \tag{4.73}$$

Here, H_0 is the limiting value of the quantity $H = (\eta + p)e^{2\psi}$, with η and p corresponding to density and pressure, and $D > 0$ is the constant of integration. The weak energy condition is then equivalent to the statement that $H_0 > 0$, which in turn ensures, from the above geodesics equation, that *families* of null geodesics, as opposed to single isolated curves, emerge from the naked singularity at $t = 0$, $r = 0$, which is a node in the (t, r) plane.

It should be noted that the Einstein equations do not require or impose an energy condition on matter distributions. It is a criterion motivated purely on physical grounds. This then suggests another possibility, namely that if somehow in the later stages of a collapse the energy conditions are violated through whatever agency, then there may be a hope to preserve the cosmic censorship. In the equation for null geodesics, if the energy conditions are violated, this would correspond to a negative value of H_0, then there are no outgoing null geodesics families from the singularity, and the censorship is essentially preserved.

2. The collapse must develop from regular initial data.

This is one of the most important physical constraints necessary for any possible version of a cosmic censorship statement. In general, the regularity conditions on the initial data for the collapse can come in many forms. If realistic collapse scenarios of matter clouds such as gravitationally collapsing massive stars are to be modeled, then the densities, pressures, and other physical quantities must be finite and regular at the initial spacelike surface from which the collapse develops. That is, the initial surface should not admit any density or curvature singularities in the initial data so as to represent a collapse from regular matter distribution.

Generally, this is ensured by imposing the usual differentiability conditions on the functions involved, together with requirements of finiteness and regularity. It is known from the gravitational collapse analysis so far, that regular distributions of initial densities and pressures (for example, they should be finite and suitably differentiable on the initial surface) do give rise to both naked singularities and blackholes, depending on the nature of the regular initial data from which the collapse evolves, as discussed in Chapter 3. It turns out that given such matter initial data, there are still sufficient numbers of free functions available to choose in the Einstein equations, subject to the weak energy condition and suitable matching to the exterior of the collapsing cloud, so that the evolution can end in either the blackhole or naked singularity outcomes as desired.

At times, more stringent requirements are imposed on the initial data, for example, a complete smoothness of densities, pressures, and the metric functions could be asked for. Usually, there are two motivations for this. One could be the requirements while calculating the analytic or numerical evolutions where smoothness (which is the same as demanding the analyticity of these functions) simplifies the analysis considerably. At other times, it is argued that astrophysically reasonable initial data must be analytic. In the case of the collapse of a dust cloud, this amounts to demanding analyticity of the density function. The initial density $\rho(r)$ then must contain no odd powers in r, and

$$\rho(r) = \rho_0 + \rho_2 r^2 + \rho_4 r^4 + \cdots \tag{4.74}$$

at the initial surface $t = t_i$, which gives an analytic density profile. It is known, however, that even in the case of smooth density profiles with only the even terms being non-vanishing, the marginally bound dust evolution can end in a naked singularity (for example, when $\rho_2 \neq 0$) that is gravitationally strong. That is, sufficiently fast divergence of curvatures does take place in the limit of approach to the singularity.

3. Singularities from realistic collapses must be gravitationally strong.

This has been one of the most useful physical requirements, which was explored rather thoroughly in order to develop a formulation for the cosmic censorship conjecture. The idea has been that any singularity that will develop from a realistic collapse has got to be physically serious in various aspects, including powerful divergences in all important physical quantities such as densities, pressures, curvatures and others, at least at the classical level. A typical condition for the singularity to be gravitationally strong is, in addition to the divergences such as above, the gravitational tidal forces must diverge and all physical volumes are crushed to zero size in the limit of approach to the naked singularity. A sufficient condition for this to happen is

$$R_{ij} V^i V^j \propto \frac{1}{k^2}, \tag{4.75}$$

where k is the affine parameter along the non-spacelike geodesics energing from the singularity, with $k = 0$ at the singularity, and V^i is the tangent vector to these curves emanating from the naked singularity. The above curvature condition corresponds to that developed through the analysis of Clarke, Królak, and Tipler (see Joshi, 1993, and Nolan, 1999, for a further discussion).

The singularity developing within the blackhole formed out of the standard dust cloud collapse as investigated by Oppenheimer and Snyder (1939) is gravitationally strong in the above sense. Now, if it could be established that whenever naked singularities form in the gravitational collapse, they are always gravitationally weak, in the sense of important divergences such as these above not being present in the limit of approach to the singularity, then such singularities could be removable from the spacetime, and it may be possible to extend the spacetime. Such removable naked singularities should no longer be regarded as physically genuine, and progress has then been made towards the cosmic censorship in some form, such as that the naked singularity could develop in the gravitational collapse; however, they would be gravitationally weak and removable always.

This possibility has been investigated thoroughly, and it is known now that gravitationally powerfully strong naked singularities actually do result from the collapse from the regular initial data (including smooth analytic density profiles), for several reasonable forms of matter such as dust, perfect fluids, Vaidya radiation collapses, and several other forms of matter that satisfy suitable energy conditions. In the next chapter, one such example, in the case of the Szekeres quasi-spherical collapse models will be discussed. At such naked singularities, the densities, curvature scalars such as the Kretschmann scalar, and gravitational tidal forces diverge most powerfully as characterized above. This is as powerfully strong as the divergences observed at physical singularities such as the big-bang in cosmology.

4. The matter fields must be sufficiently general.

If a naked singularity formed in the collapse for certain special forms of matter, such as dust or collapsing radiations only, that would not be of much interest. For example, the role of pressure cannot be underestimated in a realistic collapse and so it would be nice to know if matter with pressure would necessarily give rise to a blackhole only on undergoing gravitational collapse. If this were the case, matter fields giving rise to a naked singularity could then be ruled out as special or unphysical, in formulating the censorship, even if they satisfied an energy condition or the collapse developed from regular initial data.

It is now known, however, as discussed in Chapter 3, that naked singularities are not special to any particular form of matter field, and several physically important equations of state are also included. The collapse can be studied for a general form of matter, the so-called type I matter fields (all the known physical forms of matter, such as dust, perfect fluids, and massless scalar fields are included in this class), subject to the weak energy condition. The result is, given an arbitrary but regular distribution of matter on the initial surface, that there are always evolutions available from this initial data that would either result in a blackhole or naked singularity, depending on the allowed choice of free functions available from the Einstein equations. More specifically, in a spherically symmetric collapse with a type I general matter field, given the distribution of density and the radial and tangential pressure profiles on the initial surface from which the collapse develops, it is then possible to choose the free function describing the velocities of the in-falling shells in such a manner so as to have a blackhole or a naked singularity as the final end product, depending on this choice.

5. The collapsing cloud must obey a realistic equation of state.

It is conjectured at times that even though naked singularities may develop for general matter fields, they must go away once a physically reasonable and realistic equation of state is chosen for the collapsing cloud.

This is a very difficult argument to formulate. First, naked singularities *do* form in the collapse of several well-known equations of state, such as dust, perfect fluids, or in-flowing radiation shells. Second, it is quite difficult to make any guesses as to the state of the matter, or the realistic equation of state within a collapsing body such as a massive star that is in its advanced stages of collapse. Third, the collapsing cloud may not have a single equation of state, which might actually be changing as the collapse evolves. There have been suggestions that strange quark matter may be a good approximation to the collapsing star in its final stages, and the collapse was then examined in a Vaidya geometry, which again results in blackhole or naked singularity phases as usual. Therefore, such a choice of equation of state does not remove naked singularities. At the other extreme, as mentioned earlier, there are also arguments such as those given by Hagerdorn and Penrose, that the equation of state, in the very final stages of the collapse, closely approximates that of dust. In other words, at higher and higher densities, matter may behave more and more like dust. The point is, if pressures are not negative, then they may also contribute positively to the collapse just to add to the dust effect, and may not alter qualitatively the conclusions arrived at in the dust case. In such a case, the dust collapse situation, which has been investigated rather thoroughly, would imply that both blackholes and naked singularity phases would develop in the gravitational collapse, depending on the initial density and velocity distributions.

An interesting feature is that, while there are several widely used and familiar equations of state available that result in the formation of naked singularities as the final fate of the collapse, there is still not a single equation of state available so far that ensures necessarily that the end product will be a blackhole only. Under this situation, it is quite possible that the physics that causes collapse endstates may not be directly related to the equation of state or the form of matter collapsing.

A more promising alternative could be to work with a rather different representation of the matter, such as the one given by the Einstein–Vlasov statistical description (see for example, Rendall, 2005).

6. All radiations from a naked singularity must be infinitely red-shifted.

In certain subcases of dust collapses resulting in a naked singularity, it is seen that the red-shift along the null geodesics emerging from the naked singularity diverges in the limit of approach to the naked singularity. This has given rise to the possibility that, even if a naked singularity forms in the collapse, no energy could escape. In this sense, the naked singularity may be invisible to any external observers for all practical purposes. Of course, it has to be noted that even if true generally, this does not help the cosmic censorship hypothesis in the actual sense as, basically, cosmic censorship is about the question of principle in gravitation theory, namely whether singularities forming in the gravitational collapse are causally connected to an external observer, or not, via non-spacelike trajectories.

In any case, it is useful to explore such a possibility, because it may give some information on the structure of the naked singularity, at least in certain special models, and if true generally, then it will provide some kind of a physical formulation for the cosmic censorship. It is, however, possible that establishing in general that no energy can come out from a naked singularity can turn out to be very difficult. There could be several reasons for this. First, it is very tricky to apply the conventional definition of red-shift, which corresponds to a regular source and observer, to emissions from a naked singularity. A singularity, by definition, is not a point of the spacetime. Second, even if there was no escape of energy along the null geodesics, the possibility of mass emission via timelike or non-spacelike non-geodesic families of paths emerging from the naked singularity remains open in the very late stages of the collapse in a burst-like fashion. In the case of such a violent event being visible, particles escaping with ultra-relativistic velocities cannot be ruled out from the *neighborhood* of the naked singularity.

It may also be noted that the classical possibilities regarding the probable light or particle emission, or otherwise, from a naked singularity may not possibly offer a serious physical alternative, one way or the other. The reason is that the classical general relativity may break down eventually in the very late stages of the collapse, once the densities and curvatures are sufficiently high so that quantum gravity effects become important in the process of a continual collapse. Such quantum effects would come into play much before the actual formation of the classical naked singularity, which itself may possibly be smeared out by the quantum effects.

The key question then is that of the possible visibility, or otherwise, of these extreme strong gravity regions, which do develop in the vicinity of the classical naked singularity. It is then the causal structure, that is, the communicability, or otherwise, of these extreme strong gravity regions where quantum gravity should prevail, which would make the essential difference as far as the physical consequences of a naked singularity are concerned, rather than various purely classical aspects such as red-shift.

As seen above, the physical conditions are not able effectively to rule out naked singularities, which in turn may lead to some possible formulation of the cosmic censorship conjecture, either a physical or a mathematical one. With each of the above conditions, there are counter-examples that obey such a physical constraint, but which produce a naked singularity as the endstate of a dynamical collapse.

There are three further possibilities that are under active investigation today towards a possible formulation of the censorship conjecture, and which may offer a better hope for a cosmic censorship hypothesis. These are now briefly discussed below.

7. Will quantum gravity remove naked singularities?

It is sometimes argued that after all the occurrence of singularities is a classical phenomena, and that whether they are naked or covered should not be relevant – quantum gravity, a final theory in which there should be no singularities will remove them all anyway. But this is missing the real issue, it would appear. It is possible that in a suitable quantum gravity theory, the singularities will be smeared out (although this has not been realized so far), and there are indications that, in quantum gravity also, the singularities may not fully go away. However, it appears that the real issue is whether the extreme strong gravity regions formed due to gravitational collapse are visible to faraway observers or not. The collapse certainly proceeds classically, until quantum gravity starts governing the situation at scales of the order of the Planck length, that is, until the extreme gravity configurations have developed due to collapse. It is the visibility, or otherwise, of such regions that is under discussion.

The point is, classical gravity implies necessarily the existence of strong gravity regions, where both classical and quantum gravity come into their own. In fact, if naked singularities do develop in the gravitational collapse, then, in a literal sense, one comes face-to-face with the laws of quantum

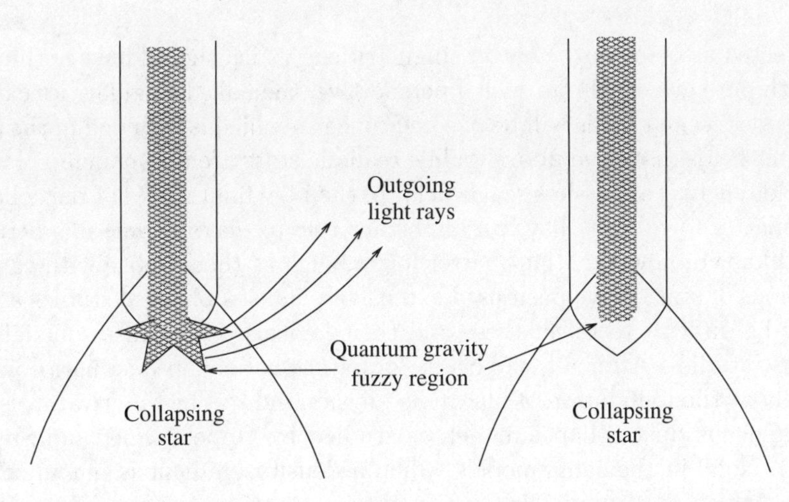

Fig. 4.6 Quantum effects may resolve the spacetime singularity, either covered or naked. If the collapse outcome is a naked singularity, there may be the opportunity to observe the quantum gravity effects taking place in the ultra-strong gravity regions in the universe.

gravity whenever such an event occurs in the universe (see for example, Vaz and Witten, 1994, 1995, and Wald, 1997). Then, the gravitational collapse phenomena could provide a possibility of actually testing the laws of quantum gravity, and every time a massive star collapses in the universe, there is potentially a laboratory to test the laws of quantum gravity (see Fig. 4.6). For an earlier discussion on quantum effects near singularities, see Kodama (1979) and Hiscock, Williams, and Eardley (1982).

In the case of a blackhole developing in the collapse of a finite sized object such as a massive star, such strong gravity regions have got to be necessarily hidden behind an event horizon of gravity, which would be well before the physical conditions became extreme. Then the quantum effects, even if they caused qualitative changes closer to the singularity, will be of no physical consequence. This is because no causal communications are then allowed from within such horizons. On the other hand, if the causal structure were that of a naked singularity, then communications from such a quantum gravity dominated extreme curvature ball would be visible in principle, either directly or via secondary effects such as shocks produced in the surrounding medium. Some of these issues will be discussed further in the next chapter.

8. Should all naked singularities produced by matter fields be considered to be unphysical?

There has been a suggestion that all naked singularities, whenever they are produced by matter fields such as dust, perfect fluids, and such others, should be rejected as being only 'matter singularities,' which should have nothing to do with pure gravity. From such a perspective, the naked singularities caused by massless scalar fields will be of some concern, which is included in the type I matter fields discussed above. While realistic stars are not made up of matter fields such as a massless scalar field, in the very final stages of the collapse such matter forms may have an important role to play in some manner.

It is known, however, that matter forms, such as those above with various equations of state, have been used extensively in astrophysical studies and it would be difficult to reject these and their logical consequences outright in collapse studies. After all, the classic gravitational collapse scenario, which is really at the foundation of blackhole physics and its chief motivator, is the homogeneous dust collapse model, as studied by Oppenheimer and Snyder (1939). Now, in the same models, when a density gradient is put in at the center, which is a physically more realistic situation, then a naked singularity rather than a blackhole results. The structure of the event and the apparent horizons then change drastically so as to expose the singularity to an external observer. All realistic stars will typically have a higher density at the center, falling off at some rate moving away from the center. In this sense, one may want to regard the naked singularity developing due to this density gradient, to be at least as physical as the blackhole. In general relativity, there have been very many studies with dust, perfect fluids, and other forms of matter for several decades, and one might want to accept the logical outcomes available within those collapse scenarios for further studies.

Again, if arguments, such as those given earlier, are considered in favor of the equations of state such as dust in the final phases of the collapse, the outcomes of such a collapse could be taken physically more seriously.

9. Are naked singularities stable and generic?

It would appear that this is the key issue on which any possible future formulation and proof of the censorship has to crucially depend. Even if naked singularities do develop in collapse models, if they were not generic and stable in some suitably well-defined sense, that would make a good case for censorship. For example, most of the current classes of naked singularities discussed here are within the framework of a spherically symmetric collapse. While there are some indications that naked singularities do develop in non-spherical collapses as well, as discussed in the next chapter, such a non-spherical collapse remains largely uncharted territory, and it would be essential to examine it rather thoroughly.

The key question to be resolved here is, whilst it is known that physically reasonable initial data do give rise to naked singularities, will the initial data subspace, which gives rise to the naked singularity as the collapse endstate, have a non-zero measure in a suitably defined sense? As is well-known, however, the formulation of the stability concept in general relativity is a rather complicated issue, as there are no well-defined formulations or criteria to test stability. Before the censorship can be tested, a satisfactory formulation for the stability criterion has to be arrived at within the framework of general relativity. Also, the issue of what is a suitable measure in the initial data space can be a complicated one. Only after making some reasonable progress here could testing these questions on the naked singularity formation start. While discussing stability and genericity, great care has to be taken on the criterion used to test them, as sometimes a criterion that also makes blackholes unstable, while trying to show the instability of naked singularities can be used.

In the absence of such well-defined criteria against which to test the available naked singularity models, various attempts have been made to examine if they would be stable to some kinds of perturbations. These attempts include perturbing the density profiles to include pressures, trying to see how the density gradients at various levels affect the global versus the local visibility of the naked singularity, imposing symmetry conditions such as self-similarity and then seeing how the conclusions change on relaxing the self-similarity condition, studying how certain perturbations grow in the limit of approach to the Cauchy horizon, which is the first ray coming out of the naked singularity, and such others. While these attempts do not provide any definitive conclusions regarding the stability, or otherwise, of the naked singularity, they certainly provide a good insight into the phenomena of the blackhole and naked singularity phases to show what is possible in gravitational collapse.

Given the complexity of the Einstein field equations, if a phenomenon occurs so widely in spherical symmetry, it is not unlikely at all that the same would be repeated in more general situations as well. In fact, before the advent of well-known singularity theorems in general relativity, it was widely believed that the singularities found in more symmetric situations such as the Schwarzschild or Friedmann–Robertson–Walker cosmological models will go away once the spacetimes are general enough. As is well-known, the singularity theorems then established that spacetime singularities occur in rather general spacetime settings without symmetry assumptions, and under a broad set of physical conditions. Therefore, the singularities that manifested earlier in symmetric situations were indicative of a deeper phenomena. Such a possibility cannot again be ruled out in the case of the occurrence of naked singularities.

Clarified above were the basic philosophy and motivation for cosmic censorship and the crucial role they play in blackhole physics. Some of the approaches that have been tried out so far to formulate or prove the censorship were then outlined. It turns out that none of the physical constraints or natural looking physical conditions devised so far are really able to ensure the validity of the cosmic censorship hypothesis. In fact, one tends to conclude that naked singularities can actually develop in physically realistic gravitational collapse situations. It then follows that more radical options, some of which were discussed above, must be formulated and tried out if the censorship conjecture is to be preserved. In fact, it would appear that only one of these, namely the one involving the stability and genericity of naked singularities, can be a potentially promising alternative as far as any possible proof of the censorship hypothesis is concerned.

As it is attempted to work towards censorship along one of these, or other, paths, it would in fact be important and quite interesting to really understand why naked singularities do actually develop in a gravitational collapse. As pointed out above, several important physical constraints on collapsing clouds do not appear to work towards helping the censorship hypothesis. It then becomes an intriguing question as to what is the physical agency that seems to be causing a naked singularity in the gravitational collapse in a rather natural manner within the framework of general relativity. Some work has been carried out in that direction, and it turns out that while the gravitational collapse proceeds, the shearing effects and inhomogeneity within the cloud could play a basic role in delaying the formation of trapped surfaces and the apparent horizon in a natural manner. This in turn exposes the singularity to outside observers, depending on the rate of growth of the shear in the limit of approach to the center. While shear can be one physical agency to delay trapping, there can be a naked singularity in shear-free collapse as well. When looked at from such a perspective, it may be thought that both blackholes and naked singularities are rather natural consequences of a gravitational collapse in classical general relativity.

It is then possible that the cosmic censorship conjecture does not hold classically, but may hold quantum mechanically in some sense that is yet to be figured out. It may be possible then that, for a star going into the final state of a naked singularity configuration, the quantum gravity induced particle creation may take over to create a thunderbolt-like burst of energy, thus clearing up the naked singularity. Such a scenario would be of physical interest because a naked singularity may have theoretical and observational properties quite different from a blackhole endstate, and it might make the communications from extreme strong gravity regions dominated by quantum gravity possible.

4.6 Genericity and stability

Are naked singularities developing in a gravitational collapse stable and generic? This is one of the most important questions as far as any possible proof of the cosmic censorship is concerned. If naked singularities formed in the collapse are not stable or generic in a suitable sense, they may not be taken seriously. The key difficulty to addressing this question is that there are no well-defined criteria or formalisms available in the general theory of relativity to test the stability and genericity. As opposed to Newtonian theory, there is no well-defined notion for stability available in Einstein's theory. Under this situation, there are a variety of ways in which the above question can be asked and treated, and there is no unique answer available in a general manner. Many times, it is the physics of the situation that guides the path adopted, as pointed out above. Nevertheless, this is an important and basic issue to try to answer.

The singularity theorems establish the existence of spacetime singularities in the form of incomplete non-spacelike geodesics, both for gravitational collapse and cosmology. These theorems, however, give no information on the nature of the singularities, such as whether they occur in the past or future, the possible growth of curvature in the limit of approach to the singularity, and whether they will be covered by an event horizon hidden from all outside observers, or whether they may be causally connected to external observers in the universe. Hence, while dealing with gravitational collapse scenarios it becomes inevitable to assume the cosmic censorship in the form of a future asymptotic predictability of the spacetime, in order to make any progress in the theory and applications of blackholes. It has to be ensured, through the censorship conjecture, that the singularities of the collapse are necessarily hidden inside an event horizon. This assumption essentially relates to the causal structure in the later stages of a gravitational collapse.

Here, the occurrence of naked singularities arising in the gravitational collapse of massive matter clouds such as stars have been discussed. The examination of such dynamical collapse scenarios, concerned with the evolution of regular initial data from a well-behaved initial value surface, imply the following basic conclusions. First, such a naked singularity forms in the dynamical evolution of several forms of matter, such as the collapse of inflowing radiation, dust, or perfect fluids. Second, a non-zero measure set of non-spacelike trajectories, in the form of families of non-spacelike curves, is emitted from the naked singularity, as opposed to a single null geodesic escaping, which corresponds to a single wave front emerging. Finally, such a singularity is physically significant in the sense that it is a powerfully strong curvature singularity as the curvatures diverge rapidly along all the trajectories meeting the naked singularity in the past. It can also be noted that in dust collapse (see Fig. 4.7), and in similar models, these are seen to arise from a non-zero measure set of initial data sets (Saraykar and Gate, 1999; Mena, Takavol, and Joshi, 2000).

How seriously such a naked singularity is to be taken, is it generic and stable, and what are the implications towards the formulation and proof of the cosmic censorship hypothesis? It may be noted that the gravitational collapse situations investigated so far have been spherically symmetric. Is it

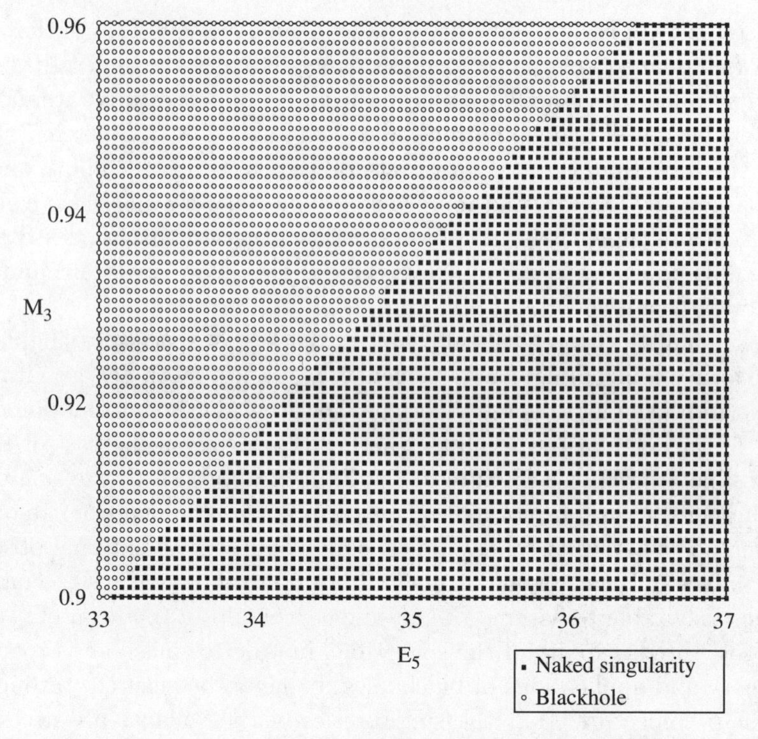

Fig. 4.7 The blackhole and naked singularity phases in a dust collapse, in terms of the initial density and velocity profiles (from Mena, Tavakol, and Joshi, 2000).

possible that the naked singularities occurring are artifacts of this assumed symmetry? As discussed later, there are results that show that naked singularities occur in some non-spherical collapse models as well, and so their existence need not be due to the assumed spherical symmetry only. Furthermore, as was shown by the singularity theorems, the singularities developing in spherical situations still persist, even when small perturbations are taken into account. It is possible that a similar situation may arise here also, and so the detailed investigation of the spherically symmetric scenario becomes quite important.

In any case, if the cosmic censorship is generically correct, it has to hold in spherical symmetry as well. Hence the results here on the spherical collapse show that only a substantially fine-tuned version of the censorship holds if one exists.

4.6.1 Topology on the space of metrics

The stability or genericity of a property in spacetimes can be characterized in the following way. If a spacetime (M, g) possesses a certain property, then for this property to be stable, it would be expected that all the nearby spacetimes would have the same property. For example, one says that the spacetime is causal, and not admitting any closed non-spacelike curves is a stable property, if for all nearby metrics $\bar{g}$ on the same manifold, the spacetime $(M, \bar{g})$ also remains causal. Similarly, for a spacetime having naked singularities, one would like to check if these naked singularities are stable. If (M, g) has naked singularities, but all nearby $(M, \bar{g})$ have none, then these are not stable.

The issue then is to make this notion of 'nearby' metrics more precise, and for that one has to have a topology on the space of all metrics $\bar{g}$ for a given spacetime M. One way to do this is to define a C^0-topology, by defining the nearby metrics to be those that are nearby in their values at the spacetime coordinates, and the open balls can be defined by the requirement that for a given $\epsilon > 0$, the ball consists of all metrics $\bar{g}$, such that $|g_{ij} - \bar{g}_{ij}| < \epsilon$.

The issue then would be, should it be required only for the values to be nearby, or must the derivatives also be nearby? In general relativity, the metric tensor is required to be at least C^2. If the first derivatives are also required to be close, then a C^1-topology should be used, and if the second derivatives are required to be close, then a C^2-topology has to be used. A concerned property, such as the spacetime having, or not having, a naked singularity should then be examined in such a context with the appropriate topology defining the nearby metrics. There is no uniqueness in defining such a topology on the space of all metrics, and a C^∞-topology could be used on the space of all metrics. The main difficulty in a stability and genericity analysis in general relativity is then that of identifying a suitable physically relevant topology for the space of all metrics, and then it has to be examined if there are open sets with non-zero measures with a given property.

As there is no uniqueness in the above procedures, the typical approach has been to examine perturbations of a given situation, as directed by the physical requirements of the model. For example, in a gravitational collapse, for the homogeneous dust collapse, one may try and see if the final state of the collapse remains unchanged if perturbations that are inhomogeneities in a given homogeneous profile are introduced, or one may want to perturb the original solution to a new solution that allows for small non-zero pressures for a given initial data from which the collapse evolves. Very many such analyzes have been carried out for gravitational collapses, as indicated in the previous chapter and in the discussion here.

4.6.2 Censorship and genericity issues

The cosmic censorship hypothesis, as suggested by Penrose (1969, 1979), emphasizes that the criteria of stability and genericity must be satisfied in some well-defined sense. For example, it could be required that spacetimes admit no naked singularities with respect to the changes in the initial data, or the equation of state. Therefore, the strong censorship hypothesis can be stated as saying that stable spacetimes must be globally hyperbolic, or that they do not admit locally naked singularities. A similar statement for the weak censorship would be that stable spacetimes do not admit globally naked singularity. Hence, one would like to know if the gravitational collapse scenarios discussed in the previous chapter, and the naked singularities forming are stable in a suitable sense, say under small departures from the spherical symmetry or changes in the equation of state.

As pointed out above, a general analysis on the question of stability is, however, a rather complicated issue because the stability theory in general relativity is a largely uncharted domain on which little is known. For example, an implication of the strong censorship principle, as stated above, would be that singularities that are spacelike in nature must remain spacelike after a small perturbation in the spacetime. However, some care has to be taken in formulating such a statement. The reason is that the Schwarzschild singularity could be thought of as being unstable in this sense, because the addition of even a small amount of charge or angular momentum changes the character of the singularity and the nature of the solution in a basic way. In this case, the solution changes to the Reissner–Nordström or Kerr spacetime where the singularity is not necessarily spacelike. The addition of a small charge changes the model so that the singularity is locally naked. It is only when a *generic* type of perturbation is introduced in the spacetime that the spacelike nature of the singularity may be retained. It is not clear, however, what such a generic perturbation would mean in general, and basically a well-defined stability theory in the framework of general relativity is needed here.

A possible approach to this, as indicated earlier, is to examine the stability of the Cauchy horizons that must form whenever the strong cosmic censorship is violated, that is, whenever global hyperbolicity of the spacetime breaks down. The Reissner–Nordström case provides an important clue here. In this case, the future Cauchy horizon extends all the way to spatial infinity, and one way of specifying this could be the following. Given any partial Cauchy surface S, the Cauchy horizon $H^+(S)$ associated with it has the property that, given any point $p \in H^+(S)$, the set $\overline{I^-(p)} \cap S$ is non-compact. In such a case, even a small perturbation in the initial data on the partial Cauchy surface will grow and diverge at $H^+(S)$, causing a blue-shift instability. The reason for this is that the signals from faraway regions on S are infinitely blue-shifted at $H^+(S)$, and in fact, weak field perturbations would diverge along $H^+(S)$ (Simpson and Penrose, 1973; McNamara, 1978; Chandrasekhar and Hartle, 1983). In such a situation, a curvature singularity, rather than a Cauchy horizon may really occur. This would imply that $H^+(S)$ is not stable against sufficiently small perturbations in the initial data. This analysis of the Reissner–Nordström case provides an indicator that the Cauchy horizons, when they form, could be unstable. In fact, such an instability is seen to be occurring for the wider class of spacetimes of the Kerr–Newman family, of which the Reissner–Nordström situation discussed above is a special case.

Care has to be taken in formulating this, because this is a different type of strong censorship violation compared with that occurring in the dynamically developing collapse geometries such as the Vaidya model or the Tolman–Bondi–Lemaître case, where the set discussed above is always compact, for any $p \in H^+(S)$. It could then be concluded that the best hope for the censorship lies in analyzing the genericity and stability properties of the currently known classes of collapse models that lead to the formation of naked singularities, rather than blackholes as the final state of collapse.

The point of view that emerges then is, as far as the occurrence is concerned, that both blackholes and naked singularities appear to be basic properties of the gravitational collapse, as a consequence of the dynamics of the Einstein equations, and emerge in a natural manner as a logical consequence of the general theory of relativity. However, the crucial issue is that of the genericity and stability of the naked singularities. It would appear that the real hope for the censorship lies in investigating in detail the stability properties of the collapse models that develop into a naked singularity.

Discussed below in some detail are some of the issues, such as those mentioned above, that are related to the genericity, with the help of a higher-dimensional model. This is related to the question of the validity of the cosmic censorship in a higher-dimensional collapse scenario. As it is possible to work out explicitly when the families of non-spacelike geodesics can come out from the singularity, or otherwise, such genericity issues can now be discussed in some detail.

The purpose has been to examine if it is possible to recover the cosmic censorship when a transition is made to a higher-dimensional spacetime, by studying the spherically symmetric dust collapse in an arbitrary higher spacetime dimension. Earlier, it was shown how certain classes of naked singularities are removed once the transition to a higher-dimensional spacetime is made. It would be of interest to see to what extent such a result is generalized in the dust collapse towards the naked singularity removal, while commenting on certain genericity aspects.

If only blackholes are to result as the endstate of a continual gravitational collapse, several conditions must be imposed on the collapsing configuration, some of which may appear to be restrictive. This needs to be studied carefully if these can be suitably motivated physically in a realistic collapse scenario. The approach developed here generalizes and unifies the earlier available results on a higher-dimensional dust collapse. Furthermore, the dependence of a blackhole or a naked singularity as collapse outcomes on the nature of the initial data from which the collapse develops, comes out explicitly.

The method used allows the genericity and stability aspects related to the occurrence of naked singularities in a gravitational collapse to be considered. The main motivation for studying a higher-dimensional collapse is that, while the censorship may fail in the four-dimensional manifold of general relativity, it can possibly be restored due to the extra physical effects arising from the transition itself to a higher-dimensional spacetime continuum.

There have been several investigations in recent years on the spherically symmetric collapse of dust in higher-dimensions. The recent revival of interest in this problem is motivated, to an extent, by various higher-dimensional theories, including the string paradigm to unify the forces of nature, and the brane-world scenarios. Different specialized subcases of the general problem of dust collapse in higher spacetime dimensions have been considered. For example, the marginally bound case in a general spacetime dimension was studied by Ghosh and Beesham (2001). The same case was also studied with an added and physically motivated assumption on initial density profiles, that the first derivatives of the initial density distribution for the collapsing cloud must be vanishing (Sil and Chaterjee, 1994; Patil, Ghate, and Saraykar, 2001; Banerjee, Debnath, and Chakraborty, 2003). Also, the non-marginally bound case with the geometric assumption that spacetime is self-similar, was examined by Ghosh and Banerjee (2003) for a five-dimensional model.

These studies do provide an idea of what is possible in a gravitational collapse as far as its endstate spectrum is concerned. It is obvious from the discussion so far that any possible proof of the censorship must be inspired by additional physical inputs into the current framework of thinking, with one of these being a possible transition to higher spacetime dimensions. Any such alternatives would be worth exploring due to the fundamental significance of the censorship in blackhole physics. If naked singularities did develop in realistic gravitational collapses of massive objects, they may have properties that would be rather different from those of blackholes, both theoretically as well as observationally, and a comparison of these two cases may prove quite interesting.

The effect of dimensions on the final fate of the evolution of the matter cloud that collapses from a given regular initial data is now examined. A spherically symmetric dust collapse is considered with $N \geq 4$ dimensions. There have been suggestions that if the in-falling velocity of the matter shells is so high that the effects of pressures are negligible, then dust may be a good approximation in the final stages of a collapse. Dust collapse is worth investigating in any case, as it has continued to serve as a basic paradigm in blackhole physics.

To focus the discussion, consider a model initial density profile given by

$$\rho(t_i, r) = \rho_0 + r\rho_1 + r^2\frac{\rho_2}{2!} + r^3\frac{\rho_3}{3!} + \cdots, \tag{4.76}$$

and the function $\mathcal{M}(r)$ is written as

$$\mathcal{M}(r) = \sum_{n=0}^{\infty} \mathcal{M}_n r^n, \quad \mathcal{M}_n = \frac{2\rho_n}{(N-2)(N+n-1)n!}, \tag{4.77}$$

along with an energy profile as specified by

$$b(r) = b_0 + rb_1 + r^2 \frac{b_2}{2!} + \cdots. \tag{4.78}$$

First, consider the marginally bound class of collapse models for a transparent understanding of the problem. This is the case when the energy function $b(r)$ above vanishes identically for the collapsing shells. In this case, the first non-vanishing coefficient $\mathcal{X}_n(0)$, where $n > 0$, could be worked out as discussed in Chapter 3. This is given by

$$\mathcal{X}_n(0) = -\frac{n!}{N-1}\left(\frac{\mathcal{M}_n}{\mathcal{M}_0^{3/2}}\right). \tag{4.79}$$

Now, it is evident that whenever $\rho_1 < 0$, there will be a naked singularity *in all dimensions*, whereas $\rho_1 > 0$ always results in a blackhole. The case $\rho_1 < 0$ corresponds to the physical situation when the density decreases with increasing comoving radius r. This is physically realistic as the density would typically be expected to be highest at the center and then gradually decrease outwards in any realistic configuration such as a massive star. Furthermore, note that the above conclusion is not dependent on the magnitude of ρ_1, but only on its sign, that is, the density should decrease away from the center with the density gradient being non-zero. Therefore, it becomes clear that it is the *density inhomogeneity* that delays the formation of the trapped surfaces, thus causing a naked singularity. This is closely connected to the non-vanishing spacetime shear, and in the next chapter it will be discussed how the inhomogeneities and related shear distort the geometry of the trapped surfaces.

Now assume that the initial density distribution has all odd terms in r vanishing, that is, it admits no 'cusps' at the center and that it is either sufficiently differentiable, or is a smooth and analytic function of r. In this case, $\rho_1 = 0$. Then, from (3.71) in the neighborhood of the singularity, the behavior of v is given by

$$\lim_{t \to t_s} \lim_{r \to 0} v = \left[\frac{N-1}{4}\sqrt{\mathcal{M}_0}\mathcal{X}_2(0)\right]^{2/(N-1)} r^{4/(N-1)}. \tag{4.80}$$

Also, in the same limit, the function F/R^{N-3} has the form

$$\lim_{t\to t_s}\lim_{r\to 0}\frac{F}{R^{N-3}}=\frac{r^2\mathcal{M}_0}{v^{(N-3)}}. \tag{4.81}$$

Therefore, it is clear from (4.80) and (4.81) that if $N>5$, then for the $\lim_{t\to t_s}, \lim_{r\to 0}$, $F/R\to\infty$ and thus the endstate of collapse will always be a blackhole, as discussed previously. It follows that for a marginally bound dust collapse, with $\rho_1=0$, that is, when the initial density profile is sufficiently differentiable and smooth, the cosmic censorship holds in a higher-dimensional spacetime with $N=6$, or higher.

In the above, the spacetime dimension was taken to be six or higher. Now consider the case when the spacetime dimension is five, but still with an analytic initial density profile. In the case of a five-dimensional marginally bound collapse with $\rho_1=0$, the tangent to the outgoing radial null geodesic at the singularity in the (R,u) plane can be written as

$$x_0^2=\sqrt{\mathcal{M}_0}\mathcal{X}_2(0)\frac{\left[1-\sqrt{F/R^2}\right]}{\left[1+\sqrt{F/R^2}\right]}. \tag{4.82}$$

The sufficient condition for the existence of an outgoing null geodesic from the singularity is that $x_0>0$, which in the above case amounts to

$$\xi\equiv\frac{\mathcal{M}_2}{\mathcal{M}_0^2}<-2. \tag{4.83}$$

But again, the outgoing null geodesic should be within the spacetime, that is, the slope of the geodesic must be less than that of the singularity curve,

$$\lim_{t\to t_s}\lim_{r\to 0}\left(\frac{dt}{dr}\right)_{\text{null}}\leq\left(\frac{dt}{dr}\right)_{\text{sing}}. \tag{4.84}$$

From the above equation,

$$\xi=\frac{\mathcal{M}_2}{\mathcal{M}_0^2}\leq -8. \tag{4.85}$$

Therefore, from (4.83) and (4.85), it can be seen that, for an outgoing null geodesic from the singularity to exist, $\xi\leq\xi_c=-8$, in which case the result is a naked singularity, otherwise a blackhole results as the collapse endstate.

Note that this situation has an interesting parallel to the four-dimensional collapse scenario, where a similar critical value exists. However, it is for the coefficient ρ_3, when both ρ_1 and ρ_2 are vanishing. Therefore, with the increase of the spacetime dimension by one, the criticality separating the blackhole and naked singularity phases shifts at the level of the second density derivative from the earlier third density derivative.

An interesting observation that could be made here is that for $\xi < -2$ there is an increasing apparent horizon at the singularity. The apparent horizon is given by $R = F$, so it initiates at the central singularity $r = R = 0$ and, in the above case, it is increasing in time (as opposed to the Oppenheimer–Snyder case of a homogeneous dust collapse). Therefore, for the range $-2 > \xi > -8$, no trapped surface is formed before the singularity epoch, but there is still a blackhole as the collapse endstate. This confirms that the absence of a trapped surface until before the singularity is necessary, but not a sufficient condition for the formation of a naked singularity. This is relevant, especially for numerical collapse simulations, where the criterion for a naked singularity formation is often taken to be just the absence of trapped surfaces on an evolving sequence of spacelike surfaces, in a particular slicing of the spacetime. The above example shows that the mere absence of trapped surfaces cannot be taken as proof that the collapse terminates in a naked singularity.

It is useful to note here that the well-known Oppenheimer–Snyder class of collapse solutions is a special case of a marginally bound dust collapse in four dimensions, in which case the initial density profile is homogeneous, that is, $\mathcal{M}_n(n > 0) = 0$ for all n. The point is, if the initial density is homogeneous, but if the collapse is *not* marginally bound, then the non-zero energy function f could inhomogenize the collapse at later epochs. In the present case, as $f = 0$, at all later epochs the density also remains a function of time only, that is, it is homogeneous at all later times as well, and it can clearly be seen that the final outcome of this class of collapse is always a blackhole. Furthermore, from (3.68) and (3.53) it can be seen that

$$t_{\mathrm{s}}(r) = t_{\mathrm{s}_0}, \quad v(t, r) = v(t). \tag{4.86}$$

As the scale function v is independent of r, all the shells collapse simultaneously to the singularity. The time taken to reach the singularity is given as

$$t_{\mathrm{s}_0} = \frac{2}{(N-1)\mathcal{M}_0^{1/2}}. \tag{4.87}$$

Therefore, in going to higher-dimensions, for a given density, the time taken to reach the singularity will reduce.

It is, however, interesting to note that even if an initially homogeneous density profile is started with, but if non-zero initial radial and tangential pressures of the form

$$p_r(t, r) = 1, \quad p_{\theta_0}(r) = 1 + p_{\theta_2} r^2 + p_{\theta_3} r^3 + \cdots, \tag{4.88}$$

are allowed for, then,

$$\mathcal{X}(0) = -\frac{1}{3}\int_0^1 \frac{v^{(N+5)/2}(p_{\theta_3})}{v^{(N-1)}\left(p_{\theta_2} - \frac{1}{3}\right) + \frac{2}{3}}. \tag{4.89}$$

Therefore, it is seen that a negative p_{θ_3} coefficient does lead to a naked singularity.

This is similar to the case of a collapse that is not marginally bound, where an initially homogeneous density profile can turn inhomogeneous at later epochs due to the non-vanishing shell velocities (Joshi and Dwivedi, 1993a). In the same way, in the case above, the non-vanishing pressures could also inhomogenize the initially homogeneous density distribution at later epochs to cause eventually a naked singularity as the collapse final state. It has to be noted, all the same, that the equation of state in situations such as above could be considered to be somewhat peculiar, although the matter is fully normal, which satisfies the positivity of the energy condition, and the regularity conditions for the collapse are fully satisfied. To state this differently, it can be argued that the models where only purely tangential pressures are taken to be non-vanishing may not be considered to be physically realistic. If the equation of state is chosen to be, say $p = k\rho, k > 0$, or any homogeneous equation of state, then, when the initial density profile is taken to be homogeneous, then so will the initial pressures, and then the collapse will end up in a blackhole only, and no naked singularity will arise.

Basically, the purpose of the above discussion was to point out how both blackholes and naked singularity final states arise rather naturally in a gravitational collapse. While one or the other of these can be created or avoided by means of one or the other conditions on the collapse, these conditions are typically somewhat restrictive, and generically both these outcomes seem to arise rather naturally as the collapse endstates. Discussed here were various special subcases of a higher-dimensional collapse scenario that result either in a blackhole or a naked singularity, depending on the values and behavior of the parameters involved.

It is necessary, however, to look at the situation in a collective manner if any insight on the genericity and stability aspects connected to the naked singularities forming in the gravitational collapse are to be gained. There may be different kinds of stabilities involved. For example, it can be asked here if the conclusions will be stable to non-spherical perturbations, or when will forms of matter more general than dust be considered, and so on. Such issues are worth a detailed investigation, and will be crucial in the important problem of collapse endstates.

At a somewhat different, but still quite interesting level, the stability of these endstates with respect to the perturbations in the initial data space that determines the final outcome of collapse can be investigated. As pointed out, this is a function space consisting of all possible mass functions F and energy functions f. It is worth knowing how, for example, a naked singularity endstate would be affected when one moves from a given density and energy profile, which gave rise to this state, to a nearby density or energy profile in this space of all initial data. The issue of how given density and energy distributions determine the final collapse state has been discussed quite extensively in the usual four-dimensional dust collapse models, although a somewhat different methodology was used. These results were completed to give a full and general treatment of the four-dimensional case, and the typical result is that given any density profile, one could choose the energy profile (and vice versa), so that the collapse endstate would be either a blackhole or a naked singularity, depending on this choice.

As can be seen from the considerations here, these results are generalized to the case of a higher-dimensional collapse situation, and the method allows a more definite statement on the genericity of naked singularity formation to be made. As seen from the discussion above and in Chapter 3, the quantity $\mathcal{X}(0)$ is fully determined from the initial data functions and their first derivatives. Once it is positive, the collapse ends in a naked singularity and a negative value gives the blackhole final state. It follows by continuity that, given a density profile, if the energy profile chosen is such that the collapse ends in a naked singularity, that is if $\mathcal{X}(0) > 0$, then there is a whole family of nearby velocities such that this will continue to be the case, and then the naked singularity forms an open subspace in the initial data space. The same of course holds for blackhole formations, and both these are neatly separated open regions in the initial data space. But if, on physical grounds, it is taken that both ρ_1 and b_1 must vanish, then the dust treatment gives $\mathcal{X}(0) = 0$, and the cosmic censor may be restored.

It may be argued that if all the assumptions such as those discussed above can be suitably motivated physically, then it may be possible to restore cosmic censorship in a higher-dimensional spacetime for the gravitational collapse of dust. These conditions will now be discussed in some detail. That the equation of state must be dust-like in the final phases of the collapse is a strong assumption, but it is not a possibility that can quite be ruled out, as discussed earlier. After all, very little is known on the equations of state, especially what it would be like in the advanced stages of the collapse. Also, it is not ruled out that, in the late stages of the collapse, the configuration is like a marginally bound one, especially in the vicinity of the singularity. The introduction of pressures may, or may not, change such a scenario.

In this case then, it may be possible to recover the censorship if one moves to a higher-dimensional spacetime arena. This is subject to the validity of several extra physical inputs, as described above. On the other hand, once more general situations of either a non-marginally bound case, or with a more general form of matter, or without restrictive extra assumptions on the nature of the initial density profiles are moved to, then generically both the blackhole and naked singularity phases could result as endstates of the collapse in a higher-dimensional spacetime scenario as well. In this way, a dynamical collapse in general relativity offers a rich spectrum of possibilities to investigate.

4.6.3 Scalar field collapses

If it is accepted that naked singularities do occur for a wide range of collapse models, the cosmic censorship requirement could then be interpreted as a question, namely whether the forms of matter, such as the dust, perfect fluid, or in-flowing radiation and others, must break down and cease to be good approximations in the very late stages of the collapse. In fact, these may not be regarded as fundamental forms of matter even at the classical level, and are only approximations to the more basic entities such as a massless or massive scalar field (in the eikonal approximation). Therefore, the question whether naked singularities occur for a scalar field coupled to gravity, or for similar matter fields other than dust, perfect fluids, or collapsing radiation could be asked.

Much attention has been given in past years to analyzing the collapse of a scalar field both analytically (see Christodoulou, 1986, 1994, 1999; Roberts, 1989; Traschen, 1994; Brady, 1995a, 1995b, and references therein), and also numerically (Abraham and Evans, 1993; Choptuik, 1993; Evans and Coleman, 1994; Gundlach, 1995, 1999). This is a model problem of a single massless scalar field that is minimally coupled to a gravitational field, and it provides possibly one of the simplest scenarios for investigating the nonlinearity effects of general relativity. On the analytic side, the results of Christodoulou show that when the scalar field is sufficiently weak, a regular solution, or global evolution for an arbitrary long time of the coupled Einstein and scalar field equations exists. During the collapse, there is a convergence towards the origin, and after a bounce the field disperses to infinity. For strong enough fields, the collapse is expected to result in a blackhole. For self-similar collapse, the results show that the collapse will result in a naked singularity. However, the initial conditions that led to the formation of a naked singularity are a set of measure zero here, and hence the naked singularity formation may be a non-generic phenomenon in these models.

Such an approach helps one to study the cosmic censorship problem as the evolution problem in the sense of examining the global Cauchy development of a self-gravitating system outside an event horizon. A dynamical version of the cosmic censorship can be suggested that, given reasonable initial data that is asymptotically flat, and assuming some suitable and reasonable energy conditions, a global Cauchy evolution of the system outside the event horizon exists in the sense that the solution exists for arbitrary large times for an asymptotic observer. For a discussion of such an approach in the context of self-gravitating scalar fields, see Malec (1995). The problem of the global existence of solutions is discussed by Malec, and an explicit example of an initial configuration that results in a naked singularity is found at the center of symmetry.

Scalar field collapse has also been numerically studied, as mentioned above. A family of scalar field solutions was considered where a parameter p characterized the strength of the field. The numerical calculations showed that, for blackhole formation, there is a critical limit $p \rightarrow p^*$ and the mass of the resulting blackholes satisfy a power law $M_{\mathrm{bh}} \propto (p - p^*)^\gamma$, where the critical exponent γ has a value of about 0.37. It was then conjectured that such a critical behavior may be a general property of a gravitational collapse, because similar behavior was found in some other cases, including imploding axisymmetric gravitational waves. Also, the case of the collapse of radiation with an equation of state $p = \rho/3$ was considered, assuming self-similarity for the solutions. It is still not clear if the critical parameter γ will have the same value for all forms of matter chosen, and further investigations may be required to determine this issue. As the parameter p moves from the weak to the strong range, very small mass blackholes can form. This has relevance to the censorship because, in such a case, one can probe and receive messages from arbitrarily near to the singularity, and this is naked singularity like behavior. Attempts have also been made to construct models analytically that may reproduce such a critical behavior assuming self-similarity, and solutions were constructed that have dispersal, together with solutions with blackholes or naked singularities.

4.6.4 Families of non-spacelike curves from a singularity

The purpose here has been to discuss the genericity and stability properties related to cosmic censorship and naked singularities. From such a perspective, it is relevant to ask about the nature of trajectories coming out from the singularity when it is visible. As such, in Chapter 3, a wide variety of situations in which a naked singularity forms in a gravitational collapse was discussed. However, the existence of only a radial null geodesic coming out from the singularity towards showing its visibility was demonstrated. If only a single null geodesic emerged from the singularity, it may not be called generic enough, as in the case of only a single photon escaping, it will be non-visible to an external observer for all practical purposes. On the other hand, the existence of families of future directed non-spacelike paths could make the singularity visible to outside observers. Also, any material particles would escape from the vicinity of the singularity only if there are timelike curves escaping from these ultra-dense regions.

The causal structure of spacetime near a singularity and the nature of trajectories emerging from it are analyzed below in some detail. If null and other timelike paths also emerge from the singularity, then in principle, particle and energy emission from such ultra-dense regions is allowed. Such emissions are basically governed by the nature of non-spacelike paths near the singularity. These trajectories are examined and it is shown that if a null geodesic emerges, then families of future directed non-spacelike curves that also necessarily escape from the naked singularity exist. The existence of such families is crucial to the physical visibility of the ultra-dense regions.

Here, no underlying symmetries are assumed for the spacetime, and some earlier considerations on the nature of causal trajectories emerging from a naked singularity are generalized and clarified. Singularities are the regions where the physical conditions such as densities and curvatures are at their extreme. While the big-bang singularity of cosmology is visible in principle, and gave rise to the universe as a whole, it cannot actually be seen. On the other hand, when a massive star dies and collapses continually under gravity, the eventual spacetime singularity can be either hidden within an event horizon of a blackhole, or it could be visible to outside observers, depending on how the collapse of the cloud evolves. While a naked singularity forming in the collapse could provide an opportunity for the physical effects taking place in these extreme regions to be observable to outside observers in the universe, the actual visibility of such extreme gravity regions will depend on the nature and structure of non-spacelike paths emerging from the singularity.

If a continual collapse leads to a naked singularity formation then, even if quantum gravity resolves it eventually, the point is that the causal structure of spacetime in the vicinity of the ultra-dense regions allows them to be seen by an external observer. Therefore, the quantum effects taking place in the regions with arbitrarily high matter densities and curvatures can be seen by the external observers. Any physical effects emerging will be again governed by the existence of families of non-spacelike paths from the vicinity of the singularity. It is therefore important to understand the structure of such families within a gravitational collapse framework.

The important physical issue then is whether such a naked singularity forming in the gravitational collapse could radiate away energy and particles. This depends crucially on the existence and structure of families of non-spacelike trajectories emerging from its vicinity. Also, the actual physical appearance and size of the singularity will be determined by the non-radial null trajectories, and the energy emission, if any, will be governed by the timelike curves and other non-spacelike trajectories escaping from the singularity. For this reason, several authors have considered the possibility of non-radial null geodesics emerging from a naked singularity in the context of spherically symmetric dust collapse models (Deshingkar and Joshi, 2001;

Mena and Nolan, 2001, 2002; Deshingkar, Joshi, and Dwivedi, 2002). Also, families of non-spacelike and timelike geodesics have been worked out in self-similar perfect fluid collapses (Joshi and Dwivedi, 1992, 1993b), and for Vaidya radiation collapse models (Dwivedi and Joshi, 1989, 1991). Most of these considerations have been in the framework of spherically symmetric spacetimes, at times together with other symmetry conditions, such as self-similarity of the models imposed, and within the framework of a specific matter model.

A general consideration of the nature of non-spacelike trajectories near a naked singularity will be of much interest from such a perspective. Here, the non-spacelike trajectories from a naked singularity in general are examined, and it is shown that if a radial null geodesic emerges, then large families of non-spacelike curves also necessarily emerge from the singularity. It is thus seen that the existence of a radial null geodesic is sufficient to ensure the existence of families of timelike and non-spacelike trajectories escaping, and in this sense a single photon escaping in a radial direction from the singularity is never an isolated phenomenon. This generalizes and clarifies earlier considerations in this direction, without assuming any symmetry conditions on the underlying spacetime or assuming a specific matter model for the collapse.

When the collapse ends in a naked singularity, the causal structure near the singularity is such that a null geodesic trajectory γ emerges from it, as shown earlier. Specifically, γ is future directed, which in the past terminates at the singularity, and is therefore a past incomplete null geodesic. To examine in general the possible existence and nature of non-spacelike curves emerging from this naked singularity, the causal boundary construction developed by Geroch, Kronheimer, and Penrose (1972), where the spacetime M is taken to satisfy a suitable causality condition such as strong causality, which rules out the existence of closed timelike curves, is used here. In this procedure, a boundary is attached to the regular spacetime manifold, which includes spacetime singularities as well as the points at infinity.

Note that a boundary attachment to the spacetime manifold is essential to treat the regular spacetime events, together with its singularities and points at infinity in a unified manner. There are different ways to attach a boundary to the spacetime, and they do not necessarily all give the same result. Here the rather basic approach, given by Geroch, Kronheimer, and Penrose (1972), is used as it depends essentially only on the causal structure of the spacetime, which is much more fundamental than, for example, the differential structure of the spacetime manifold. Also, from a physical point of view, each ideal point here is directly associated with the region of spacetime that it can influence, or that it would be influenced by.

An open set W in the spacetime is called a *future set* if it contains its own future, that is, $I^+(W) \subset W$. Furthermore, a future set W is called an *indecomposable future set* (IF) if it cannot be expressed as the union of two proper subsets that are themselves future sets. Indecomposable past sets (IPs) are similarly defined. The idea of the causal boundary construction is to divide the collection of IFs and IPs into two classes, namely the one representing regular points of the spacetime, and the other class giving all its boundary points or the ideal points, which include spacetime singularities as well as points at infinity. The collection of IFs (or IPs) can be divided into the two parts as follows. For a set W that is an IF, if an event in the spacetime $p \in M$ exists such that $W = I^+(p)$, then W is called a *proper* IF or a PIF. All other IFs are called *terminal* IFs or TIFs, and represent spacetime singularities and the points at infinity.

Consider now the set $I^+(\gamma)$ where γ is any null geodesic curve emerging from the singularity. It is then a future set, because for any $p \in I^+(\gamma)$, $I^+(p) \subset I^+(\gamma)$. It can be seen that $I^+(\gamma)$ is an indecomposable future set, or an IF. To show this, a somewhat modified version of the proof of Theorem 2.1 of Geroch, Kronheimer, and Penrose (1972) is used. Suppose $I^+(\gamma) = A \cup B$ with A and B both being future sets. If neither A is fully contained in B nor vice versa, then two events x and y can be found such that $x \in A - B$ and $y \in B - A$. As $x, y \in I^+(\gamma)$, there are points $x', y' \in \gamma$ such that $x \in I^+(x')$ and $y \in I^+(y')$. But, x' and y' are causally related, so suppose now that x' is in the past of y' on γ. Then there is a null geodesic from x' to y', and there is a timelike curve from y' to y, as above. This implies that there must be a timelike curve from x' to y (see for example, Hawking and Ellis, 1973, p. 183). It follows that $y \in I^+(x')$. As $x \in I^+(x')$, this implies that $x, y \in I^+(x')$. Therefore, x' lies in the intersection of the sets $I^-(x)$ and $I^-(y)$, which is an open set and so contains a neighborhood N of x'. Let z be an event in $I^+(x') \cap N$, then $z \in I^+(\gamma)$ and so has to be in one of the future sets A or B. Suppose it is in A, then since there are future directed timelike curves from z to both x and y, it follows that both $x, y \in A$, which is a contradiction. Hence, it follows that $I^+(\gamma)$ has to be an IF. Since γ is a past incomplete null geodesic, there is no regular point $p \in M$ such that $I^+(p) = I^+(\gamma)$, and hence it follows that $I^+(\gamma)$ is necessarily a TIF.

The TIF set $I^+(\gamma)$ here represents a boundary point of the spacetime that is the naked singularity. While the naked singularity formation as the endstate of a continual gravitational collapse has been investigated extensively in the last decade or so (especially within the framework of spherically symmetric collapses and for certain non-spherical examples), the main technique there has been to show that a radial null geodesic coming out in the future and terminating in the past at the spacetime singularity exists (see for example, Joshi and Dwivedi, 1993a; Joshi and Goswami, 2004). These results basically show that the gravitational collapse from regular initial matter profiles could result in either the blackhole or naked singularity endstates, depending on the nature of the initial data from which it evolves and the dynamical evolutions of the collapsing cloud, as allowed by the Einstein equations.

However, as remarked above, if the visibility and other related physical characteristics of a naked singularity that formed in the gravitational collapse are to be explored, then it is important to examine and understand the structure of the families of non-spacelike curves from the singularity. Again, if non-spacelike curves emerge from the singularity, but do not go out of the boundary of the collapsing cloud, then the singularity will be only locally visible, and outside observers would not be able to see it. It is necessary therefore to understand the structure of non-spacelike curves from a naked singularity in general.

It is now possible to do this. Since the set $S = I^+(\gamma)$ is a TIF, it follows from Geroch, Kronheimer, and Penrose (1972) that in this case, a past inextendible timelike curve λ must exist, such that $S = I^+(\lambda)$. In the case of the collapse ending in a naked singularity and a radial null geodesic γ escaping from it, it can thus be seen that the set $I^+(\gamma)$ *is* a TIF, and so, by the above result, there is a timelike curve λ generating this TIF, in the sense that $S = I^+(\lambda)$. Since both the non-spacelike trajectories γ and λ represent the same ideal or boundary point of the spacetime that is the naked singularity, and since $I^+(\gamma) = I^+(\lambda)$ by definition, it follows that the future directed timelike curve λ must terminate in the past at the naked singularity. In other words, it has been shown that a timelike curve λ, which escapes away to the future, and which terminates in the past at the naked singularity exists.

It follows that if $p \in \lambda$ and q is any other event such that $q \in I^+(p)$, then there are timelike curves from the naked singularity to q. This proves the existence of families of infinitely many future directed non-spacelike trajectories escaping away from the naked singularity. In general, if λ' is any other future directed non-spacelike curve such that $I^+(\lambda') = I^+(\lambda)$, then it follows that they all represent the same TIF, which is the naked singularity, and that λ' terminates in the past at this singularity.

Therefore, it can be seen that there is an infinity of future going non-spacelike curves that emerge from the singularity, if a single null geodesic has emerged. These include timelike curves as well as non-radial non-spacelike geodesics. It is seen that the usual method employed to show the existence of a naked singularity in the collapse, which establishes the existence of a radial null geodesic escaping away, is sufficient to lead to the existence of infinite families of future going non-spacelike curves from the naked singularity, as shown here. In the present consideration, it is no longer required to have any special symmetry assumptions on the spacetime, such as spherical symmetry, self-similarity, or others, or any specific form of matter model such as the dust equation of state, which are usually assumed in such discussions.

In particular, this also clarifies and generalizes the earlier results on dust collapses and other models mentioned above, which have focused on non-spacelike null geodesics. The null geodesics of the spacetime have, of course, a special role to play as far as the visibility of the singularity is concerned. From such a perspective, the existence of radial versus the non-radial families of null geodesics from a naked singularity will be briefly discussed. Suppose a radial null geodesic emerges from the naked singularity S developing in a continual collapse. In this case, as seen above, a timelike curve λ generating the TIF set $I^+(\lambda)$ that represents the boundary point S exists. All other future directed non-spacelike curves γ that satisfy $I^+(\lambda) = I^+(\gamma)$ generate the same TIF representing the boundary point S, and they give the families of particle or photon trajectories escaping away from the naked singularity. The boundary of this future set, which is a TIF, is a three-dimensional null hypersurface that is ruled by the radial, as well as non-radial, null geodesics generators γ, which are all incomplete when extended in the past, and which all have the property that $I^+(\lambda) = I^+(\gamma)$. This shows that the existence of a radial null geodesic is sufficient also to give families of non-radial null geodesics emerging from the singularity. This generalizes the earlier results on the existence of non-radial null geodesics from the singularity for the spherically symmetric dust collapse, when a radial null geodesic emerges from the naked singularity.

It is seen that once a singularity is naked, it gives rise to infinitely many null as well as timelike curves to escape away from. In this sense, the emission of paths representing particle or photon trajectories from the naked singularity is a generic phenomena. This is essential and is a necessary condition for the naked singularity to give rise to any physical effects that may possibly be observed by external observers. In the present consideration, the global visibility of the singularity, that is, the situation in which once the families of non-spacelike curves emerge from the naked singularity when they will actually cross the boundary of the cloud to escape to an outside observer, has not been discussed. It is known, however, in several cases including spherical dust collapses, that whenever a singularity is locally naked, then the rest of the free functions in the model can be chosen so as to make it globally visible.

It is known, for example, in the case of dust collapses, that once the singularity is locally naked, the choice of a suitable behavior of the mass function (which is a free function, subject only to some physical conditions such as an energy condition and regularity of the initial data) away from the center, allows the null rays to emerge from the boundary of the cloud. It may also be noted that in various classes of self-similar collapses, once the singularity is locally naked it becomes necessarily globally visible (Joshi and Dwivedi, 1992, 1993b). In any case, as there is no scale in the problem, once the singularity is locally visible, an observer within a large enough blackhole will still be able to see it for a long enough time. In such a scenario, the escape of rays outside the boundary of the cloud would not be crucial. A discussion on global visibility in a more general context of perfect fluids has recently been given by Giambo (2006).

5
Final fate of a massive star

The considerations on gravitational collapse so far have been with the motivation to address the physical questions such as the role of collapse in astrophysics and cosmology. Many of the cosmic processes, such as the birth of stars, the formation of galaxies, and others, are not well understood today, but it is clear that gravitational collapse will play a major role there. Hence, understanding the dynamics of the collapse is important, as has been attempted here in various cases.

The important question of the final fate of massive stars at the end of their life cycle, when they have used all their nuclear fuel, and when gravity becomes the sole and key governing force, has drawn much attention for many decades. The importance of this issue was highlighted by Chandrasekhar (1934), who pointed out that the life history of a star of small mass must be essentially different from that of a star of large mass, and that while a small mass star can pass into a white dwarf stage, a star of large mass *cannot* go to this state, and one is left speculating on other possibilities. The question as to what happens when a massive star, heavier than a few solar masses, collapses under its own gravity has been a fundamental key problem in astronomy and astrophysics. If the star is sufficiently massive, beyond the white dwarf or neutron star mass limits, then a continued gravitational collapse must ensue without achieving any equilibrium state, when the star has exhausted its nuclear fuel. To understand the possible endstates of such a continual gravitational collapse, the dynamical collapse scenarios must be studied within the framework of a gravitation theory such as Einstein's theory. The theory of singularities discussed earlier then implies that, under rather general physical conditions, a spacetime singularity must develop.

Considerations here show that, according to Einstein's theory of gravity, such a star in a continual collapse can end up either in a blackhole final state, from which no communications in the form of light or particles would come out, or it can go to a naked singularity, where the collapse proceeds but the trapped surfaces do not form early enough to cover the singularity. In such a case, the extreme gravity regions can communicate and send out physical effects to the external universe, and it is also suggested that a huge amount of energy could possibly be released, in principle, during the final stages of the collapse from the regions close to the classical singularity (Joshi, Dadhich, and Maartens, 2000). According to these suggestions, an enormous amount of energy may probably be generated, either by some kind of a quantum gravity mechanism, or by means of an astrophysical process where the region could simply turn into a fireball-like situation, creating shocks into the surrounding medium. One way to study the structure of these extreme regions is to examine the complete spectrum of non-spacelike geodesics through which this energy could escape. Even if a fraction of the energy so generated is able to escape to a distant observer, it cannot be ruled out that an observational signature may be generated. It therefore becomes important to look into these possibilities in some detail, and to consider likely observable signatures of each of the blackhole and naked singularity final states of the collapse, from the perspective of a faraway observer.

The theoretical properties and possible observational signatures of a blackhole and naked singularity would be significantly different from each other, and this could be of potential interest from the perspective of astrophysical observations. An immediate distinction is, for example, in the case where the collapse ends in a blackhole, an event horizon develops well before the occurrence of the singularity, and thus the regions of extreme physical conditions are always hidden from the outside world. But if the collapse developed into a globally naked singularity, then the energy of the region neighbouring the singularity can escape via the available non-spacelike geodesics paths or via other non-geodetic, non-spacelike trajectories to a distant observer.

In this chapter, various aspects on gravitational collapse that have emerged from the analysis so far are considered. An important issue is the final fate of a non-spherical collapse. In earlier treatment, it was shown that strong curvature naked singularities arise in a variety of situations involving dust, perfect fluids, and other forms of matter. There are many interesting questions that are under active investigation at the moment. For example, could naked singularities generate bursts of gravity waves? What kind of quantum effects will take place near a naked singularity? What will be the generic outcome for the case of a non-spherical collapse? Many of these issues could have interesting physical implications. The possibility that the ultra-high energy astrophysical phenomena, such as the gamma ray bursts, may have a strong connection to the physics and dynamics of the gravitational collapse of massive stars cannot be ruled out. In fact, many of the current gamma ray burst models involve a collapsar, emphasizing the role of the massive star collapse. Another intriguing possibility is that a naked singularity may possibly provide some kind of observational signatures for the quantum gravity effects taking place in ultra-strong gravity regions. This would then be an exciting prospect in view of the current lack of knowledge on quantum gravity.

In Section 5.1, the life cycle of massive stars is discussed. The large mass stars, several times heavier than the Sun, follow a characteristically different life cycle from stars of about one solar mass. Such massive stars typically live a much shorter life than small mass stars, and undergo a catastrophic gravitational collapse at the end of their life cycle. In Section 5.2, how a physically realistic collapse must evolve and the different dynamical forces at work to govern the final fate of such a collapse are discussed. To give an explicit example, it is shown how the spacetime shearing forces, caused by the inhomogeneities in the matter distribution of the star, delay the formation of trapped surfaces during the evolution of the collapse. This allows the ultra-strong regions of gravity and the spacetime singularity to become visible to faraway observers, and it is seen how a naked singularity forms rather naturally as a collapse endstate.

For any realistic consideration of gravitational collapse, departures from spherical symmetry have to be taken into account. While this problem is much more complex and not much work is still carried out on this, the Szekeres collapse models are discussed in Section 5.3. These are not spherically symmetric, and have no Killing vectors. It is shown that these again admit both blackhole and naked singularity final states. While much analytic and numerical work remains to be carried out on non-spherical collapse, this model explicitly illustrates and shows that the naked singularity final state for the collapse is not necessarily limited to spherical symmetry, and that these can result in a non-spherical collapse as well.

If both blackhole and naked singularity final states do occur in physically realistic gravitational collapse scenarios, the main issue would then be the outcome that nature may prefer for the final state of a collapsing star. While, mathematically, many evolutions may be possible, in physically realistic situations only certain evolutions may be actually realized. From such a perspective, in Section 5.4 the paradoxes that are associated with the blackhole formation as the collapse final state, when a massive star undergoes gravitational collapse at the end of its life cycle, are discussed. Finally, in Section 5.5 it is discussed how the quantum effects, which will be prominent in the late stages of the collapse, can play a crucial role in creating a massive emission of mass and energy in the very late stages of the collapse. It could then be asked if a 'quantum star' comes into being in the very final stages of the collapse, which allows for a powerful emission of mass and energy. Such a burst-like emission can have observational features on the one hand and, on the other, it would help resolve the naked singularity formation, thus allowing the classical singularity to be removed.

5.1 Life cycle of massive stars

Here, the life of a star that comes into existence by gravitational contraction and collapse within a dense interstellar gas cloud is outlined. This helps in understanding the final fate of a massive star that has exhausted its nuclear fuel that provided the internal pressure against the inwards pull of gravity.

As the dense interstellar gas compresses, possibly due to a variety of reasons, in the process the central temperatures of the material rises to ignite a nuclear fuel burning cycle. The major bulk of the cloud is made up of hydrogen, and this now starts burning into helium. This produces energy and internal pressure to counter the internal gravity pull, and the gravitational contraction is halted. The star enters a quasi-static period when it supports itself against gravity by means of thermal and radiation pressures. Such a phase may continue for billions of years, depending on the original mass of the star. If $M < M_\odot$ (where $M_\odot$ denotes the mass of the Sun, $\sim 2 \times 10^{30}$ kg), this period would typically be longer than 10^{10} years. However, such a lifetime decreases and becomes much shorter if $M > 10M_\odot$, and it has to be less than 2×10^7 years. That is, much more massive stars burn out their nuclear fuel much faster. (For a review, see for example, Blandford and Thorne, 1979; Blandford, 1987, and references therein).

The star spends this larger portion of its life on the main sequence. This evolution of the star is actually a balance between the nuclear burning in its interior and gravitational collapse, and, once much of the hydrogen in the core is exhausted providing no further pressure against the inwards pull of gravity, then the collapse must continue further if the star is still sufficiently massive. In such a process, the core temperatures rise again to initiate thermonuclear reactions converting, in the second phase, the helium that formed earlier into carbon, and the core stabilizes again. For a heavy enough star, this process repeats itself and finally a large core of stable nuclei, such as iron and nickel, is built up.

The outcome is that the final state for such an evolution is either an equilibrium star, or a state of continual gravitational collapse if there are no internal forces available that are strong enough to build up a pressure within to resist the pull of gravity of the massive star. The key factor towards deciding the possibility of such a stability is the equation of state for the cool matter of the star in its ground state, that is when all possible nuclear reactions have taken place within the star and no further energy can be derived from such a burning. The support against the pull of gravity in such a case must then come either from electron degeneracy pressure, or from neutron degeneracy pressure. If the electron degeneracy pressure can balance the gravity, a *white dwarf* comes into being. If somewhat higher forces are required, these can be still provided by the neutron degeneracy pressure, and a *neutron star* forms.

The equation of state for an ideal electron Fermi gas was approximated by Chandrasekhar (1931, 1934), who showed that there is a maximum mass limit for the mass of a spherical, non-rotating star to achieve a white dwarf stable state. This is given by

$$M_c \sim 1.4 \left(\frac{2}{\mu_e}\right)^2 M_\odot, \tag{5.1}$$

where μ_e is the constant mean molecular weight per electron. Subsequently, considerable work has been carried out on equations of state for matter at nuclear densities (see for example, Arnett and Bowers, 1977) and it is seen that the maximum mass for non-rotating white dwarfs lies in the range $1.0M_\odot$–$1.5M_\odot$, depending on the composition of matter. Similar considerations can be made for neutron stars, and this gives the mass range as $1.3M_\odot$–$2.7M_\odot$.

The radius of a typical white dwarf is $\sim 10^4$ km and it has a central density of $\sim 10^3\,\mathrm{kg\,cm^{-3}}$. For a neutron star, these numbers are ~ 10 km and $\sim 10^{12}\,\mathrm{kg\,cm^{-3}}$ respectively. The maximum mass upper limits stated above are raised somewhat when either the rotation or differential rotation of the star is taken into account. However, under very general circumstances, a firm upper limit of about $5M_\odot$ is obtained, beyond which the degeneracy pressures cannot support the star in any case (Harrison *et al.*, 1965; Hartle, 1978). By now very many examples of white dwarfs are known to exist in the universe. The discovery of pulsars provided strong support for the existence of neutron stars that must be rotating with periods of fractions of a second in order to produce the observed pulsar signals.

It follows that if a star has a mass higher than, or about, $5M_\odot$, it must then enter a state of perpetual gravitational collapse and contraction once it has exhausted all its nuclear fuel, and no equilibrium configurations such as those given above are possible. Of course, there is a possible escape from the continual collapse if the star manages to throw away most of its mass by some process during this evolution, and settles again below the neutron star or the white dwarf limits. In fact, mass ejection is observed in a *supernova* explosion for the star. During the process of the collapse, the iron core of the star would convert to the formation of a neutron core and a neutron star is born, at least momentarily. Then, the neutron degeneracy pressure develops, which provides a balancing force against further collapse. Then, the core collapse is halted or at least slowed down at nuclear densities, and a shock wave is generated, which propagates outwards in the envelope of the star. In this case, while the inner core remains a neutron star, the outer parts are then driven away by the shock, thus releasing enormous mass and energy, which is believed to be a supernova explosion. However, the theory for such ejection of matter is not still well-understood.

In any case, it does not seem likely that all such massive stars would be able to throw away almost all, or a very major part, of their mass in such a process. That is because, for stars having tens of solar masses, this would amount to throwing away almost ninety percent of the mass of the star. In a supernova, typically only the outer layers of the star are blown off, and no suitable mechanism that could achieve such a high degree of efficiency for the mass ejection from the star is envisaged as yet. In the case of a massive star, in the course of its normal life of nuclear burning, not a very large portion of its mass can be removed through radiation. Then, once the gravitational collapse initiates, which is a catastrophic process, a star that has lived millions of years collapses gravitationally in a matter of seconds. Now, during such a catastrophic collapse, if the shock that is produced could not blow off almost all the outer layers, these layers would fall on the newborn neutron star that has momentarily formed at the core in the process of the collapse, and the further collapse continues again as the pull of gravity would exceed the balancing neutron degeneracy pressures.

It is thus seen that the evolution of a massive star, when it exhausts its nuclear fuel, causes an inevitable continual collapse at the end point of a stellar evolution. The general relativity theory then implies that a spacetime singularity must form of necessity in such a scenario. The basic ingredients of such spacetime singularities were discussed earlier. The cosmic censorship conjecture then asserts that any such singularities forming in gravitational collapse must be covered necessarily within an event horizon of gravity, invisible to any external observer, and that in general the final state of such an evolution must then be a blackhole. This requires an appropriate formulation of the concepts of a blackhole as well as the cosmic censorship, within the framework of an appropriate spacetime geometry. To decide on this issue, the dynamical gravitational collapse has to be analyzed to see how the structure of the horizons evolve. It is within such a perspective that the analysis here on gravitational collapse has taken place, and it can be seen that the dynamical gravitational collapse processes are entirely fundamental in this way to the basics and applications of blackhole physics.

5.2 Evolution of a physically realistic collapse

It was shown earlier that a continued collapse of a massive matter cloud could terminate either in a blackhole or a naked singularity, depending on the matter initial data and the allowed evolutions as decided by the Einstein equations. These results were obtained under physical conditions such as the validity of the energy condition, regularity of the initial data, and such others. However, the actual outcome of a physically realistic collapse still remains to be investigated. While, mathematically, many models may be permitted by general relativity, nature might select only a few of these for the actual evolution of a massive star. Therefore, further examination of the physical forces operating within a star as the collapse proceeds is required.

This issue is discussed here in the context of a spherical collapse by means of an explicit example. This also throws some light on why naked singularities form at all in a gravitational collapse. The key physical features that possibly cause the development of a naked singularity, rather than a blackhole, as the endstate of a gravitational collapse are investigated, and it is seen that sufficiently strong shearing effects near the singularity can delay the formation of the apparent horizon. This exposes the singularity to an external observer, in contrast to a blackhole, where it is hidden behind an event horizon due to the early formation of the trapped surfaces. The final outcome of a gravitational collapse in general relativity is an issue of much importance from the perspective of blackhole physics and its astrophysical implications, and one needs to understand the key physical characteristics and dynamical features in the collapse that give rise to a naked singularity, rather than a blackhole. Here, the treatment of Joshi, Dadhich, and Maartens (2002) is followed.

It is seen that it is the inhomogeneity and related shearing effects within the cloud that, if sufficiently strong near the central worldline of the collapsing cloud, would delay the formation of the apparent horizon so that the singularity becomes visible, and communication from the extreme strong gravity regions to outside observers becomes possible. When the inhomogeneity and related shear forces are weak, or for the extreme case of no shear in a fully homogeneous collapse, the collapse necessarily ends in a blackhole, because an early formation of the apparent horizon leads to the singularity being hidden inside an event horizon.

For the spherical gravitational collapse of a massive matter cloud, the interior metric in comoving coordinates is

$$ds^2 = -e^{2\nu(t,r)}dt^2 + e^{2\psi(t,r)}dr^2 + R^2(t,\,r)d\Omega^2. \tag{5.2}$$

The matter shear is

$$\sigma_{ab} = e^{-\nu}\left(\frac{\dot{R}}{R} - \dot{\psi}\right)\left(\tfrac{1}{3}h_{ab} - n_a n_b\right), \tag{5.3}$$

where $h_{ab} = g_{ab} + u_a u_b$ is the induced metric on the three-surfaces orthogonal to the fluid four-velocities u^a, and n^a is a unit radial vector. The initial data for collapse are the values on $t = t_i$ of the three metric functions, the density, the pressures, and the mass function that arises from integrating the Einstein equations, as discussed earlier,

$$F(t_i, r) = \int \rho(t_i, r) r^2 dr, \tag{5.4}$$

where $4\pi F(t_i, r_b) = M$, which is the total mass of the collapsing cloud, and where $r > r_b$ is a Schwarzschild spacetime. As earlier, the rescaling freedom in r is used to set

$$R(t_i, r) = r\,, \tag{5.5}$$

so the physical area radius R increases monotonically in r, and with $R'_i = 1$, there are no shell-crossings on the initial surface from which the collapse develops. Of interest here is the central shell-focusing singularity at $R = 0, r = 0$, which is a gravitationally strong singularity, as opposed to the shell-crossing singularities that are weak, and through which the spacetime may sometimes be extended.

The evolution of the density and radial pressure are given by

$$\rho = \frac{F'}{R^2 R'}, \quad p_r = \frac{\dot{F}}{R^2 \dot{R}}\,. \tag{5.6}$$

The central singularity at $r = 0$, where density and curvature are infinite, is naked if there are outgoing non-spacelike geodesics that reach outside observers in the future and terminate at the singularity in the past. Outgoing radial null geodesics of the metric (5.2) are given by

$$\frac{dt}{dr} = e^{\psi - \nu}. \tag{5.7}$$

First consider the case of a homogeneous density collapse, $\rho = \rho(t)$. Writing $f = e^{-2\psi} R'^2 - 1$, the Einstein equations give $f - e^{-2\nu}\,\dot{R}^2 = -F/R$. Then, as discussed in Chapter 3, the geodesic equations can be written as

$$\frac{dR}{du} = \left(1 - \sqrt{\frac{f + F/R}{1 + f}}\right) \frac{R'}{\alpha r^{\alpha - 1}}\,, \tag{5.8}$$

where $u = r^\alpha$ $(\alpha > 1)$. If there are outgoing radial null geodesics terminating in the past at the singularity with a definite tangent, then at the singularity, $dR/du > 0$. Now in the case of homogeneous density, the entire mass of the cloud collapses to the singularity simultaneously at the event $(t = t_s, r = 0)$, so that $F/R \to \infty$. Then in that case, from (5.8), $dR/du \to -\infty$, so that no radial null geodesics can emerge from the central singularity. It can be shown that all the later epochs $t > t_s$ are similarly covered.

It has thus been shown that for a spherical gravitational collapse with homogeneous density, the final outcome is necessarily a blackhole. The pressures, however, can be arbitrary on which no conditions are imposed. This conclusion does not require homogeneity of the pressures p_r and p_θ, and is independent of their behavior. This generalizes the well-known Oppenheimer–Snyder–Datt result of the special case of dust, where the homogeneous cloud always collapses to form a blackhole. An immediate consequence is that if the final outcome of a spherical gravitational collapse is not a blackhole, then the density must be inhomogeneous necessarily. In any physically realistic scenario, the density will typically be higher at the center, so that generically the collapse is inhomogeneous. An inhomogeneous density profile is thus a necessary condition for a naked singularity to develop as the collapse endstate.

To understand more clearly the role that the inhomogeneities play in delaying the trapped surfaces during the collapse, consider now a collapsing inhomogeneous dust cloud ($p = 0$), with density that is higher at the center. The metric is the TBL geometry, given by (5.2) with $\nu = 0$, $e^{2\psi} = R'^2/(1+f)$, and

$$\dot{R}^2 = f(r) + \frac{F(r)}{R}. \tag{5.9}$$

These models are fully characterized by the initial data, specified on an initial surface $t = t_i$ from which the collapse develops, which consist of two free functions that are the initial density $\rho_i(r) = \rho(t_i, r)$ (or equivalently, the mass function $F(r)$), and $f(r)$, which describes the initial velocities of the collapsing matter shells. At the onset of the collapse, the spacetime is singularity-free, so that from (5.6),

$$F(r) = r^3 \bar{F}(r)\,, \quad 0 < \bar{F}(0) < \infty. \tag{5.10}$$

The initial density $\rho_i(r)$ is

$$\rho_i(r) = r^{-2} F'(r). \tag{5.11}$$

The shell-focusing singularity appears along the curve $t = t_{\rm s}(r)$, and is defined by

$$R(t_{\rm s}(r), r) = 0. \tag{5.12}$$

As the density grows without bound, trapped surfaces develop within the collapsing cloud. These can be traced explicitly via the outgoing null geodesics, and the equation of the apparent horizon, $t = t_{\rm ah}(r)$, which marks the boundary of the trapped region and is given by (see also the discussion on apparent horizons in Section 3.6)

$$R(t_{\rm ah}(r), r) = F(r). \tag{5.13}$$

If the apparent horizon starts developing earlier than the epoch of the singularity formation, then the event horizon can fully cover the strong gravity regions including the final singularity, which will thus be hidden within a blackhole. On the other hand, if trapped surfaces form sufficiently later during the evolution of the collapse, then it is possible for the singularity to communicate with outside observers.

For the sake of clarity, consider a marginally bound collapse, $f = 0$, although the conclusions can be generalized to hold for the general case. Then, (5.9) can be integrated to give

$$R^{3/2}(t,r) = r^{3/2} - \frac{3}{2}(t - t_i)F^{1/2}(r), \tag{5.14}$$

and (5.12) and (5.13) lead to

$$t_{\rm s}(r) = t_i + \frac{2}{3}\left[\frac{r^3}{F(r)}\right]^{1/2}, \tag{5.15}$$

$$t_{\rm ah}(r) = t_{\rm s}(r) - \frac{2}{3}F(r). \tag{5.16}$$

The central singularity at $r = 0$ appears at the time

$$t_0 = t_{\rm s}(0) = t_i + \frac{2}{\sqrt{3\rho_{\rm c}}}, \tag{5.17}$$

where $\rho_{\rm c} = \rho_i(0)$. Unlike the homogeneous dust case of Oppenheimer and Snyder, the collapse is not simultaneous in comoving coordinates, and the singularity is described by a curve, the first point being $(t = t_0, r = 0)$.

For inhomogeneous dust, (5.3) and (5.14) give

$$\sigma^2 \equiv \frac{1}{2}\sigma_{ab}\sigma^{ab} = \frac{r}{6R^4 R'^2 F}\left(3F - rF'\right)^2. \tag{5.18}$$

Here, a generic inhomogeneous mass profile can be chosen to have the form

$$F(r) = F_0 r^3 + F_1 r^4 + F_2 r^5 + \cdots \tag{5.19}$$

near $r = 0$, where $F_0 = \rho_{\rm c}/3$. The homogeneous Oppenheimer–Snyder dust collapse has $F_n = 0$ for all $n > 0$, and (5.18) then implies $\sigma = 0$. The converse is also true in this case, namely, if a vanishing shear $\sigma = 0$ is imposed, then $F_n = 0$ necessarily. Whenever there is a negative density gradient, that is, when there is a higher density at the center, then $F_n \neq 0$ for some $n > 0$, and it follows from (5.18) that the shear is then necessarily non-zero. Here, note that if the density profile is required to be analytic, all the odd terms F_{2n-1} in the mass function can be set to be zero. This, however, is not required by the analysis here, which is independent of any assumptions on F_n.

It is therefore important to evaluate the effect of such a shear on the development and time evolution of the trapped surfaces. In other words, there is a need to determine the behavior of the apparent horizon in the vicinity of the central singularity at $R = 0, r = 0$. For this purpose, let the first non-vanishing derivative of the density at $r = 0$ be the nth one ($n > 0$), that is,

$$F(r) = F_0 r^3 + F_n r^{n+3} + \cdots, \quad F_n < 0, \tag{5.20}$$

near the center. By (5.18) and (5.16),

$$\sigma^2(t,r) = \frac{n^2 {F_n}^2}{6F_0}\left[1 - 3F_0^{1/2}(t-t_i) + \frac{9}{4}F_0(t-t_i)^2\right] r^{2n} + O\big(r^{2n+1}\big), \tag{5.21}$$

$$t_{\rm ah}(r) = t_0 - \frac{2}{3}F_0 r^3 - \frac{F_n}{3F_0^{3/2}} r^n + O\big(r^{n+1}\big). \tag{5.22}$$

The time dependent factor in square brackets on the right of (5.21) decreases monotonically from 1 at $t = t_i$ to 0 at $t = t_0$. Therefore, the qualitative role of the shear in singularity formation can be seen by looking at the initial shear. The initial shear $\sigma_i = \sigma(t_i, r)$ on the surface $t = t_i$ grows as r^n, $n \geq 1$, near $r = 0$. A dimensionless and covariant measure of the shear is the relative shear, $|\sigma/\Theta|$, where

$$\Theta = 2\frac{\dot{R}}{R} + \frac{\dot{R}'}{R'} \tag{5.23}$$

is the volume expansion. It follows that

$$\left|\frac{\sigma}{\Theta}\right|_i = \frac{-nF_n}{3\sqrt{6}F_0}\, r^n\, [1 + O(r)]. \tag{5.24}$$

It can now be seen how such an initial shear distribution determines the growth and evolution of the trapped surfaces, as prescribed by the apparent horizon curve $t_{\rm ah}(r)$, given by (5.22). If the initial density profile is assumed to be smooth at the center, then $\rho_i(r) = \rho_{\rm c} + \rho_2 r^2 + \cdots$, with $\rho_2 \leq 0$, which corresponds to $F(r) = F_0 r^3 + F_3 r^5 + \cdots$, with $F_2 \leq 0$. Now suppose that ρ_2 (and hence F_2) is non-zero. Then, (5.22) implies that the apparent horizon curve initiates at $r = 0$ at the epoch t_0, and increases near $r = 0$ with increasing r, moving to the future (see Fig. 5.1). Note that as soon as F_2 is non-zero, even with a very small magnitude, the behavior of the apparent horizon changes qualitatively. Rather than going back into the past from the center, as would happen in the homogeneous collapse case with $F_2 = 0$, it is now future pointed. This leads to a locally naked singularity as the collapse endstate. The singularity may be globally naked and visible to faraway observers, depending on the nature of the density function at large values of r.

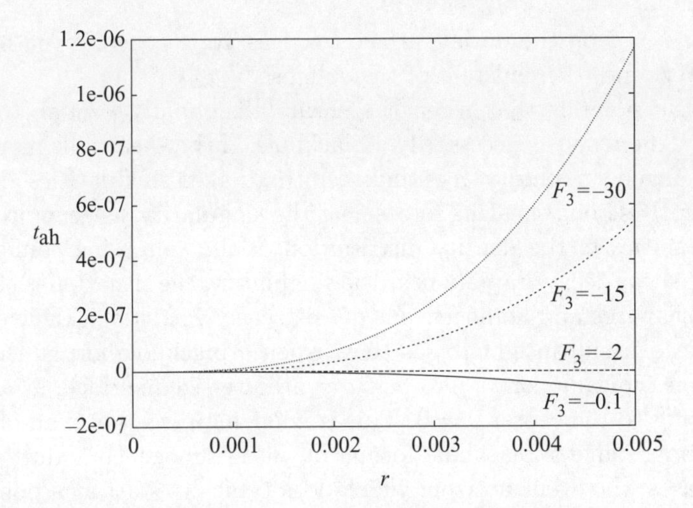

Fig. 5.1 The apparent horizon behavior.

A naked singularity occurs when a comoving observer at fixed r does not encounter any trapped surfaces until the time of the singularity formation, whereas, for a blackhole, the trapped surfaces form before the singularity. Therefore, for a blackhole to form,

$$t_{\rm ah}(r) \leq t_0 \text{ for } r > 0, \quad \text{near } r = 0. \tag{5.25}$$

In the general case, where there is not a necessarily smooth initial density, this condition is violated for $n = 1, 2$, as follows from (5.22). The apparent horizon curve initiates at the singularity $r = 0$ at the epoch t_0, and increases with increasing r, moving to the future, i.e. $t_{\rm ah} > t_0$ for $r > 0$ near the center. The behavior of the outgoing families of null geodesics has been analyzed in detail in these cases, and it is known that the geodesics terminate at the singularity in the past, which results in a naked singularity, as discussed in Section 3.6. In such cases, the extreme strong gravity regions can communicate with outside observers. For the case where $n = 3$, (5.25) shows that it is possible to have a blackhole if $F_3 \geq -2F_0^{5/2}$, or a naked singularity if $F_3 < -2F_0^{5/2}$. For $n \geq 4$, (5.25) is always satisfied, and a blackhole forms.

When the dust density is homogeneous, the apparent horizon starts developing earlier than the epoch of the singularity formation, and the singularity then is fully hidden within a blackhole. There is no density gradient at the center, and no shear present within the cloud. On the other hand, if a density gradient is present at the center, then the trapped surface development is delayed due to the presence of the shear, as seen above, and, depending on the 'strength' of the density gradient and shear at the center, this may allow the singularity to be visible. It is the rate of decrease of the shear, as the center $r = 0$ on the initial surface $t = t_i$ is approached, given by (5.24), which determines the endstate of the collapse.

The basic point is that when the shear falls rapidly enough to zero at the center, the result is necessarily a blackhole. If the shear falls more slowly, there is a naked singularity. It is thus seen that naked singularities are caused by sufficiently strong shearing forces near the singularity, as generated by the inhomogeneities in the density distribution of the collapsing configuration. When the shear decays rapidly near the singularity, the situation is effectively like the shear-free and homogeneous density case, with a blackhole endstate. It provides a useful insight to see that, when a blackhole forms, the apparent horizon typically comes into being as a finite-sized surface, at a finite r, then moving to the center $r = 0$. This is what happens, for example, in the Oppenheimer–Snyder blackhole formation in a homogeneous dust collapse. In such cases, the event horizon, which does typically start at a point, could have formed earlier than the apparent horizon. On the other hand, in the case of a naked singularity, it follows from (5.16) and (5.22), that the apparent horizon starts at $r = 0$, and then is future directed in time, that is, $t_{\rm ah}$ grows with increasing coordinate radius r along the apparent horizon curve $R = F$. These two behaviors of the apparent horizon curve are very different and are governed by the shearing effects. A comoving observer will not encounter any trapped surfaces until the time of the singularity formation in the naked singularity case, whereas, in the blackhole case, the apparent horizon typically develops *before* the epoch of the singularity formation. This is what is meant by the delayed formation of the apparent horizon, caused by the shearing effects or inhomogeneity present within the cloud.

The relation between density gradients and shear may be understood via the non-local or free gravitational field. Density gradients act as a source for the electric Weyl tensor as given by Maartens and Bassett (1998)

$$D^b E_{ab} = \tfrac{1}{3} D_a \rho \,, \tag{5.26}$$

where D_a is the covariant spatial derivative. Note that the magnetic Weyl tensor vanishes for spherical symmetry. In turn, the gravito-electric field is a source for shear, or equivalently, the shear is a gravito-electric potential,

$$u^c \nabla_c \sigma_{ab} + \frac{2}{3} \Theta \sigma_{ab} + \sigma_{ac} \sigma^c{}_b - \frac{2}{3} \sigma^2 h_{ab} = -E_{ab} \,. \tag{5.27}$$

Therefore, density gradients may be directly related to shear as

$$D_a \rho = -4\sigma D_a \sigma - 2\Theta D^b \sigma_{ab} - 3D^b \left(u^c \nabla_c \sigma_{ab}\right) - 3\sigma_a{}^b D^c \sigma_{bc} - 3D^b \left(\sigma_{ac} \sigma^c{}_b\right), \tag{5.28}$$

where the shear constraint $D^b\sigma_{ab} = 2/3 D_a\Theta$ has been used. Equation (5.28) explicitly links the behavior of the density gradients and the shear near the center, which was discussed above. The free gravitational field that mediates this link can also provide a covariant characterization of the singularity formation. By (5.24) and (5.27), the relative gravito-electric field E/Θ^2 (where $E^2 = 1/2E^{ab}E_{ab}$) near $r = 0$ is given at $t = t_i$ by

$$\left(\frac{E}{\Theta^2}\right)_i = \frac{-7nF_n}{18\sqrt{6}F_0} r^n \left[1 + O(r)\right]. \tag{5.29}$$

Therefore, naked singularities in a spherical dust collapse are signaled by a less rapid fall-off of the relative gravito-electric field as the singularity is approached. Equations (5.24) and (5.29) provide two equivalent ways of expressing this result. This specifies how much shear is sufficient to create a locally naked singularity. It is sometimes asked, in the case of a continual collapse when the densities and curvatures grow without bound, and when the gravitational fields grow to extreme, how the trapping of light can be delayed and how such extreme gravity regions can be visible. The above consideration provides an insight into this, namely that the shearing effects due to inhomogeneity, which are purely general relativistic effects, cause the delay of trapped surfaces by distorting the geometry of the trapped surfaces and the apparent horizon, thus allowing the light to escape, even from extreme gravity regions.

For the case of a dust collapse, the role of inhomogeneity or shear in deciding the endstate of the collapse is fairly transparent. To understand how shear affects the formation of the apparent horizon for general matter fields with pressures included is much more complicated, in particular because $F = F(t, r)$ in a general case, whereas $\dot{F} = 0$ for dust. In fact, even in certain general classes of non-dust models with non-zero pressures, it is possible to characterize the collapse covariantly. In the above, it was shown that *homogeneous density* implies a blackhole endstate. The next logical step would be to consider models for which the *initial* density is homogeneous. For example, if the mass function is

$$F(t, r) = f(r) - R^3(t, r), \quad f(r) = 2r^3, \tag{5.30}$$

then (5.6) shows that ρ_i and $(p_r)_i$ are constants. The density and pressure may however develop inhomogeneities as the collapse proceeds, depending on the choice of the remaining functions, including in particular the initial velocities of the collapsing shells. The collapse may then end up in either a blackhole or a naked singularity (for a discussion on this for the case of a dust collapse, see Joshi and Dwivedi, 1993a). In fact, it can be shown that zero shear implies a blackhole for these models. By (5.3), (5.6), and (5.30), the shear-free condition leads to $R'/R = 1/r$, and (5.6) then shows that $\rho = \rho(t)$, i.e. the density evolution is necessarily homogeneous. As shown above, the collapse thus necessarily ends in a blackhole. For the class of models given by (5.30), whenever the collapse ends in a naked singularity, the shear must necessarily be non-vanishing. Although this class of models is somewhat special, the result indicates that the behavior of the shear remains a crucial factor even when pressures are non-vanishing.

It would appear that the only way a singularity becomes visible is when suitable modifications occur in the geometry of trapped surfaces and the apparent horizon, and a delay in the trapped surface formation is necessary. As has been shown here, the shear provides a rather natural explanation for the occurrence of locally naked singularities in certain collapse models. Sufficiently strong shearing forces in spherical collapsing dust do affect the delay in the formation of the apparent horizon, thereby exposing the strong gravity regions to the outside world, and leading to a naked singularity formation. When shear decays rapidly near the singularity, the situation is effectively like the shear-free case, with a blackhole endstate. The important point is that naked singularities can develop in quite a natural manner, very much within the standard framework of general relativity, as governed by shearing effects.

In the case of a spherical dust collapse, shear and density inhomogeneity are equivalent, that is, one implies the other. Although shear contributes positively to the focusing effect via the Raychaudhuri equation

$$\dot{\Theta} + \frac{1}{3}\Theta^2 = -\frac{1}{2}\rho - 2\sigma^2, \tag{5.31}$$

its dynamical action can make the collapse incoherent and dispersive. Depending on the rate of fall-off of shear near the singularity, its dispersive effect can play the critical role of delaying the formation of the apparent horizon, without directly hampering the process of collapse. The dispersive effect of shear always tends to delay the formation of the apparent horizon, but it is able to expose the singularity only when the shear is strong enough near the singularity.

Here, the effects of inhomogeneities and shear are analyzed for certain collapse models. The implication is certainly not that it is only the shear that can cause the changes in the trapped surfaces' geometry. The important point to emphasize is that modifications in the trapped surfaces' geometry can occur rather naturally in the process of a dynamical collapse, due to physical agencies such as inhomogeneities and spacetime shear. In general, the phenomena such as trapped surfaces formation and apparent horizons are independent of any spacetime symmetries, and it would seem clear that a naked singularity will not develop in a more general gravitational collapse as well, unless there is a suitable delay of the apparent horizon. This suggests that the forces such as shear will continue to be pivotal in determining the final fate of physically realistic gravitational collapse scenarios, independent of any spacetime symmetries. In any case, the main purpose here has been to try to understand and find the physical mechanism that leads the collapse to the development of a naked singularity rather than a blackhole, in some of the well-known classes exhibiting such behavior. It is found that the shear and inhomogeneity provide a covariant dynamical explanation of the phenomenon of naked singularity formation in a spherical gravitational collapse.

The basic question is: what governs the geometry of the trapped surfaces, or the formation, or otherwise, of the naked singularities in gravitational collapse? In other words, what is it that causes the naked singularity rather than a blackhole to develop as the final end product of the collapse? It turns out from the above that physical agencies such as inhomogeneities in matter profiles, as well as spacetime shear, play a key role in distorting the trapped surface geometry, and could delay the trapped surface formation during the collapse, thus giving rise to a naked singularity, rather than a blackhole, as the collapse final state. This, in a way, provides the physical understanding of the phenomena of blackhole and naked singularity final states in the collapse. It is seen that in a spherical collapse, these phases are generic, and are seen to be determined by the nature of the initial data from which the collapse develops, and in terms of the allowed dynamical evolutions. Physical agencies such as inhomogeneities and shear may cause them, and given the initial data, there are non-zero measure classes of evolutions that evolve into either of these outcomes.

5.3 Non-spherical models

In terms of analytic calculations, not much is known about the non-spherical gravitational collapse. There are indications, however, in view of the properties of the Kerr geometry, which is a solution to the Einstein equations for a rotating particle, that to avoid naked singularity formation the object should not be rotating very fast. In Kerr geometry, the metric is characterized in terms of the mass of the particle and its angular momentum. If the angular momentum is larger than the mass, then a naked singularity naturally forms in the spacetime. It should be noted, however, that there is no interior metric yet known for a Kerr exterior solution. The ideal situation would be the existence of a non-spherical collapsing cloud, the exterior of which is given by the Kerr solution, and then to investigate the final state of the collapse.

A spherically symmetric homogeneous dust cloud collapse goes through the phases of implosion and the subsequent formation of a horizon and a spacetime singularity completely hidden within the horizon and a blackhole.

The consequences when the collapse is no longer homogeneous, and of different equations of state for the matter, are investigated within the spherically symmetric framework in the previous treatment.

Small perturbations over the spherically symmetric situation were taken into account in the work of Doroshkevich, Zel'dovich, and Novikov (1966), de la Cruz, Chase, and Israel (1970), and in the perturbation calculations of Price (1972). The main outcome of these works is that the basic result of the spherically symmetric collapse situation remains unchanged, at least in the sense that an event horizon will continue to form in the advanced stages of the collapse. It was indicated by the work of Doroshkevich, Zel'dovich, and Novikov (1966) that, as the collapse progresses and the star reaches the horizon, the perturbations remain small and there are no forces arising to destroy the horizon. This analysis contained the basic idea of the no-hair theorems for blackholes that were developed later, as it is pointed out that, from the point of view of an outside observer, non-spherical perturbations in the geometry and the electromagnetic field die down as the star approaches the horizon. These findings were supported by the numerical calculations of de la Cruz, Chase, and Israel (1970) and Price (1972), where linearly perturbed collapse models were integrated to show that the perturbations died out with time as far as they could be followed. It may be noted, however, as pointed out by Israel (1986a), that a result showing that a small change in the initial data perturbs the solution only slightly would be desired. The above works lead to such a result for bounded time intervals only, whereas the existence of a horizon depends on the entire future behavior of the solution. Detailed numerical models for stellar collapse may help reach the desired solution.

One would then like to inquire further, namely, do horizons still form when the fluctuations from the spherical symmetry are large, and when the collapse is highly non-spherical? This issue is again related basically to the nature and evolution of the trapped surface geometry discussed earlier, as the collapse develops, but now in the case of a non-spherical geometry. It is known, for example, that when there is no spherical symmetry, the collapse of infinite cylinders do give rise to naked singularities in general relativity, which are not covered by horizons (Thorne, 1972; Misner, Thorne, and Wheeler, 1973; see also Herrera and Santos, 2005; Nakao *et al.*, 2007). However, the situation of greater physical interest would be that of finite systems, possibly in an asymptotically flat spacetime. Not much is known about this, except the *hoop conjecture* of Thorne (1972), which characterizes the final fate of a non-spherical collapse in the following statement. The horizons of gravity form when and only when a mass M gets compacted in a region whose circumference in *every* direction obeys

$$\mathcal{C} \leq 2\pi(2GM/c^2). \tag{5.32}$$

Therefore, unlike the cosmic censorship conjecture, the hoop conjecture now no longer rules out *all* the naked singularities, but only makes a definite assertion on the occurrence of the event horizons in the gravitational collapse.

It is thus seen that the hoop conjecture allows for the occurrence of naked singularities in general relativity, at least when the collapse is sufficiently aspherical, and especially when one or two dimensions are sufficiently larger than the others. A known example of this is given by Lin, Mestel, and Shu (1965), which discusses gravitational collapse for uniform spheroidal objects from the perspective of instability in Newtonian gravity (see also, Thorne, 1972; Shapiro and Teukolsky, 1991, 1992, for a discussion). Here, a non-rotating homogeneous spheroid collapses, maintaining its homogeneity and spheroidicity, but its deformations grow as the collapse progresses. If the initial condition is that of a slightly oblate spheroid, the collapse results in a pancake singularity through which the evolution could proceed and continue. However, for a slightly prolate spheroidal configuration, the matter collapses to a thin thread that ultimately results in a spindle singularity. This is more serious in nature in that the gravitational potential and force and the tidal forces blow up, as opposed to only the densities blowing up, in a mere shell-crossing singularity. Even in the case of an oblate collapse, the passing of matter through the pancake causes prolateness, and subsequently a spindle singularity again results without the formation of any horizon.

It was indicated by the numerical calculations of Shapiro and Teukolsky (1991) that a similar situation is maintained in general relativity, also in conformity with the hoop conjecture. They evolved collissionless gas spheroids in full general relativity that collapse in all cases to singularities. When the spheroid is sufficiently compact, a blackhole that contains the singularity forms, but when the semi-major axis of the spheroid is sufficiently large, a spindle singularity results without an apparent horizon forming. These have to be treated as numerical results, as opposed to a full analytic treatment, which need not be in contradiction to a suitably formulated version of the cosmic censorship. The definition of naked singularity here was basically in terms of the non-occurrence of trapped surfaces in a certain family of non-spacelike surfaces. Sometimes this may, or may not, indicate a genuine naked singularity, and a blackhole may still form. However, this gives rise to the possibility of the occurrence of naked singularities in the collapse of finite systems in asymptotically flat spacetimes, which could be in violation of the weak cosmic censorship, but possibly in conformity with the hoop conjecture.

A somewhat broader statement in a similar spirit to the hoop conjecture is the *event horizon conjecture* of Israel (1984, 1986a, 1986b). A general statement for this conjecture is given by the requirement that an event horizon must form whenever a matter distribution (satisfying appropriate energy conditions) has passed a certain critical point in its gravitational collapse, namely the formation of a closed trapped surface. A strong motivation for believing in this conjecture is that, unlike the cosmic censorship hypothesis, no counter-examples have been found so far. For example, the inequality $A < 4\pi(2M)^2$ must hold if the event horizon conjecture is true, for the area A of a closed trapped surface that forms at time $t = 0$ (say) in the gravitational collapse of a mass M. The inequality depends only on the initial data on the spacelike surface $t = 0$ and hence there is no need to trace future evolution of the system in order to verify it. Clearly, further investigations in a non-spherical collapse are needed to probe such a conjecture.

In the case of the exact spherical symmetry holding for the spacetime, it is known (see for example, Leibovitz and Israel, 1970) that an event horizon must form when a star collapses to a sufficiently small radius, and when the positivity of energy is satisfied. The only alternative to this is that the star must radiate away all of its mass in the process of the collapse. As discussed in Section 5.1 above, this appears very difficult, at least using purely classical or astrophysical processes, for the stars having tens of solar masses. Even for a non-spherical collapse, which involves only small perturbations from the spherical symmetry, the numerical calculations mentioned above seem to indicate that a non-singular event horizon must develop during the collapse. As noted by Jang and Wald (1977), without an event horizon to stop the flow of outgoing radiation, the gravitational mass of the star must become negative.

In fact, for all the classes of spacetimes discussed in Chapter 3, an event horizon always forms in the process of the collapse, even though it fails to cover the singularity completely. This results in the formation of naked singularities in the spacetime. Note that merely the formation of an event horizon is not a sufficient condition for the avoidance of the naked singularity. If the trapped surfaces and event horizon form late enough as the collapse develops, a strong curvature singularity can still be visible, as discussed here. Not only should the event horizon develop, but it must develop early enough, as is the case for the example in the Oppenheimer–Snyder collapse, in order to create a blackhole as the gravitational collapse final state.

Of course, in such a case, when the event horizons form accompanied by the naked singularities, the usual interpretation of blackhole physics becomes unclear due to the back reaction on the metric as a result of the possible emission from the naked singularities. For example, in the detailed analysis for the case of radiation collapse using the Vaidya metric, the horizon could be fully covered by the emissions from the strong curvature naked singularity, some of which may fall back inwards, thus affecting the area and structure of the horizons.

As pointed out by Israel (1984), there are two main aspects of the formulation of such an event horizon conjecture. First, one has to characterize the formation of a trapped surface in terms of the initial data on a spacelike surface in the form of a statement, such as a closed two-surface S will be trapped if the gravitational mass interior to S exceeds a certain critical value (defined suitably in terms of the geometry of S). The works such as those of Schoen and Yau (1983) and Ludvigsen and Vickers (1983) would be relevant here. Whereas this could be carried out for the spherical symmetric case, the situation could be fairly complicated in general, because one would need to characterize the concept of 'mass interior to S' much more precisely. The other aspect is that of time evolution of such a trapped surface, which should extend for a finite distance in the future to generate a spacelike three-cylinder, the sections of which are trapped surfaces with bounded dimensions. Assuming energy conditions, the regularity of the energy-momentum tensor would be desired on this three-cylinder. If these bounds and extension depend only on the initial geometry and the mass within the trapped surface, then, as pointed out by Israel (1986a, 1986b), the three-cylinder could be extended infinitely into the future, provided it encounters no singularities in its future development.

Now, the occurrence and nature of naked singularities forming in a class of non-spherical models known as the Szekeres spacetimes are examined. These represent irrotational dust collapses, they have no Killing vectors, and are generalizations of the Tolman–Bondi–Lemaitre spacetimes. It is seen that in these spacetimes, naked singularities exist that satisfy the limiting focusing condition and the strong limiting focusing condition. This generally ensures that various classes of naked singularities forming in the Szekeres models and also in the TBL collapse spacetimes, as discussed earlier, are strong curvature naked singularities.

A future incomplete (or past incomplete) causal geodesic terminates in a strong curvature singularity in the future (or past) if, for every point $q \in \lambda$, the expansion θ of the congruence of the future directed (or past directed) causal geodesics originating from q and infinitesimally neighboring λ diverges. If the strong curvature condition on a spacetime is defined as holding if all future and past incomplete null geodesics generating an achronal set terminate in a strong curvature singularity, then Królak (1983) tried to prove that, under a strong curvature condition on the spacetime, the cosmic censorship should hold. Such attempts, however, did not succeed and it turned out that such a conjecture does not hold, as shown by further work. Also, the theorems proved by Królak (1983, 1986) and later by Królak (1999) required an extra restrictive assumption on the causal structure of spacetimes. Furthermore, explicit examples of naked strong curvature singularities were found in Królak's sense (Eardley and Smarr, 1979; Christodoulou, 1984; Newman, 1986), and also in Tipler's sense (Tipler, Clarke, and Ellis, 1980), and by Dwivedi and Joshi (1989, 1991) and Joshi and Dwivedi (1993a). These demonstrated the occurrence of strong curvature naked singularities in the TBL spacetimes representing a spherically symmetric inhomogeneous collapse of dust, and also in the Vaidya radiation collapse models.

The TBL dust collapse spacetimes, even though they generalize the homogeneity assumption, are special in some ways. They are spherically symmetric and they have matter in the form of irrotational pressureless dust. It is interesting to know whether naked strong curvature singularities occur in more general situations than this. Here, it is discussed how naked strong curvature singularities occur in the Szekeres spacetimes that do not have any Killing vectors (Joshi and Królak, 1986). This result shows that naked strong curvature singularities do not arise necessarily as a result of the spherical symmetry. Nevertheless, the Szekeres spacetimes have the same special form of matter as the TBL spacetimes, which is again irrotational pressureless dust. Moreover, Szekeres spacetimes are also special because they can be matched to the Schwarzschild spacetime and they cannot contain any gravitational radiation (Bonnor, 1976).

The Szekeres spacetime (Szekeres, 1975) is a solution of the Einstein equations representing irrotational dust, where

$$G_{ab} = T_{ab} = \rho u_a u_b, \quad u_a u^a = 1, \tag{5.33}$$

and the units have been chosen so that $c = 8\pi G = 1$. The metric has the diagonal form given by

$$ds^2 = dt^2 - S^2 dr^2 - Y^2(dx^2 + dy^2), \tag{5.34}$$

where (r, x, y) are comoving spatial coordinates. The solution is given below for the case $Y' = \partial Y/\partial r \neq 0$,

$$Y = \frac{R(t,r)}{P(r,x,y)}, \qquad S = \frac{P(r,x,y)Y'(t,r)}{\sqrt{1+f(r)}}, \tag{5.35}$$

where $f(r) > -1$ and

$$P = a(r)(x^2 + y^2) + 2b_1(r)x + 2b_2(r)y + c(r), \tag{5.36}$$

$$ac - b_1^2 - b_2^2 = \frac{1}{4}, \tag{5.37}$$

$$\dot{R}^2 = f + \frac{F(r)}{R}. \tag{5.38}$$

Here $F(r)$ is again an arbitrary function of r as in the TBL collapse case, R denotes the physical radius for the cloud, and the dot denotes the partial derivative with respect to the time coordinate t.

As earlier, to ensure physical reasonability of the collapse models, some regularity conditions are imposed, and a class of Szekeres collapse spacetimes is dealt with. First, the metric is taken to be C^1 everywhere in the spacetime. Then, the function P must be non-zero everywhere, and its derivative with respect to r must be continuous and vanishing at $r = 0$. Also, take the metric to be locally Euclidean at $r = 0$. Then, it is necessary to set

$$f(0) = 0. \tag{5.39}$$

The function $R_0(r) = R(r, 0)$, which again indicates the physical radius of the shells, is a monotonically increasing function of r. The freedom in the choice of the radial coordinate r can be used to do the scaling as

$$R_0(r) = r. \tag{5.40}$$

The dust density ρ is given by

$$\rho = \frac{PF' - 3FP'}{P^2 R^2 Y'}. \tag{5.41}$$

Although for $P > 0$, the surfaces $r =$ const., $t =$ const. are spheres, the solution is not spherically symmetric here because the spheres are no longer concentric in this case, and their centers are given by $(-a^{-1}b_1, -a^{-1}b_2)$. Szekeres has also analyzed the singularities and their causal structure in these spacetimes. When $R = 0$, the singularity is of the *first kind*, and when $Y' = 0$ the singularity is of the *second kind*. The singularities of the second kind are familiar shell-crossing singularities that also occur in TBL spacetimes. As in the TBL spacetimes, the shell-crossing singularities in Szekeres spaces can also be both locally and globally naked (Szekeres, 1975). However, they are generally believed to be mild and they will not be considered here. These singularies can be eliminated by imposing a regularity condition,

$$Y' > 0. \tag{5.42}$$

This is no loss of generality, as the main interest here is in examining the properties of the shell-focusing singularities in terms of visibility, or otherwise. Szekeres (1975) has also shown that whenever $r > 0$, the shell of the dust always crosses the apparent horizon before collapsing to a singularity, and therefore for $r > 0$, the singularity cannot be naked, and all these points hide behind the event horizon. Therefore, the singularity of the first kind can be naked only when $r = 0$, which is called the central singularity. This situation is again analogous to the TBL case. It will be shown that, as in TBL spacetimes, naked strong curvature singularities do occur in Szekeres spacetimes. Consider the case of a gravitational collapse, that is, $\dot{R} < 0$. For simplicity, only the class of marginally bound collapse models is considered, that is,

$$f(r) = 0, \tag{5.43}$$

as mentioned above. Then, the function $R(r,t)$ is given by

$$R = r\left(1 - \frac{3}{2}\sqrt{\frac{F}{r^3}}t\right)^{2/3}. \tag{5.44}$$

The analysis here can again be similar to that in the TBL collapse models case, as given by Joshi and Dwivedi (1993a). A set of new functions can be introduced,

$$X = \frac{R}{r^\alpha}, \quad \eta = r\frac{F'}{F}, \quad \Lambda = \frac{F}{r^\alpha},$$
$$\Theta = \frac{1 - \frac{1}{3}\eta}{r^{3(\alpha-1)/2}}, \quad \mathcal{L} = r\frac{P'}{P}, \tag{5.45}$$

where $\alpha \geq 1$, and the unique value of the constant α is determined by the condition that $\Theta/\sqrt{X}$ does not vanish or does not go to infinity identically as $r \to 0$ in the limit of approach to the central singularity along any $X =$ const. direction. It will be assumed that the above functions are at least C^2. Partial derivatives R' and $\dot{R}'$ that are useful in the analysis of the singularity are then given by

$$\dot{R} = -\sqrt{\frac{\Lambda}{X}}, \quad R' = r^{\alpha-1}H, \quad \dot{R}' = -\frac{N}{r}, \tag{5.46}$$

where

$$H = \frac{1}{3}\eta X + \frac{\Theta}{\sqrt{X}}, \tag{5.47}$$

$$N = -\frac{\sqrt{\Lambda}}{2X^2}\left(\Theta - \frac{2}{3}\eta X^{3/2}\right). \tag{5.48}$$

The tangents $K^a = dx^a/dk$ for the outgoing radial ($x =$ const., $y =$ const.) null geodesics can be written as

$$K^t = \frac{dt}{dk} = \frac{\mathcal{P}}{Y}, \tag{5.49}$$

$$K^r = \frac{dr}{dk} = \frac{\mathcal{P}}{PYY'}, \tag{5.50}$$

$$K^x = \frac{dx}{dk} = 0, \tag{5.51}$$

$$K^y = \frac{dy}{dk} = 0, \tag{5.52}$$

where $\mathcal{P}$ satisfies the differential equation

$$\frac{d\mathcal{P}}{dk} + \mathcal{P}^2 \left(\frac{\dot{Y}'}{YY'} - \frac{\dot{Y}}{Y^2} - \frac{1}{PY^2} \right) = 0. \tag{5.53}$$

The parameter k is an affine parameter along the null geodesics.

If the future directed outgoing null geodesics are to terminate in the past at the central singularity at $r = 0$, which occurs at some time $t = t_0$ at which $R(t_0, 0) = 0$, then along such geodesics we must have $R \to 0$ as $r \to 0$. The following equation is satisfied along the null geodesics

$$\frac{dR}{du} = \frac{1}{\alpha r^{\alpha-1}} \left[\dot{R}\frac{dt}{dr} + R' \right] \tag{5.54}$$

$$= \left(1 - \sqrt{\frac{\Lambda}{X}} \right) \frac{H(X,u)}{\alpha} + \frac{\sqrt{X\Lambda}}{\alpha} \mathcal{L} \equiv U(X,u), \tag{5.55}$$

where $u = r^\alpha$. Note that when one writes the quantity U, one is essentially working out the necessary condition for the null geodesics to emerge from the singularity at $R = 0$, $r = 0$, with a well-defined tangent given by dR/dt. Therefore, if the trajectories *are* emerging with a well-defined tangent from the singularity, then the quantity $U(x, u)$ is well-defined by definition. For more details, see, for example, Joshi and Dwivedi (1993a). Consider the limit X_0 of the function X along the null geodesic terminating at the singularity at $R = 0, u = 0$. Using the l'Hospital rule,

$$X_0 = \lim_{R\to 0,\, u\to 0} \frac{R}{u} = \lim_{R\to 0,\, u\to 0} \frac{dR}{du} = \lim_{R\to 0,\, u\to 0} U(X,u) = U(X_0, 0). \tag{5.56}$$

The necessary condition for the existence of the null geodesic outgoing from the central singularity is the existence of the positive real root X_0 of the equation

$$V(X) \equiv U(X,0) - X = 0. \tag{5.57}$$

By the regularity conditions, $\lim_{r\to 0} \mathcal{L} = 0$. Consequently, the necessary condition for the existence of the naked singularity in the marginally bound case of the Szekeres spacetime is the existence of the positive real root of the equation

$$\left(1 + \sqrt{\frac{\Lambda_0}{X}} \right) \frac{H(X,0)}{\alpha} - X = 0, \tag{5.58}$$

where

$$\eta_0 = \eta(0),\ \Lambda_0 = \Lambda(0),\ \Theta_0 = \Theta(0). \tag{5.59}$$

This is exactly the same equation as in the marginally bound TBL case. Consequently, the same analysis as by the TBL case given earlier by Joshi and Dwivedi (1993a) applies here. Here, only a few key results will be summarized. To show that the singularity is naked, there is a need to prove that a solution of the geodesic equation exists such that the tangent X_0 is realized at the singularity. It can be proved that there is always at least a single null geodesic outgoing from the central singularity. Therefore, the existence of the real and positive root of the (5.58) is both a necessary and sufficient condition for the existence of a naked singularity.

Some further remarks will now be made on the strength of the singularity, which will be investigated in some detail. Let M be the spacetime manifold, and let $J(k)$ be the quantity along a null geodesic in the spacetime as given below, with $\lambda : (k_0, 0] \to M$, being parametrized by an affine parameter k,

$$J(k) = \int_{k_0}^{0} R_{ab} K^a K^b dk'. \tag{5.60}$$

Say that the *limiting focusing condition* holds if $J(k)$ is unbounded in the interval $(k_0, 0]$, and say that the *strong limiting focusing condition* holds if $J(k)$ is non-integrable on an interval $(k_0, 0]$. It is proved in Clarke and Królak (1986), that the limiting focusing condition implies that λ terminates in a strong curvature singularity in the sense of Królak, in the future, whereas the strong limiting focusing condition implies that λ terminates in Tipler's strong curvature singularity in the future.

To find out whether the naked singularity satisfies the strong limiting focusing condition, the limit $\lim_{k\to 0} k^2 R_{ab} K^a K^b$ along the future directed null geodesics emerging from the singularity will be investigated. Using the l'Hospital rule, regularity conditions, and (5.53),

$$\lim_{k\to 0} k^2 R_{ab} K^a K^b = \lim_{k\to 0} \frac{k^2 \left(F' - 3F\frac{P'}{P}\right)(K^t)^2}{R^2\left(R' - R\frac{P'}{P}\right)} = \frac{\eta_0 \Lambda_0 H_0}{X_0^2(\alpha - N_0)^2}, \tag{5.61}$$

where $N_0 = N(X_0, 0)$. Therefore, if $\Lambda_0 \neq 0$ the naked singularity satisfies the strong limiting focusing condition.

Suppose then that $\Lambda_0 = 0$. Equation (5.53) can be written in the form

$$\frac{d(\ln K^t)}{dk} = \left(N - \sqrt{\frac{\Lambda}{X}}\mathcal{L}\right)\frac{1}{r}\frac{dr}{dk}. \tag{5.62}$$

Since as $k \to 0$, $N \to 0$ because $\Lambda_0 = 0$ and $\mathcal{L} \to 0$, by the regularity conditions, the right-hand side of the above equation is integrable on the interval $(k_0, 0]$ with respect to k. Therefore the limit $\lim_{k\to 0} K^t$ exists, and

$$K^r = \frac{K^t}{r^{\alpha-1}(H - X\mathcal{L})}. \tag{5.63}$$

Therefore if $\alpha = 1$ the limit $\lim_{k\to 0} K^r$ also exists. Suppose that $\alpha = 1$, and consider the limit $\lim_{k\to 0} kR_{ab}K^aK^b$. Applying the l'Hospital rule twice and using (5.53),

$$\lim_{k\to 0} kR_{ab}K^aK^b = \frac{\eta_0 \Lambda_0' H_0}{X_0^2(\alpha - N_0)^2} \lim_{k\to 0} K^r. \tag{5.64}$$

Therefore when $\alpha = 1$ the above limit is finite and the naked singularity just satisfies the limiting focusing condition ($R_{ab}K^aK^b$ diverges logarithmically). If $\alpha > 1$ the above limit diverges and the naked singularity also satisfies the limiting focusing condition (but not the strong focusing condition, unless $\Lambda_0 \neq 0$).

The above results show that under the regularity conditions, and for a marginally bound collapse, any central naked singularity in the Szekeres spacetime is always a strong curvature singularity in Królak's sense. Furthermore, as the considerations above reduce to the TBL case under various appropriate limits, the above also implies that many classes of naked singularities occurring in the TBL dust collapse models are strong curvature naked singularities necessarily. The significance of this is that these are physically serious curvature singularities, which are not removable from the spacetime. As discussed earlier in Chapter 4, all volume forms along non-spacelike geodesics falling into such a singularity go to a vanishing value. In other words, all physical objects are crushed to a zero volume as they fall into a strong curvature singularity.

5.4 Blackhole paradoxes

While the Einstein gravity has been a highly successful theory of gravitation, it generically admits the occurrence of spacetime singularities under fairly general physical conditions. These are extreme gravity regions in the spacetime, where the matter densities and spacetime curvatures, as well as various physical quantities, typically blow up and the theory must break down. Such singularities may develop in cosmology, indicating a beginning for the universe, or in a gravitational collapse that ensues whenever a massive star exhausts its nuclear fuel and undergoes the process of a continual collapse. It is expected that possibly a future theory of quantum gravity may resolve these singularities, where all known laws of physics break down. There have been very many attempts over past decades in this direction, by trying to construct a quantum theory of gravity that includes the string theory and the loop quantum gravity formalisms. All the same, it would appear that the goal of achieving a fully consistent and complete quantum theory of gravity is yet to be reached. The possibility of singularity resolution through quantum gravity is clearly linked to such efforts, and much work may be needed before it can be fully realized.

The singularity theorems predicting the occurrence of the spacetime singularities, however, contain three main assumptions under which the existence of a singularity is predicted in the form of geodesic incompleteness in the spacetime. These are in the form of a typical causality condition that ensures a suitable and physically reasonable global structure of the spacetime, an energy condition that requires the positivity of energy density at the classical level as seen by a local observer, and finally a condition demanding that trapped surfaces must exist in the dynamical evolution of the universe, or in the later stages of a continual gravitational collapse. A trapped region in the spacetime consists of trapped surfaces, which are two-surfaces, such that both in going as well as outgoing wavefronts normal to it must converge. Such trapped surfaces then necessarily give rise to a spacetime singularity either in a gravitational collapse or in cosmology.

For the same reason, the process of trapped surface formation in a gravitational collapse is also central to blackhole physics. The role of such a trapping of light and matter within the framework of Einstein's theory of gravitation was highlighted by Datt (1938) and Oppenheimer and Snyder (1939), within the context of the continual collapse of a massive matter cloud. They studied the collapse of pressureless dust clouds using general relativity, and showed that it leads to the formation of an *event horizon*, and a *blackhole* as the collapse endstate, assuming that the spatial density distribution within the star was strictly homogeneous, that is, $\rho = \rho(t)$ only. In the later stages of the collapse, an *apparent horizon* develops that is the boundary of the trapped region, thus giving rise to an event horizon. Once the collapsing star has entered the event horizon, the causal structure of the spacetime there would then imply that no non-spacelike curves from that region would escape away, and the star is cut off from the faraway observers, thus giving rise to a blackhole in the spacetime.

While blackhole physics has led to several interesting theoretical, as well as observational, developments and discussions, it is necessary, however, to study more realistic models of the gravitational collapse in order to put blackhole physics on a sound footing. This is because the Oppenheimer–Snyder–Datt scenario is rather idealized and pressures would play an important role in the dynamics of any realistic collapsing star. Also, density distributions within the star could not be completely homogeneous in any physically realistic model, but would be higher at the center of the cloud, with a typical negative gradient as one moved away from the center. The study of a dynamical collapse within Einstein's gravity is, however, a difficult subject because of the non-linearity of the Einstein equations. It was hence proposed in 1969 by Penrose that any physically realistic continual gravitational collapse must *necessarily* end in a blackhole final state only. This is the cosmic censorship conjecture as discussed earlier. Although cosmic censorship has played a rather crucial role as a basic assumption in all the physics and astrophysical applications of blackholes so far, no proof or any mathematically rigorous formulation for it is available as yet, despite many attempts. Hence, this has been widely recognized as the single most important problem in the theory of blackholes, and the gravitation physics of today. In order to make any progress on the censorship hypothesis, or to understand the final endstate of a massive collapsing star it is necessary to study realistic gravitational collapse scenarios within the framework of Einstein's gravity theory (Joshi, 1993).

Such a study of gravitational collapse is also warranted because of several deep paradoxes that are associated with the blackholes, which have been widely discussed. First, all the matter entering a blackhole, must, of necessity, collapse into a spacetime singularity of infinite density and curvatures at the center of the cloud, where all known laws of physics must break down. It is not clear how such a model can be stable at the classical level. Second, there is a need to formulate mathematically and prove the cosmic censor hypothesis, that a generic gravitational collapse gives rise to a blackhole only. As is clear from the discussions so far, this is turning into a formidably difficult problem, and gravitational collapse studies so far have pointed to several detailed collapse models that show that the final fate of a collapsing star could be either a blackhole or a naked singularity, depending on the nature of the initial data from which the collapse evolves, and the possible evolutions allowed by the Einstein equations, as discussed here. As opposed to a blackhole final state, the naked singularity is a scenario where ultra-strong density and curvature regions of spacetime, forming as result of the collapse, would be visible to faraway observers in the universe, in violation of the cosmic censorship. Finally, it is well-known that a blackhole would create information loss, violating the unitarity principle, thus creating a basic contradiction with fundamental principles of quantum theory.

Within the context of such a scenario, it may be worth investigating the possibility that the singularity problem in general relativity, as well as the blackhole paradoxes, can be resolved, by possibly avoiding the trapped surface formation in the spacetime during the process of a dynamical gravitational collapse. Note that within a cosmological context, singularity free solutions have been discussed for example, (see for example, Senovilla, 2006, and references therein). It would be possible to construct classes of collapsing solutions for the Einstein equations, where the trapped surface formation could be delayed or avoided during the collapse. This could then offer a somewhat natural resolution to the issues such as those above. The main issue would then be that of finding a mechanism, either at a classical or quantum level, so that much of the matter of the collapsing star can escape away and would be thrown out during the final stages of the gravitational collapse, thus resolving the problems such as those above, and also that of the infinite density spacetime singularity. Towards realizing such a scenario, the usual physical reasonability conditions could be imposed, and while regularity of the initial data could be required, as well as the weak energy condition, the pressure could be allowed to be negative.

Such a possibility at a quantum level is discussed in some detail in the next section, where the role of quantum effects towards generating an outwards flow of matter and radiation from the collapsing cloud in the very late stages of the collapse is explored. In any case, the occurrence of a spacetime singularity indicates the breakdown of the Einstein gravitation theory in these extreme regimes. The singularity problem and various blackhole paradoxes at classical and quantum levels were mentioned above. Note that quantum corrections could generate a strong negative pressure in the interior of the cloud in the very late stages of the collapse, where the classical theory should break down. The collapsing star could then radiate away most of its matter as the process of the gravitational collapse evolves, so as to avoid the formation of trapped surfaces and the spacetime singularity.

5.5 Resolution of a naked singularity

While a naked singularity may form as the endstate of a continual gravitational collapse within the framework of classical general relativity, the very final stages of the collapse should be dominated by the quantum effects. Therefore, these must be taken into account when determining the final fate of the collapse in any physically realistic scenario. Here, some possibilities in this direction, especially from the perspective of whether the singularity can be resolved once the quantum gravity effects are taken into consideration, are explored.

A class of collapsing scalar field models with a non-zero potential, where the weak energy condition is satisfied by the collapsing configuration are constructed and studied. It is seen that the endstate of the collapse at the classical level can be either a blackhole or a naked singularity, and that physically it is the rate of collapse that governs these outcomes of the dynamical evolution. This feature is similar to the Vaidya radiation collapse models (see Joshi, 1993, for details and references), where again it is the rate of collapse that determines either the blackhole or naked singularity outcomes.

There have been considerations of models of four-dimensional gravity coupled to a scalar field with potential $V(\phi)$, and modifications and numerical studies of such a scenario have been discussed in some detail (Alcubierre *et al.*, 2004; Dafermos, 2004; Garfinkle, 2004; Gutperle and Krauss, 2004; Hertog, Horowitz, and Maeda, 2004; Hubeny *et al.*, 2004). Some of these models satisfy the positive energy theorem, but would violate the energy conditions within an asymptotically anti-de Sitter framework. While the cosmic censorship violation is not clear, these investigations provide information on some aspects of self-interacting scalar fields. Though the basic question of cosmic censorship in scalar field collapse models remains open to further analysis, the case of a massless scalar field has been analyzed to see that a naked singularity develops, but not generically (Christodoulou, 1994, 1999).

Considered below is the scalar field collapse with the potential to examine in some detail the final state for such a collapse. Some implications of the loop quantum gravity formalism, when applied to and considered within such a collapsing cloud in the later stages of the collapse, especially to examine whether these quantum effects help resolve the classical naked singularity will be discussed. It is possible that the occurrence of a naked singularity as the outcome of the collapse may offer the possibility of observing quantum gravity effects taking place in such visible ultra-strong gravity regions. No such possibilities exist if the collapse necessarily always ends in a blackhole. Whereas the quantum effects would be certainly important in the vicinities of the ultra-strong gravity regions that a spacetime singularity creates, these will be of no observational consequence if the singularity is always covered within a blackhole, where all physical effects are hidden within the event horizon and cannot be seen by any external observer. On the other hand, if the collapse terminates in a naked singularity, the ultra-strong gravity regions have a causal connection to the outside universe, and there is a possibility of observing the quantum gravity effects taking place in these extreme gravity situations, which are created by the very late stages of the collapse.

As discussed earlier, the generic conclusion in the studies on the gravitational collapse of various matter fields such as dust, perfect fluids, radiation collapse and other forms of matter is, depending on the nature of the regular initial data in terms of the initial distributions of density and pressures of the matter and other collapse parameters from which the collapse develops, that the final outcome could be either a blackhole or a naked singularity. Special importance is, however, sometimes attached to the investigation of the collapse of a scalar field, because one would like to know if the cosmic censorship is preserved in the collapse for fundamental matter fields that are derived from a suitable Lagrangian. Also, in the later stages of the collapse it is not certain what the equation of state for the matter would be, and it is possible that the matter at these epochs could be of a fundamental form, such as a Klein–Gordon or Maxwell field.

From such a perspective, a class of continual collapse models of a scalar field with potential are constructed and studied here (see also Goswami and Joshi, 2004c). It is required that the weak energy condition is preserved throughout the collapse, although the pressures are allowed to be negative closer to the singularity. The interior collapsing sphere is matched, using the Israel–Darmois conditions, with a generalized Vaidya exterior to complete the model. It is found that there are classes of collapse models included here for which no trapped surfaces form in the spacetime as the collapse evolves, and the singularity that develops as the collapse endstate is visible. In these classes, it can be seen that naked singularities are created through a scalar field collapse with potential from generic initial conditions that violate the cosmic censorship. It is seen that physically it is the rate of collapse that governs the formation of either a blackhole or a naked singularity as the collapse endstate.

A spherically symmetric homogeneous scalar field, $\Phi = \Phi(t)$, with a potential $V(\Phi)$ is discussed. This ensures that the interior spacetime must have a Friedmann–Robertson–Walker (FRW) metric. Furthermore, the marginally bound $(k=0)$ case is chosen. Then, the interior metric is of the form

$$ds^2 = -dt^2 + a^2(t)\left[dr^2 + r^2 d\Omega^2\right], \tag{5.65}$$

where $d\Omega^2$ is the line element on a two-sphere. In this comoving frame, the energy-momentum tensor of the scalar field is given as

$$T^t{}_t = -\rho(t) = -\left[\frac{1}{2}\dot{\Phi}^2 + V(\Phi)\right], \tag{5.66}$$

$$T^r{}_r = T^\theta{}_\theta = T^\phi{}_\phi = p(t) = \left[\frac{1}{2}\dot{\Phi}^2 - V(\Phi)\right], \tag{5.67}$$

with all other off-diagonal terms being zero.

It may be noted that the comoving coordinate system chosen has a particular physical significance as compared with an arbitrary system, and the quantities ρ and p are interpreted as the density and pressure of the scalar field respectively. It is then easily seen that the scalar field behaves like a perfect fluid, as the radial and tangential pressures are equal. Take the scalar field to satisfy the weak energy condition, that is, the energy density as measured by any local observer is non-negative, and for any timelike vector V^i, $T_{ik}V^iV^k \geq 0$. This amounts to $\rho \geq 0$, $\rho + p \geq 0$.

The dynamical evolution of the system is now determined by the Einstein equations, which for the metric (5.65) become (in the units $8\pi G = c = 1$),

$$\rho = \frac{F'}{R^2 R'} \quad p = \frac{-\dot{F}}{R^2 \dot{R}}, \tag{5.68}$$

$$\dot{R}^2 = \frac{F}{R}. \tag{5.69}$$

For general spherically symmetric spacetimes, $F(t, r)$ has the interpretation of the mass function for the cloud, with $F \geq 0$. The quantity $R(t, r) = ra(t)$ is the area radius for a shell labeled by the comoving coordinate r. In order to preserve the regularity of the initial data, $F(t_i, 0) = 0$, that is, the mass function should vanish at the center of the cloud. From (3.28) it is evident that on any regular epoch $F \approx r^3$ near the center, in order for the initial data to be singularity free.

The Klein–Gordon equation for the scalar field is given by

$$\frac{d}{dt}\left[a^3 \Phi\right] = -a^3 V(\Phi)_{,\Phi}. \tag{5.70}$$

Since the aim here is to construct a continual collapse model, the class with $\dot{a} < 0$, which is the collapse condition implying that the area radius of a shell at a constant value of comoving radius r decreases monotonically, is considered. In general, there may be classes of solutions where a scalar field may disperse also (see for example, Choptuik, 1993). The objective, however, is to examine whether the singularities forming in the scalar field collapse could be naked, or would be necessarily covered within a blackhole, and if so under what conditions. The singularity resulting from a continual collapse is given by $a = 0$, that is, when the scale factor vanishes and the area radius for all the collapsing shells becomes zero. At the singularity $\rho \to \infty$.

The key factor that decides the visibility, or otherwise, of the singularity is the geometry of the trapped surfaces that may form as the collapse evolves, that is the two-surfaces in the spacetime from which both outgoing and in going wavefronts necessarily converge. The boundary of the trapped region in a spherically symmetric spacetime is given by the equation $F = R$, which describes the apparent horizon for the spacetime. The spacetime region where the mass function F satisfies $F < R$ is not trapped, while $F > R$ describes a trapped region.

In terms of the scale variable a, the mass function can be written as

$$F = r^3 \left[\frac{1}{3} a^3 \left\{\frac{1}{2} \Phi(a)^2{}_{,a} \dot{a}^2 + V(\Phi(a))\right\}\right]. \tag{5.71}$$

This is because from the above Einstein equation for density, the mass function can be solved as

$$F = \frac{1}{3}\rho(t)R^3. \tag{5.72}$$

From (5.72), it can be seen that

$$\frac{F}{R} = \frac{1}{3}\rho(t)r^2a^2. \tag{5.73}$$

The above relation decides the trapping, or otherwise, in the spacetime as the collapse develops. The classes of collapse solutions for a scalar field with potential, and the trapping, or otherwise, as the collapse develops will be constructed and investigated. This is relevant from the perspective of the cosmic censorship conjecture in order to understand the development of a blackhole or a naked singularity as the collapse outcomes.

Towards such a purpose, consider the class of models where, near the singularity, the divergence of the density is given by $\rho(t) \approx 1/a(t)$. Then, using (5.71), the above condition implies, near the singularity,

$$\frac{1}{2}\Phi(a)_{,a}^2\dot{a}^2 + V(\Phi(a)) = \frac{1}{a}. \tag{5.74}$$

Now, solving the equation of motion (5.69),

$$\dot{a} = -\frac{\sqrt{a}}{\sqrt{3}}. \tag{5.75}$$

The negative sign implies a collapse scenario where $\dot{a} < 0$. Using (5.75) and (5.74) in the latter part of (5.68),Φ can be solved as

$$\Phi(a) = -\ln a. \tag{5.76}$$

Note that as the singularity is approached, $a \to 0$ implies $\Phi \to \infty$, that is, the scalar field blows up at the singularity. Finally, using (5.75) and (5.76) in (5.70), the potential V can be solved as

$$V(\Phi) = \frac{5}{6}e^{\Phi}. \tag{5.77}$$

Therefore near the singularity,

$$\rho(t) \approx \frac{1}{a(t)}, \quad p(t) \approx -\frac{2}{3a(t)}. \tag{5.78}$$

It is seen that in the limit of approach to the singular epoch ($t = t_s$) $F/R = 0$ for all shells and there is no trapped surface developing in the spacetime.

In the model above the weak energy condition is satisfied as $\rho > 0$ and $\rho + p > 0$, although the pressure would be negative. Also, it should be noted that the pressure does not have to be negative from the initial epoch, because the specific behavior of ρ in (5.78) has been required only near the singularity. It is possible to choose a $V(\Phi)$ such that at the initial epoch $1/2\dot{\Phi}^2 > V(\Phi)$, and then the pressure would be positive. But, near the singularity, $V(\Phi)$ would behave according to (5.78), and hence the pressure should decrease monotonically from the initial epoch and tend to $-\infty$ at the singularity.

If, from an epoch $t = t^*$ (or equivalently for some $a = a^*$) the density starts growing as a^{-1}, then integrating (5.75), the singular epoch can be obtained as

$$t_s = t^* + 2\sqrt{3}a^*. \tag{5.79}$$

Therefore, the collapse reaches the singularity in a finite comoving time, where the matter energy density as well as the Kretschman scalar $\kappa = R^{ijkl}R_{ijkl}$ diverge. Note that from the equation of motion (3.31) it follows that the metric function a is given by

$$a(t) = \left[\sqrt{a^*} - \frac{1}{2\sqrt{3}}(t - t^*)\right]^2. \tag{5.80}$$

This completes the interior solution within the collapsing cloud, thus giving the required construction.

It can be seen from the above considerations that the absence, or otherwise, of the trapped surfaces and the behavior of the pressure crucially depend on the rate of divergence of the density ρ near the singularity. To examine this more carefully, near the singularity set $\rho = a^{-n}$ with $n > 0$, as it is known that $\rho(t)$ must diverge as $a(t)$ goes to zero in the limit of approach to the singularity. In this case, solving the Einstein equations gives

$$p = \frac{(n-3)}{3}a^{-n}. \tag{5.81}$$

The corresponding values of Φ and $V(\Phi)$ are

$$\Phi = -\sqrt{n}\ln(a), \quad V(\Phi) = \left(1 - \frac{n}{6}\right)e^{\sqrt{n}\Phi}. \tag{5.82}$$

Again, calculating F/R in this general case,

$$\frac{F}{R} = \frac{1}{3}a^{2-n}. \tag{5.83}$$

Therefore, it can be seen that for low enough divergences ($0 < n < 2$) no trapped surfaces form and there are negative pressures near the singularity. For $2 \leq n < 3$, trapped surfaces do form, but the pressure still remains negative at the singularity. For $n \geq 3$, $p \geq 0$ and there are trapped surfaces forming in the spacetime as the collapse advances in time. Conversely, a non-negative pressure always ensures trapped surfaces in a homogeneous scalar field collapse. Therefore, the role of the chosen potential $V(\Phi)$ is to control the divergence of the density near the singularity, which in turn governs the development, or otherwise, of trapped surfaces. Equation (5.82) shows the behavior of the functions $V(\Phi)$ with respect to Φ, for different values of n. It is seen that the naked singularity arises from a non-zero measure open set of initial conditions ($n < 2$), whereas the rest of the initial data set produces a blackhole as the final endstate of the collapse.

In the above class of solutions, the behavior is described mainly near the singularity. However, note that an exact solution is given, and the analysis holds even if right from initial epoch the density behavior is as above. This gives a global exact solution where the weak energy condition is satisfied. As such, the treatment holds for any general class of solutions where near singularity behavior is as described.

To complete the model, this interior spacetime needs to be matched to an exterior spacetime. For the required matching, the Israel–Darmois conditions are used, where the first and second fundamental forms (the metric coefficients and the extrinsic curvature respectively) are matched at the boundary of the cloud. Whereas the procedures used below are standard, the particular case treated here is described in some detail in order to give the picture of the emerging overall collapse scenario. A useful fact is that since the second fundamental form K_{ij} is matched, there is *no* surface stress–energy or surface tension at the boundary (Israel, 1966a, 1966b; Mazur and Mottola, 2004).

The spherical ball of a collapsing scalar field is matched to a generalized Vaidya exterior geometry (Wang and Wu, 1999; Joshi and Dwivedi, 1999) at the boundary hypersurface Σ given by $r = r_{\rm b}$. Then the metric just inside Σ is

$$ds_-^2 = -dt^2 + a^2(t)\left[dr^2 + r_{\rm b}^2\, d\Omega^2\right], \tag{5.84}$$

while the metric in the exterior of Σ is

$$ds_+^2 = -\left(1 - \frac{2M(r_{\rm v}, v)}{r_{\rm v}}\right) dv^2 - 2\, dv\, dr_{\rm v} + r_{\rm v}^2\, d\Omega^2, \tag{5.85}$$

where v is the retarded null coordinate and $r_{\rm v}$ is the Vaidya radius. Matching the area radius at the boundary,

$$r_b\, a(t) = r_{\rm v}(v). \tag{5.86}$$

Then on the hypersurface Σ, the interior and exterior metrics are given by

$$ds_{\Sigma-}^2 = -dt^2 + a^2(t) r_{\rm b}^2\, d\Omega^2 \tag{5.87}$$

and

$$ds^2_{\Sigma+} = -\left(1 - \frac{2M(r_{\rm v}, v)}{r_{\rm v}} + 2\frac{dr_{\rm v}}{dv}\right) dv^2 + r_{\rm v}^2\, d\Omega^2. \tag{5.88}$$

Matching the first fundamental form on this hypersurface,

$$\left(\frac{dv}{dt}\right)_{\Sigma} = \frac{1}{\sqrt{1 - \frac{2M(r_{\rm v},v)}{r_{\rm v}} + 2\frac{dr_{\rm v}}{dv}}}, \quad (r_{\rm v})_{\Sigma} = r_{\rm b}\, a(t). \tag{5.89}$$

To match the second fundamental form (extrinsic curvature) for interior and exterior metrics, note that the normal to the hypersurface Σ, as calculated from the interior metric, is given by $n^{\rm i}_{-} = [0, a(t)^{-1}, 0, 0]$, and the non-vanishing components of the normal derived from the generalized Vaidya metric are

$$n^{\rm v}_{+} = -\left[1 - \frac{2M(r_{\rm v}, v)}{r_{\rm v}} + 2\frac{dr_{\rm v}}{dv}\right]^{-1/2}, \tag{5.90}$$

$$n^{r_{\rm v}}_{+} = \frac{1 - \dfrac{2M(r_{\rm v}, v)}{r_{\rm v}} + \dfrac{dr_{\rm v}}{dv}}{\sqrt{1 - \dfrac{2M(r_{\rm v}, v)}{r_{\rm v}} + 2\dfrac{dr_{\rm v}}{dv}}}. \tag{5.91}$$

Here the extrinsic curvature is defined as

$$K_{ab} = \frac{1}{2}\mathcal{L}_{\boldsymbol{n}} g_{ab}, \tag{5.92}$$

that is, the second fundamental form is the Lie derivative of the metric with respect to the normal vector $\boldsymbol{n}$. The above expression is equivalent to

$$K_{ab} = \frac{1}{2}\left[g_{ab,c}n^c + g_{cb}n^c{}_{,a} + g_{ac}n^c{}_{,b}\right]. \tag{5.93}$$

Setting $\left[K^{-}_{\theta\theta} - K^{+}_{\theta\theta}\right]_{\Sigma} = 0$ on the hypersurface Σ,

$$r_{\rm b} a(t) = r_{\rm v}\frac{1 - \dfrac{2M(r_{\rm v}, v)}{r_{\rm v}} + \dfrac{dr_{\rm v}}{dv}}{\sqrt{1 - \dfrac{2M(r_{\rm v}, v)}{r_{\rm v}} + 2\dfrac{dr_{\rm v}}{dv}}}. \tag{5.94}$$

Simplifying the above using (5.89) and (5.69) on the boundary, one gets

$$F(t, r_{\rm b}) = 2M(r_{\rm v}, v). \tag{5.95}$$

Using the above equation and (5.89),

$$\left(\frac{dv}{dt}\right)_{\Sigma} = \frac{1 + r_{\mathrm{b}}\dot{a}}{1 - \dfrac{F(t, r_{\mathrm{b}})}{r_{\mathrm{b}}a(t)}} \tag{5.96}$$

Finally, setting $[K^{-}_{\tau\tau} - K^{+}_{\tau\tau}]_{\Sigma} = 0$, where τ is the proper time on Σ,

$$M(r_{\mathrm{v}}, v)_{,r_{\mathrm{v}}} = F/2r_{\mathrm{b}}a + r_{\mathrm{b}}^2 a\ddot{a}. \tag{5.97}$$

The above equations, together with (5.94), completely specify the matching at the boundary of the collapsing scalar field.

It is known (see for example, Wang and Wu, 1999), that a generalized Vaidya spacetime describes the matter that is a combination of matter fields of Type I and Type II. Specifically, the energy–momentum tensor T_{ik} for the matter can be written as a linear superposition of two tensors $T^{(n)}{}_{ik}$ and $T^{(m)}{}_{ik}$ given by

$$T^{(n)}{}_{ik} = \mu l_i l_k, \tag{5.98}$$

$$T^{(m)}{}_{ik} = (\rho + P)(l_i n_k + n_i l_k) + P g_{ik}. \tag{5.99}$$

Here, l_k and n_k are null vectors defined by

$$l_k = \delta^0{}_k, \quad n_k = \frac{1}{2}\left(1 - \frac{2M}{r_{\mathrm{v}}}\right)\delta^0{}_k + \delta^1{}_k. \tag{5.100}$$

The tensor $T^{(n)}{}_{ik}$ represents the matter field that moves along the null hypersurface $v = \text{const}$. The other tensor describes the matter moving out along timelike trajectories. The matching conditions uniquely describe the quantities μ and ρ on the matching surface, given by $\mu = (2/9\sqrt{a})(\sqrt{3}+r_{\mathrm{b}}\sqrt{a})$ and $\rho = (2/3a)$. The pressure component is determined by the specific choice of mass function satisfying the matching equations. It can be seen that these quantities are well-defined everywhere, except at the singularity, hence it can be deduced that the exterior spacetime has no source or singularities.

The above provides the full spacetime. However, one could also consider the fields that asymptotically fall off to vanishing values, rather than matching the collapsing cloud to an exterior metric such as above. In this case, suitable fall off conditions which are not discussed in details here, need to be prescribed. Although the matching used above is in a somewhat general setting, particular examples which form subcases have been discussed earlier. These serve to show that the set of all functions $M(v, r_{\mathrm{v}})$ that satisfies (3.102), is non-empty. Such examples include the charged Vaidya spacetime $M = M(v) + Q(v)/r_{\mathrm{v}}$ and also the anisotropic de-Sitter spacetime

$M = M(r_v)$, which are two different solutions of (3.102) (see for example, Joshi and Dwivedi, 1999; Wang and Wu, 1999; Giambo, 2005). These give two unique exterior spacetimes, both of which are subclasses of the generalized Vaidya geometry described above.

Now it can be seen that at the singular epoch $t = t_s$, $2M(r_v, v)/r_v \to 0$. Therefore the exterior metric around the singularity smoothly transforms to

$$ds^2 = -dv^2 - 2\,dv\,dr_v + r_v^2\,d\Omega^2. \tag{5.101}$$

This describes a Minkowski spacetime in retarded null coordinates. Hence, it can be seen that the exterior generalized Vaidya metric, together with the singular point at $(t_s, 0)$, can be smoothly extended to the Minkowski spacetime as the collapse completes. From (3.101), it can be seen that the trajectory that emerges from the singularity (before it evaporates into free space) in the generalized Vaidya geometry is a null geodesic. It follows that non-spacelike trajectories can emerge from the singularity that develops as the collapse end point. Hence, a naked singularity is produced in the collapse of the scalar field with potential for a non-zero measure set of initial conditions, and the occurrence of trapped surfaces in the spacetime is avoided.

An interesting observation to be made is that the null geodesic from the singularity lies completely in the spacetime that is exterior to the collapsing cloud. This is different, for example, from the globally naked singularities developing as collapse endstates in the TBL dust collapse models, where the null geodesic from the singularity starts within the interior spacetime and then crosses the boundary of the collapsing cloud to emerge into the exterior. In the present case, the complete interior spacetime collapses simultaneously at the epoch $t = t_s$ and no future directed non-spacelike geodesic from the singularity can lie in the interior of the cloud. Also, it should be pointed out that the proof that all simultaneous singularities in non-extendible spacetimes must be covered, does not hold in this case, as the spacetime here (together with the singularity) is extendible to a Minkowski spacetime when the collapse completes.

Quantum effects in the gravitational collapse of the scalar field model considered above, which classically leads to a naked singularity, are now discussed. It is seen that non-perturbative semi-classical modifications near the singularity, based on loop quantum gravity, give rise to a strong outward flux of energy. This leads to the dissolution of the collapsing cloud before the singularity can form. Quantum gravitational effects thus resolve the naked singularity by avoiding their formation. Furthermore, quantum gravity induced mass flux has a distinct feature that may lead to a novel observable signature in astrophysical bursts. The visible singularities, predicted by classical general relativity as gravitational collapse endstates, can, in principle, be directly observed by an external observer, unlike their black-hole siblings. Hence, it is of interest to examine these from such a perspective as there are many classes where, given the initial density and pressure profiles for a matter cloud, the collapse evolution leads to naked singularity formation, subject to an energy condition and astrophysically reasonable equations of state.

Since singularities originate in the regime where the classical general relativity is expected to be replaced by quantum gravity, whether a quantum theory of gravity could resolve their formation has remained an outstanding problem. Also, with the lack of observable signatures from the Planck regime, naked singularities could, in fact, be a boon for a quantum theory of gravity. Because, the singularity is visible, any quantum gravitational signature originating in the ultra-high curvature regime near a classical singularity can, in principle, be observed, thus providing a rare test for quantum gravity.

One of the non-perturbative quantizations of gravity is given by the loop quantum gravity formalism (see for example, Ashtekar, 1986, 1991; Ashtekar and Lewandowski, 2004). Its key predictions include the Bekenstein–Hawking entropy formula (Ashtekar *et al.*, 1998), and its application to symmetry reduced mini-superspace quantization of homogeneous spacetimes is called loop quantum cosmology (Bojowald, 2001, 2002, 2005), which has applications towards the resolution of the big-bang singularity, initial conditions for inflation (Tsujikawa, Singh, and Maartens, 2004), and possible observable signatures in cosmic microwave background radiation.

As the dynamics of a generic collapse would be complex, it is useful to work here with a collapse scenario such as that of a scalar field. It serves as a good toy model to gain insights into the role of quantum gravity effects closer to a naked singularity, at the late stages of a gravitational collapse. One of the simplest settings is to consider an initial configuration of a homogeneous and isotropic scalar field $\Phi = \Phi(t)$ with a potential $V(\Phi)$, as discussed above, and the canonical momentum P_Φ. In this case, as seen above, the fate of the singularity being naked or covered depends on the rate of the gravitational collapse. For an appropriately chosen potential, formation of trapped surfaces can be avoided even as the collapse progresses, resulting in a naked singularity with an outward energy flux, which would be, in principle, observable. Since the interior of the homogeneous scalar field collapse is described by a Friedmann–Robertson–Walker (FRW) metric, techniques of loop quantum cosmology can be used to investigate the way in which quantum gravity modifies the collapse.

Consider the classical collapse of a homogeneous scalar field $\Phi(t)$ with potential $V(\Phi)$ and the canonical momentum for the marginally bound ($k = 0$) case. The interior metric is given by

$$ds^2 = -dt^2 + a^2(t)\left[dr^2 + r^2 d\Omega^2\right], \tag{5.102}$$

with the classical energy density and pressure of the scalar field of

$$\rho(t) = \dot{\Phi}^2/2 + V(\Phi), \quad p(t) = \dot{\Phi}^2/2 - V(\Phi) \,. \tag{5.103}$$

The dynamical evolution of the system is obtained from the Einstein equations that yield

$$\dot{R}^2 R = F(t,r), \quad \rho = F_{,r}/\kappa a R^2, \quad p = -\dot{F}/\kappa R^2 \dot{R}. \tag{5.104}$$

Here, $\kappa = 8\pi G$, and $F(t,r) = (\kappa/3)\rho(t) r^3 a^3$ has the interpretation of the mass function of the collapsing cloud, with $F \geq 0$, and $R(t,r) = ra(t)$ is the area radius of a shell labeled by the comoving coordinate r. In a continual collapse, the area radius of a shell at a constant value of the comoving radius r decreases monotonically. The spacetime region is trapped, or otherwise, depending on the value of the mass function. If F is greater (or less) than R, the region is trapped (or untrapped). The boundary of the trapped region is given by $F = R$. The collapsing interior can be matched at some suitable boundary $r = r_{\text{b}}$ to a generalized Vaidya exterior geometry, as discussed earlier.

The form of the potential that leads to a naked singularity is determined by writing the energy density of the scalar field in a generic form as $\rho = l^{n-4} a^{-n}$, where $n > 0$ and l is a proportionality constant. Using the energy conservation equation, this leads to the pressure $p = [(n-3)/3]\ l^{n-4} a^{-n}$. On substituting (5.103) in these (Goswami and Joshi, 2004),

$$\Phi = -\sqrt{n/\kappa} \ln a, \quad V(\Phi) = (1 - n/6) l^{n-4} e^{\sqrt{\kappa n}\ \Phi} \,. \tag{5.105}$$

Then it is easily seen that $F/R = (\kappa/3) l^{n-4} a^{2-n} r^2$. Therefore, in the collapsing phase as $a \to 0$, whether or not the trapped surfaces form is determined by the value of n. As discussed above, for $0 < n < 2$, if no trapped surfaces exist initially then no trapped surfaces would form until the epoch $a(t) = 0$ with $a(t) = \left(1 - nt/2\sqrt{3}\right)^{2/n}$.

The absence of trapped surfaces is accompanied by a negative pressure, implying that for a constant value of the comoving coordinate r, $\dot{F}$ is negative and so the mass contained in the cloud of that radius keeps decreasing. This leads to a classical outward energy flux. As the collapse proceeds, the scale factor vanishes in finite time and the physical densities blow up, leading to a naked singularity. Since no trapped surfaces form during the collapse, the outward energy flux will, in principle, be observable. However, near the singularity, when the energy density is close to Planckian values, this classical picture has to be modified and the scenario incorporating quantum gravity modifications into classical dynamics needs to be investigated. A non-perturbative semi-classical modification, based on loop quantum gravity for the interior, was considered by Goswami, Joshi, and Singh (2006),

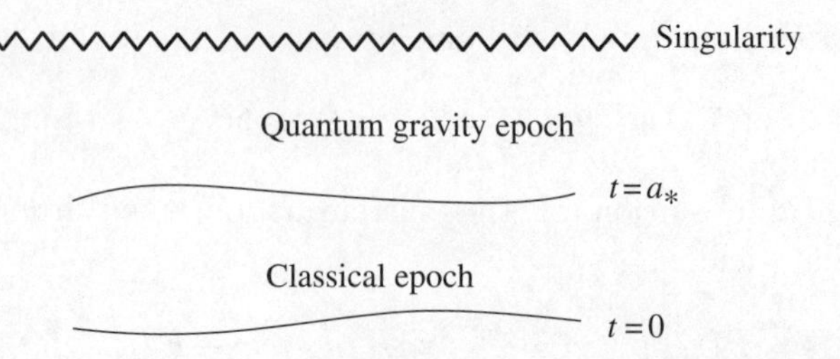

Fig. 5.2 Classical and quantum epochs in gravitational collapse.

which is discussed below in some detail. The underlying geometry for the FRW spacetime in loop quantum cosmology is discrete and both the scale factor and the inverse scale factor operators have discrete eigenvalues. In particular, a critical scale $a_* = \sqrt{j\gamma/3}\ell_{\rm P}$ exists below which the eigenvalues of the inverse scale factor become proportional to the positive powers of the scale factor (see Fig. 5.2). Here $\gamma \approx 0.2375$ is the Barbero–Immirzi parameter, $\ell_{\rm P}$ is the Planck length, and j is a half-integer free parameter that arises because the inverse scale factor operator is computed by tracing over SU(2) holonomies in an irreducible spin j representation. The value of this parameter is arbitrary and is constrained only by phenomenological considerations.

The change in the behavior of the classical geometrical density $(1/a^3)$ for scales $a \sim a_*$ can be approximated by

$$d_j(a) = D(q)\, a^{-3}, \quad q := a^2/a_*^2, \quad a_* := \sqrt{j\gamma/3}\,\ell_{\rm P}, \tag{5.106}$$

with

$$\begin{aligned} D(q) = (8/77)^6\, q^{3/2} \Big\{ 7 \Big[(q+1)^{11/4} - |q-1|^{11/4} \Big] \\ - 11q \Big[(q+1)^{7/4} - \operatorname{sgn}(q-1)|q-1|^{7/4} \Big] \Big\}^6 . \end{aligned} \tag{5.107}$$

For $a \ll a_*$, $d_j \propto (a/a_*)^{15} a^{-3}$, and for $a \gg a_*$ it behaves classically with $d_j \approx a^{-3}$. The scale at which transition to the behavior of the geometrical density takes place is determined by the parameter j.

At the fundamental level, the dynamics in the loop quantum regime is discrete, however, recent investigations pertaining to the evolution of coherent states showed that for scales $a_0 = \sqrt{\gamma}\ell_{\rm P} \sim a \sim a_* = \sqrt{j\gamma/3}\ell_{\rm P}$, dynamics can be described by modifications to the Friedmann dynamics on a continuous spacetime with the modified matter Hamiltonian,

$$\mathcal{H}_\Phi = d_j(a)\, P_\Phi^2/2 + a^3 V(\Phi), \tag{5.108}$$

and the modified Friedmann equation,

$$\dot{a}^2/a^2 = (\kappa/3)(\dot{\Phi}^2/2D + V(\Phi)), \tag{5.109}$$

which is obtained by the vanishing of the total Hamiltonian constraint and the Hamilton equations, which are $\dot{\Phi} = d_j(a)P_\Phi, \quad \dot{P}_\Phi = -a^3\, V_{,\Phi}(\Phi)$. These also lead to the modified Klein–Gordon equation,

$$\ddot{\Phi} + \left(3\dot{a}/a - \dot{D}(q)/D(q)\right)\, \dot{\Phi} + D(q)\, V_{,\Phi}(\Phi) \;= 0\;. \tag{5.110}$$

Since, at classical scales ($a \gg a_*$) $D \approx 1$, the modified dynamical equations reduce to the standard Friedmann dynamical equations. For scales $a \sim a_*$, the $\dot{\Phi}$ term acts like a frictional term for a collapsing phase.

Note that since semi-classical modifications for the inhomogeneous case are still not known, a complete quantum analysis of the interior and exterior cannot be carried out, and the exterior is assumed to remain classical. In any case, the quantum effects are supposed to be important mainly inside the cloud, which is collapsing and facing a singular fate. Also, as a continuous spacetime can be approximated to a scale factor a_0, the matching of the interior and exterior spacetimes remains valid during the semi-classical evolution.

The modified energy density and pressure of the scalar field in the semi-classical regime can be similarly obtained from the eigenvalues of the density operator, and by using the stress–energy conservation equation

$$\rho_{\rm eff} = d_j(a)\, \mathcal{H}_\Phi = \dot{\Phi}^2/2 + D(q)\, V(\Phi), \tag{5.111}$$

and

$$p_{\rm eff} = \left[1 - \frac{2}{3}\,\frac{1}{(\dot{a}/a)}\,\frac{\dot{D}(q)}{D(q)}\right]\frac{\dot{\Phi}^2}{2} - D(q)\, V(\Phi) - \frac{\dot{D}(q)}{3(\dot{a}/a)}\, V(\Phi)\;. \tag{5.112}$$

It is then straightforward to check that $p_{\rm eff}$ is generically negative for $a \sim a_*$, and for $a \ll a_*$ it becomes very strong. For example, at $a \sim a_0$, $p_{\rm eff} \approx -9\rho_{\rm eff}$. This is much stronger than its classical counterpart $p = [(n-3)/3]\;\rho$ with $0 < n < 2$. Therefore, a strong burst of outward energy flux can be expected in the semi-classical regime. Furthermore, for $a \ll a_*$, $D(q) \ll 1$ and the Klein–Gordon equation yields $\dot{\Phi} \propto a^{12}$. Hence, from (5.111) it can be easily seen that the effective density, instead of blowing up, becomes extremely small and remains finite.

The modified mass function of the collapsing cloud can be evaluated using (3.31) and (5.109),

$$F = (\kappa/3)(d_j^{-1}\,\dot{\Phi}^2/2 + a^3\, V(\Phi))\, r^3\;. \tag{5.113}$$

In the regime $a \sim a_0$, $d_j^{-1}\dot{\Phi}^2$ becomes proportional to a^{12}, the potential term becomes negligible, and thus the mass function becomes vanishingly small at small scale factors.

The picture emerging from the loop quantum modifications to the collapse is then as follows. First, before the area radius of the collapsing shell reaches $R_* = ra_*$ at $t = t_*$, the collapse proceeds as per classical dynamics, and as smaller scale factors are approached $\dot{\Phi}$ and the energy density $\rho \propto a^{-n}$ increase. The mass function is proportional to a^{n-3} and (as $0 < n < 2$) it decreases with decreasing scale factor, so there is a mass loss to the exterior, which is also explained by the existence of a negative classical pressure. Next, as the collapsing cloud reaches R_*, the geometric density classically given by a^{-3}, modifies to d_j and the dynamics is governed by the modified Friedmann and Klein–Gordon equations. The scalar field that experienced anti-friction in the classical regime, now experiences friction leading to a decrease in $\dot{\Phi}$. Finally, the slowing down of Φ decreases the rate of the collapse, and the formation of the singularity is delayed. Eventually, when the scale factor becomes less than a_0, this leads to a breakdown of the continuum spacetime approximation and semi-classical dynamics. Discrete quantum geometry emerges at this scale and the dynamics can only be described by the quantum difference equation. The naked singularity is thus avoided until the scale factor at which a continuous spacetime exists.

If the evolution of the area radius in time as the collapse proceeds is considered, the semi-classical evolution closely follows the classical trajectory until the time t_*. Within a finite time after t_*, the classical collapse leads to a vanishing R and a naked singularity. However, the area radius never vanishes in the loop modified semi-classical dynamics, and the naked singularity does not form as long as the continuum spacetime approximation holds. The evolution of the energy density in Planck units can be examined, and while the energy density would blow up classically, it remains finite and in fact decreases in the semi-classical regime.

The phenomena of delay and avoidance of the naked singularity in continuous spacetime is accompanied by a burst of matter to the exterior. If the mass function at scales $a \gg a_*$ is F_i and its difference with a mass of the cloud for $a < a_*$ is $\Delta F = F_i - F$, then the mass loss can be computed as

$$\frac{\Delta F}{F(a_i)} = \left[1 - \frac{\rho_{\text{eff}} d_j^{-1}}{l^{n-4} a_i^{3-n}}\right]. \tag{5.114}$$

For $a < a_*$, as the scale factor decreases, the energy density and mass in the interior decrease and the negative pressure increases strongly. This leads to a strong burst of matter. The absence of trapped surfaces enables the quantum gravity induced burst to propagate via the generalized Vaidya exterior to an observer at infinity.

As for the evolution of the mass function in the semi-classical regime, $\Delta F/F_i$ approaches unity very rapidly. This feature is independent of the choice of parameter j. The choice of potential causes mass loss to the exterior in the classical collapse also, but it is much smaller and, in any case, the classical description cannot be trusted at an energy density greater than the Planck value, when the quantum effects as described above must be considered. Basically, it can be seen that the loop quantum evolution leads to the dissolution of all the mass of the collapsing shell.

Interestingly, for a given collapsing configuration, the scale at which the strong outward flux initiates depends on the loop parameter j that controls a_*. If j is large, then the burst occurs at an earlier area radius and vice versa. For all such choices, $\Delta F/F_i \to 1$, but the outgoing flux profiles change. The loop quantum burst has a distinct signature in that at $a \sim a_*$ the flux decreases for a short period and then rapidly increases. Since the causal structure of classical spacetime is such that trapped surface formation is avoided, this quantum gravitational signature can be, in principle, observed by an external observer as a slight dimming and subsequent brightening of the collapsing star. This peculiar phenomena is directly related to the peak in the function $d_j(a)$, and depends solely on the value of the parameter j. If this is compared with the other phenomenological applications mentioned above, the effect could not be masked by the role of other loop quantum parameters in a more general setting. This phenomena is possibly a direct probe to measure j, and an observer can estimate this loop quantum parameter by observing the flux profile of the burst, based on this mechanism, and by measuring the variation in luminosity of the collapsing cloud.

During such a burst, most of the mass of the cloud is ejected and this may dissolve the singularity. Therefore, non-perturbative semi-classical modifications may not allow the formation of the naked singularity as the collapsing cloud evaporates away due to super-negative pressures in the late regime, as generated by the quantum effects. Such super-negative pressures would also exist for arbitrary matter configurations, which indicates that the results such as above could hold in a more general setting as well.

In this sense, loop quantum effects then imply a quantum gravitational cosmic censorship, alleviating the naked singularity problem. Note that the semi-classical effects do not show that the singularity is absent, it is only avoided to scale factor a_0, below which the semi-classical dynamics and matching may break down. If, for a given choice of initial data, semi-classical dynamics is unable to dissolve the singularity completely, the final fate of the naked singularity must be decided by using full quantum evolution. Even in such cases, there are valuable insights from semi-classical loop quantum effects, with the possibility of phenomenologically constraining the j parameter.

In the toy model considered above, it was shown that the classical outcome and evolution of the collapse is radically altered by the non-perturbative modifications to the dynamics, as induced by the quantum effects. These considerations are, of course, within the mini-superspace setting, and the general case of inhomogeneities and anisotropies remains open. However, the possibility of such observable signatures in astrophysical bursts originating from the quantum gravity regime near the spacetime singularity is indeed intriguing, indicating that gravitational collapse scenarios can be used as probes to test quantum gravity models.

References

Abraham, A. and Evans, C. R. (1993). 'Critical behavior and scaling in vacuum axisymmetric gravitational collapse'. *Phys. Rev. Lett.*, **70**, 2980.

Alcubierre, M., Gonzalez, J., Salgado, M. and Sudarsky, D. (2004). 'The cosmic censor conjecture: is it generically violated?', gr-qc/0402045.

Arnett, W. D. and Bowers, R. L. (1977). 'A microscopic interpretation of neutron star structure'. *Astrophys. J. Suppl.*, **33**, 415.

Ashtekar, A. (1986). 'New variables for classical and quantum gravity'. *Phys. Rev. Lett.*, **57**, 2244.

Ashtekar, A. (1991). 'Old problems in the light of new variables'. In *Conceptual Problems of Quantum Gravity*, ed. A. Ashtekar and J. Stachel. Boston: Birkhauser.

Ashtekar, A. and Lewandowski, J. (2004). 'Background independent quantum gravity: a status report'. *Class. Quant. Grav.*, **21**, R53.

Ashtekar, A., Pawlowski, T. and Singh, P. (2006). 'Quantum nature of the big bang: an analytical and numerical investigation', gr-qc/0604013.

Ashtekas, A., Baez, J., Corichi, A. and Krasanou, K. (1998). 'Quantum geometry and black hole entropy'. *Phys. Rev. Lett.*, **80**, 904.

Avez, A. (1963). 'Essais de geometrie Riemannienne hyperbolique globale. Applications à la relativité générale'. *Ann. Inst. Fourier (Grenoble)*, **132**, 105.

Banerjee, A., Debnath, U. and Chakraborty, S. (2003). 'Naked singularities in higher dimensional gravitational collapse'. *Int. J. Mod. Phys. D.*, **12**, 1255.

Bardeen, J. M., Carter, B. and Hawking, S. W. (1973). 'The four laws of black hole mechanics'. *Commun. Math. Phys.*, **31**, 161.

Bekenstein, J. D. (1973). 'Black holes and entropy'. *Phys. Rev. D.*, **7**, 2333.

Birkhoff, G. D. (1923). *Relativity and Modern Physics.* Cambridge, MA: Harvard University Press.

Bishop, R. L. and Critendon, R. J. (1964). *Geometry of Manifolds.* New York: Academic Press.

Blandford, R. D. (1987). 'Astrophysical black holes'. In *Three Hundred Years of Gravitation*, eds S. W. Hawking and W. Israel. Cambridge: Cambridge University Press.

Blandford, R. D. and Thorne, K. (1979). 'Black hole astrophysics'. In *General Relativity: An Einstein Centenary Survey*, eds S. W. Hawking and W. Israel. Cambridge: Cambridge University Press.

Bojowald, M. (2001). 'Absence of a singularity in loop quantum cosmology'. *Phys. Rev. Lett.*, **86**, 5227.

Bojowald, M. (2002). 'Inflation from quantum geometry'. *Phys. Rev. Lett.*, **89**, 261301.

Bojowald, M. (2005). 'Loop quantum cosmology'. *Living Rev. Rel.*, **8**, 11.

Bondi, H. (1948). 'Spherically symmetric models in general relativity'. *Mon. Not. R. Astron. Soc.*, **107**, 400.

Bondi, H., van der Burgh, M. G. J. and Metzner, R. (1962). 'Gravitational waves in general relativity VII: waves from axi-symmetric isolated systems'. *Proc. Roy. Soc. Lond. A.*, **269**, 21.

Bonnor, W. B. (1976). 'Non-radiative solutions of Einstein's equations for dust'. *Commun. Math. Phys.*, **51**, 191.

Brady, P. R. (1995a). 'Analytic example of critical behaviour in scalar field collapse'. *Class. Quant. Grav.*, **11**, 1255.

Brady, P. R. (1995b). 'Self-similar scalar field collapse: naked singularities and critical behavior'. *Phys. Rev. D.*, **51**, 4168.

Cahill, M. E. and Taub, A. H. (1971). 'Spherically symmetric similarity solutions of the Einstein field equations for a perfect fluid'. *Commun. Math. Phys.*, **21**, 1.

Carr, B. J. and Coley, A. (1999). 'Self-similarity in general relativity'. *Class. Quant. Grav.*, **16**, R31.

Carter, B. (1971). 'Causal structure in space-time'. *Gen. Relat. Grav.*, **1**, 349.

Celerier, M. and Szekeres, P. (2002). 'Timelike and null focusing singularities in spherical symmetry: a solution to the cosmological horizon problem and a challenge to the cosmic censorship hypothesis'. *Phys. Rev. D.*, **65**, 123516.

Chandrasekhar, S. (1931). 'The maximum mass of ideal white dwarfs'. *Astrophys. J.*, **74**, 81.

Chandrasekhar, S. (1934). 'Stellar configurations with degenerate cores'. *Observatory*, **57**, 373.

Chandrasekhar, S. and Hartle, J. (1983). 'On crossing the Cauchy horizon of a Reissner: Nordström black hole'. *Proc. Roy. Soc. Lond. A.*, **384**, 301.

Choptuik, M. W. (1993). 'Universality and scaling in gravitational collapse of a massless scalar field'. *Phys. Rev. Lett.*, **70**, 9.

Christodoulou, D. (1984). 'Violation of cosmic censorship in the gravitational collapse of a dust cloud'. *Commun. Math. Phys.*, **93**, 171.

Christodoulou, D. (1986). 'Global existence of generalized solutions of the spherically symmetric Einstein scalar equations in the large'. *Commun. Math. Phys.*, **106**, 587.

Christodoulou, D. (1994). 'Examples of naked singularity formation in the gravitational collapse of a scalar field'. *Ann. Maths.*, **140**, 607.

Christodoulou, D. (1999). 'The instability of naked singularities in the gravitational collapse of a scalar field'. *Ann. Maths.*, **149**, 183.

Clarke, C. J. S. (1986). 'Singularities: global and local aspects'. In *Topological Properties and Global Structure of Space-time*, eds P. G. Bergmann and V. de Sabbata. New York: Plenum Press.

Clarke, C. J. S. (1993). *The Analysis of Spacetime Singularities.* Cambridge: Cambridge University Press.

Clarke, C. J. S. and Joshi, P. S. (1988). 'On reflecting space-times'. *Class. Quant. Grav.*, **5**, 19.

Clarke, C. J. S. and Królak, A. (1986). 'Conditions for the occurrence of strong curvature singularities'. *J. Geo. Phys.*, **2**, 127.

Dafermos, M. (2004). 'On naked singularities and the collapse of self-gravitating Higgs fields', gr-qc/0403033.

Datt, B. (1938). 'Über eineklasse von lösungen der gravitationsgleichungen der relativität'. *Z. Physik.*, **108**, 314.

Debnath, U., Chakraborty, S. and Barrow, J. D. (2004). 'Quasispherical gravitational collapse in any dimension'. *Gen. Relat. Grav.*, **36**, 231.

de la Cruz, V., Chase, J. E. and Israel, W. (1970). 'Gravitational collapse with symmetries'. *Phys. Rev. Lett.*, **24**, 423.

Deshingkar, S. and Joshi, P. S. (2001). 'Structure of nonspacelike geodesics in dust collapse'. *Phys. Rev. D.*, **63**, 024007.

Deshingkar, S. S. , Joshi, P. S. and Dwivedi, I. H. (1999). 'Physical nature of the central singularity in spherical collapse'. *Phys. Rev. D.*, **59**, 044018.

Deshingkar, S., Joshi, P. S. and Dwivedi, I. H. (2002). 'Appearance of the central singularity in spherical collapse'. *Phys. Rev. D.*, **65**, 084009.

Deshingkar, S. S., Chamorro, A., Jhingan, S. and Joshi, P. S. (2001). 'Gravitational collapse and cosmological constant'. *Phys. Rev. D.*, **63**, 124005.

Deutsch, D. and Candelas, P. (1980). 'Boundary effects in quantum field theory'. *Phys. Rev. D.*, **20**, 3063.

Dixon, W. G. (1970). 'Dynamics of extended bodies in general relativity: momentum and angular momentum'. *Proc. Roy. Soc. Lond. A.*, **314**, 499.

Doroshkevich, A. G., Zel'dovich, Ya. B. and Novikov, I. (1966). 'Gravitational collapse of non-symmetric non-rotating masses'. *Sov. Phys. JETP*, **22**, 122.

Dwivedi, I. H. and Joshi, P. S. (1989). 'On the nature of naked singularities in Vaidya space-times'. *Class. Quant. Grav.*, **6**, 1599.

Dwivedi, I. H. and Joshi, P. S. (1991). 'On the nature of naked singularities in Vaidya space-times II'. *Class. Quant. Grav.*, **8**, 1339.

Dwivedi, I. H. and Joshi, P. S. (1992). 'Cosmic censorship violation in non-self-similar Tolman–Bondi models'. *Class. Quant. Grav.*, **9**, L69.

Eardley, D. M. and Smarr, L. (1979). 'Gravitational collapse of dust spheres: time coordinate conditions'. *Phys. Rev. D.*, **19**, 2239.

Ellis, G. F. R. and King, A. R. (1974). 'Was the big bang a whimper?'. *Commun. Math. Phys.*, **38**, 119.

Ellis, G. F. R. and Schmidt, B. (1977). 'Singular space-times'. *Gen. Relat. Grav.*, **8**, 915.

Evans, C. R. and Coleman, J. S. (1994). 'Critical phenomena and self-similarity in the gravitational collapse of radiation fluid'. *Phys. Rev. Lett.*, **72**, 1782.

Finkelstein, D. (1958). 'Past-future asymmetry of the gravitational field of a point particle'. *Phys. Rev.*, **110**, 965.

Fock, V. (1939). 'Sur le mouvement des masses finies d'aprés la théorie gravitation einsteinienne'. *J. Phys. USSR.*, **1**, 81.

Foglizzo, T. and Henriksen, R. (1993). 'General relativistic collapse of homothetic ideal gas spheres and planes'. *Phys. Rev. D.*, **48**, 4645.

Friedman, J., Morris, M. S., Novikov, I. D., Echeverria, F., Klinkhammer, G. and Thorne, K. S. (1990). 'Cauchy problem in space-times with closed timelike curves'. *Phys. Rev. D.*, **42**, 1915.

Garfinkle, D. (2004). 'Numerical simulation of a possible counterexample to cosmic censorship', gr-qc/0403078.

Geroch, R. (1967). 'Topology in general relativity'. *J. Math. Phys.*, **8**, 782.

Geroch, R. (1968a). 'What is a singularity in general relativity?'. *Ann. Phys.*, **48**, 526.

Geroch, R. (1968b). 'Spinor structure of space-times in general relativity I'. *J. Math. Phys.*, **9**, 1739.

Geroch, R. (1970a). 'Domain of dependence'. *J. Math. Phys.*, **11**, 437.

Geroch, R. (1970b). 'Singularities'. In *Relativity*, eds S. Fickler, M. Carmeli and L. Witten. New York: Plenum Press.

Geroch, R. (1971). 'Space-time structure from a global view point'. In *General Relativity and Cosmology, Proceedings of the International School in Physics 'Enrico Fermi'*, ed. R. K. Sachs. New York: Academic Press.

Geroch, R. and Horowitz, G. (1979). 'Global structure of space-times'. In *General Relativity: An Einstein Centenary Survey*, eds S. W. Hawking and W. Israel. Cambridge: Cambridge University Press.

Geroch, R., Kronheimer, E. and Penrose, R. (1972). 'Ideal points in space-time'. *Proc. Roy. Soc. Lond. A.*, **327**, 545.

Ghosh, S. G. and Banerjee, A. (2003). 'Non-marginally bound inhomogeneous dust collapse in higher dimensional spacetime'. *Int. J. Mod. Phys. D.*, **12**, 639.

Ghosh, S. G. and Beesham, A. (2001). 'Higher dimensional inhomogeneous dust collapse and cosmic censorship'. *Phys. Rev. D.*, **64**, 124005.

Ghosh, S. G. and Deshkar, D. W. (2003). 'Gravitational collapse of perfect fluid in self-similar higher dimensional space-times'. *Int. J. Mod. Phys. D.*, **12**, 913.

Giambo, R. (2005). 'Gravitational collapse of homogeneous scalar fields'. *Class. Quant. Grav.*, **22**, 2295.

Giambo, R. (2006). 'Global visibility of naked singularities', gr-qc/0603120.

Giambo, R., Giannoni, F., Magli, G. and Piccione, P. (2003). 'New solutions of Einstein equations in spherical symmetry: the cosmic censor to the court'. *Commun. Math. Phys.*, **235**, 545.

Giambo, R., Giannoni, F., Magli, G. and Piccione, P. (2004). 'Naked singularities formation in the gravitational collapse of barotropic spherical fluids'. *Gen. Rel. Grav.*, **36**, 1279.

Gödel, K. (1949). 'An example of a new type of cosmological solution of Einstein's field equations of gravitation'. *Rev. Mod. Phys.*, **21**, 447.

Goncalves, S. M. C. V. and Jhingan, S. (2001). 'Singularities in gravitational collapse with radial pressure'. *Gen. Rel. Grav.*, **33**, 2125.

Goswami, R. and Joshi, P. S. (2002). 'What role do pressures play in determining the final end state of gravitational collapse?'. *Class. Quant. Grav.*, **19**, 5229.

Goswami, R. and Joshi, P. S. (2004a). 'Spherical dust collapse in higher dimensions'. *Phys. Rev. D.*, **69**, 044002.

Goswami, R. and Joshi, P. S. (2004b). 'Cosmic censorship in higher dimensions'. *Phys. Rev. D.*, **69**, 104002.

Goswami, R. and Joshi, P. S. (2004c). 'Naked singularity formation in scalar field collapse', gr-qc/0410144.

Goswami, R. and Joshi, P. S. (2006). 'Spherical gravitational collapse in N-dimensions', gr-qc/0608136.

Goswami, R., Joshi, P. S. and Singh, P. (2006). 'Quantum evaporation of a naked singularity'. *Phys. Rev. Lett.*, **96**, 031302.

Gundlach, C. (1995). 'Choptuik spacetime as an eigenvalue problem'. *Phys. Rev. Lett.*, **75**, 3214.

Gundlach, C. (1999). 'Critical phenomena in gravitational collapse'. *Living Rev. Rel.*, **2**, 4.

Gutperle, M. and Krauss, P. (2004). 'Numerical study of cosmic censorship in string theory'. hep-th/0402109.

Hagerdorn, R. (1968). 'Hadronic matter near the boiling point'. *Nuovo Cimento A.*, **56**, 1027.

Harada, T. (1998). 'Final fate of the spherically symmetric collapse of a perfect fluid'. *Phys. Rev. D.*, **58**, 104015.

Harada, T. and Maeda, H. (2001). 'Convergence to a self-similar solution in general relativistic gravitational collapse'. *Phys. Rev. D.*, **63**, 084022.

Harada, T., Iguchi, H. and Nakao, K. I. (2000). 'Naked singularity explosion'. *Phys. Rev. D.*, **61**, 101502.

Harada, T., Iguchi, H. and Nakao, K. (2002). 'Physical processes in naked singularity formation'. *Prog. Theor. Phys.*, **107**, 449.

Harrison, B. K., Thorne, K. S., Wakano, M. and Wheeler, J. (1965). *Gravitation Theory and Gravitational Collapse*. Chicago: University of Chicago Press.

Hartle, J. (1978). 'Relativistic stars, gravitational collapse, and black holes'. In *Relativity, Astrophysics and Cosmology*, ed. W. Israel. Dordrecht: Reidel.

Hawking, S. W. (1971). 'Gravitational radiation from colliding black holes'. *Phys. Rev. Lett.*, **26**, 1344.

Hawking, S. W. (1975). 'Particle creation by black holes'. *Commun. Math. Phys.*, **43**, 199.

Hawking, S. W. and Ellis, G. F. R. (1973). *The Large Scale Structure of Space-time*. Cambridge: Cambridge University Press.

Hawking, S. W. and Israel, W. (1979a). *General Relativity: An Einstein Centenary Survey*. Cambridge: Cambridge University Press.

Hawking, S. W. and Israel, W. (1979b). 'An introductory survey'. In *General Relativity: An Einstein Centenary Survey*, eds S. W. Hawking and W. Israel. Cambridge: Cambridge University Press.

Hawking, S. W. and Penrose, R. (1970). 'The singularities of gravitational collapse and cosmology'. *Proc. Roy. Soc. Lond. A.*, **314**, 529.

Hellaby, J. and Lake, K. (1985). 'Shell crossings and the Tolman model'. *Astrophys. J.*, **290**, 381.

Herrera, L. and Santos, N. O. (2005). 'Cylindrical collapse and gravitational waves'. *Class. Quant. Grav.*, **22**, 2407.

Hertog, T., Horowitz, G. and Maeda, K. (2004). 'Generic cosmic-censorship violation in anti-de Sitter space'. *Phys. Rev. Lett.*, **92**, 131101.

Hicks, N. J. (1965). *Notes on Differential Geometry*. Princeton, NJ: Van Nostrand.

Hiscock, W. A., Williams, L. G. and Eardley, D. M. (1982). 'Creation of particles by shell-focusing singularities'. *Phys. Rev. D.*, **26**, 751.

Hubeny, V., Lio, X., Rangamani, M. and Shenker, S. (2004). 'Comments on cosmic censorship in AdS/CFT', hep-th/0403198.

Iguchi, H., Nakao, K. I. and Harada, T. (1998). 'Gravitational waves around a naked singularity: odd-parity perturbation of Lemaitre–Tolman–Bondi spacetime'. *Phys. Rev. D.*, **57**, 7262.

Israel, W. (1966a). 'Singular hypersurfaces and thin shells in general relativity'. *Nuovo Cimento B.*, **44**, 1.

Israel, W. (1966b). 'Singular hypersurfaces and thin shells in general relativity'. *Nuovo Cimento B.*, **48**, 463.

Israel, W. (1984). 'Does a cosmic censor exist?'. *Found. Phys.*, **14**, 1049.

Israel, W. (1986a). 'Must non-spherical collapse produce black holes? A gravitational confinement theorem'. *Phys. Rev. Lett.*, **56**, 789.

Israel, W. (1986b). 'The formation of black holes in non-spherical collapse and cosmic censorship'. *Can. J. Phys.*, **64**, 120.

Jang, P. S. and Wald, R. (1977). 'The positive energy conjecture and the cosmic censor hypothesis'. *J. Math. Phys.*, **18**, 41.

Jhingan, S. and Magli, G. (2000). 'Black holes versus naked singularities formation in collapsing Einstein clusters'. *Phys. Rev. D.*, **61**, 124006.

Joshi, P. S. (1981). 'On higher order causality violations'. *Phys. Lett. A.*, **85**, 319.

Joshi, P. S. (1993). *Global Aspects in Gravitation and Cosmology.* Oxford: Clarendon Press, Oxford University Press.

Joshi, P. S. (2000). 'Gravitational collapse: the story so far'. *Pramana*, **55**, 529.

Joshi, P. S. and Dwivedi, I. H. (1992). 'The structure of naked singularity in self-similar gravitational collapse'. *Commun. Math. Phys.*, **146**, 333.

Joshi, P. S. and Dwivedi, I. H. (1993a). 'Naked singularities in spherically symmetric inhomogeneous Tolman–Bondi dust cloud collapse'. *Phys. Rev. D.*, **47**, 5357.

Joshi, P. S. and Dwivedi, I. H. (1993b). 'The structure of naked singularity in self-similar gravitational collapse II'. *Lett. Math. Phys.*, **27**, 235.

Joshi, P. S. and Dwivedi, I. H. (1999). 'Initial data and the end state of spherically symmetric gravitational collapse'. *Class. Quant. Grav.*, **16**, 41.

Joshi, P. S. and Goswami, R. (2004). 'Role of initial data in spherical collapse'. *Phys. Rev. D.*, **69**, 064027.

Joshi, P. S. and Królak, A. (1996). 'Naked strong curvature singularities in Szekeres spacetimes'. *Class. Quant. Grav.*, **13**, 3069.

Joshi, P. S. and Narlikar, J. V. (1982). 'Black hole physics in globally hyperbolic spacetimes'. *Pramana*, **18**, 385.

Joshi, P. S. and Saraykar, R. V. (1987). 'Cosmic censorship and topology change in general relativity'. *Phys. Lett. A.*, **120**, 111.

Joshi, P. S. and Singh, T. P. (1995). 'Phase transition in gravitational collapse of inhomogeneous dust'. *Phys. Rev. D.*, **51**, 6778.

Joshi, P. S., Dadhich, N. and Maartens, R. (2000). 'Gamma-ray bursts as the birth-cries of black holes'. *Mod. Phys. Lett. A.*, **15**, 991.

Joshi, P. S., Dadhich, N. and Maartens, R. (2002). 'Why do naked singularities form in gravitational collapse?'. *Phys. Rev. D.*, **65**, 101501.

Kahana, S., Baron, E. and Cooperstein, J. (1984). 'Successful supernovae, the anatomy of shocks: neutrino emission and the adiabatic index'. In *Problem Collapse and Numerical Relativity*, eds D. Bancel and M. Signore. Dordrecht: Reidel.

Kerr, R. P. (1963), 'Gravitational field of a spinning particle as an example of algebrically special metrics'. *Phys. Rev. Lett.*, **11**, 237.

Kodama, H. (1979). 'Inevitability of a naked singularity associated with the blackhole evaporation'. *Prog. Theor. Phys.*, **62**, 1434.

Krasinski, A. (1997). *Inhomogeneous Cosmological Models.* Cambridge: Cambridge University Press.

Królak, A. (1983). 'A proof of the cosmic censorship hypothesis'. *Gen. Relat. Grav.*, **15**, 99.

Królak, A. (1986). 'Towards the proof of the cosmic censorship hypothesis'. *Class. Quant. Grav.*, **3**, 267.

Królak, A. (1999). 'Nature of singularities in gravitational collapse'. *Prog. Theor. Phys. Suppl.*, **136**, 45.

Kruskal, M. D. (1960). 'Maximal extension of Schwarzschild metric'. *Phys. Rev.*, **119**, 1743.

Lake, K. (1991). 'Naked singularities in gravitational collapse which is not self-similar'. *Phys. Rev. D.*, **43**, 1416.

Lake, K. (1992). 'Precursory singularities in spherical gravitational collapse'. *Phys. Rev. Lett.*, **68**, 3129.

Lake, K. (2000). 'Gravitational collapse of dust with a cosmological constant'. *Phys. Rev. D.*, **62**, 027301.

Lake, K. and Roeder, R. C. (1979). 'Some remarks on surfaces of discontinuity in general relativity'. *Phys. Rev. D.*, **17**, 1935.

Landau, L. and Lifshitz, E. (1975). *The Classical Theory of Fields.* New York: Pergamon Press.

Lee, C. W. (1983). 'The topology of geodesically complete space-times'. *Gen. Relat. Grav.*, **15**, 21.

Leibovitz, C. and Israel, W. (1970). 'Maximum efficiency of energy release in spherical collapse'. *Phys. Rev. D.*, **1**, 3226.

Lemaître, G. (1933). 'L'univers en expansion'. *Ann. Soc. Sci. Bruxelles I A.*, **53**, 51.

Leray, J. (1952). *Hyperbolic Differential Equations.* Princeton: Institute for Advanced Study (Preprint).

Lin, C. C., Mestel, L. and Shu, F. H. (1965). 'The gravitational collapse of a uniform spheroid'. *Astrophys. J.*, **142**, 1431.

Lindquist, R. W., Schwartz, R. A. and Misner, C. W. (1965). 'Vaidya's radiating Schwarzschild metric'. *Phys. Rev. B.*, **137**, 1364.

Lovelock, D. (1972). 'The four-dimensionality of space and the Einstein tensor'. *J. Math. Phys.*, **13**, 874.

Ludvigsen, M. and Vickers, J. A. G. (1983). 'A simple proof of the positivity of Bondi mass'. *J. Phys. A.*, **15**, L67.

Maartens, R. and Bassett, B. (1998). 'Gravito-electromagnetism'. *Class. Quant. Grav.*, **15**, 705.

Madhav Arun, T., Goswami, R. and Joshi, P. S. (2005). 'Gravitational collapse in asymptotically anti-de Sitter/de Sitter backgrounds'. *Phys. Rev. D.*, **72**, 084029.

Maeda, H. (2006). 'Final fate of spherically symmetric gravitational collapse of a dust cloud in Einstein–Gauss–Bonnet gravity'. *Phys. Rev.*, **D73**, 104004.

Mahajan, A., Goswami, R. and Joshi, P. S. (2005). 'Cosmic censorship in higher dimensions. II.'. *Phys. Rev. D.*, **72**, 024006.

Malec, E. (1995). 'Global solutions of a free boundary problem for selfgravitating scalar fields', gr-qc/9506005.

Markovic, D. and Shapiro, S. L. (2000). 'Gravitational collapse with a cosmological constant'. *Phys. Rev. D.*, **61**, 084029.

May, M. M. and White, R. H. (1966). 'Hydrodynamical calculations of general relativistic collapse'. *Phys. Rev.*, **141**, 1232.

Mazur, P. O. and Mottola, E. (2004). 'Gravitational vacuum condensate stars'. *Proc. Nat. Acad. Sci.*, **111**, 9545.

McNamara, J. M. (1978). 'Instability of black hole inner horizons'. *Proc. Roy. Soc. Lond. A.*, **358**, 499.

Mena, F. C., Tavakol, R. and Joshi, P. S. (2000). 'Initial data and spherical dust collapse'. *Phys. Rev. D.*, **62**, 044001.

Mena, F. C. and Nolan, B. (2001). 'Non-radial null geodesics in spherical dust collapse'. *Class. Quant. Grav.*, **18**, 4531.

Mena, F. C. and Nolan, B. (2002). 'Geometry and topology of singularities in spherical dust collapse'. *Class. Quant. Grav.*, **19**, 2587.

Misner, C. (1963). 'The flatter regions of Newman, Unti and Tamburino's generalized Schwarzschild space'. *J. Math. Phys.*, **4**, 924.

Misner, C. (1967). 'Taub-NUT space as a counterexample to almost anything'. In *Relativity Theory and Astrophysics I: Relativity and Cosmology*, ed. J. Ehlers. Lectures in Applied Mathematics, Vol. 8. Providence: American Mathematical Society.

Misner, C. and Sharp, D. H. (1964). 'Relativistic equations for adiabatic, spherically symmetric gravitational collapse'. *Phys. Rev. B.*, **136**, 571.

Misner, C., Thorne, K. and Wheeler, J. (1973). *Gravitation.* San Francisco: Freeman.

Morris, M. S., Thorne, K. S. and Yurtsever, U. (1988). 'Wormholes, time machines, and the weak energy condition'. *Phys. Rev. Lett.*, **61**, 1446.

Nakamura, T. and Sato, H. (1982). 'General relativistic collapse of non-rotating, axisymmetric stars'. *Prog. Theor. Phys.*, **67**, 1396.

Nakao, K., Kurita, Y., Morisawa, Y. and Harada, T. (2007). 'Relativistic gravitational collapse of cylindrical dust'. *Prog. Theor. Phys.*, **117**, 75.

Newman, R. P. A. C. (1986). 'Strengths of naked singularities in Tolman-Bondi space-times'. *Class. Quant. Grav.*, **3**, 527.

Newman, R. P. A. C. and Joshi, P. S. (1988). 'Constraints on the structure of naked singularities in classical general relativity'. *Ann. Phys.*, **182**, 112.

Nolan, B. C. (1999). 'Strengths of singularities in spherical symmetry'. *Phys. Rev. D.*, **60**, 024014.

Omer, G. C. (1965). 'Spherically symmetric distributions of matter without pressure'. *Proc. Nat. Acad. Sci.*, **53**, 1.

Oppenheimer, J. R. and Snyder, H. (1939). 'On continued gravitational contraction'. *Phys. Rev.*, **56**, 455.

Ori, A. (1992). 'Inner structure of a charged black hole: an exact mass-inflation solution'. *Phys. Rev. Lett.*, **67**, 789.

Ori, A. and Piran, T. (1987). 'Naked singularity in self-similar spherical gravitational collapse'. *Phys. Rev. Lett.*, **59**, 2137.

Ori, A. and Piran, T. (1990). 'Naked singularities and other features of self-similar general relativistic gravitational collapse'. *Phys. Rev. D.*, **42**, 1068.

Ostriker, J. P. and Steinhardt, P. J. (1995). 'The observational case for a low density universe with a nonzero cosmological constant'. *Nature*, **377**, 600.

Papapetrou, A. (1985). 'Formation of a singularity and causality'. In *A Random Walk in Relativity and Cosmology*, eds N. Dadhich, J. Krishna Rao, J. V. Narlikar, C. V. Vishveshwara. New Delhi: Wiley Eastern.

Papapetrou, A. and Hamoui, A. (1967). 'Surfaces caustiques dégénérées dans la solution de Tolman. La singularité physique en relativité générale'. In *Ann. Inst. H Poincare AVI*, p. 343.

Patil, K. D., Ghate, S. H. and Saraykar, R. V. (2001). 'Spherically symmetric inhomogeneous dust collapse in higher dimensional spacetime and cosmic censorship hypothesis'. *Pramana*, **56**, 502.

Penrose, R. (1965). 'Zero rest mass fields including gravitation: asymptotic behaviour'. *Proc. Roy. Soc. Lond. A.*, **284**, 159.

Penrose, R. (1968). 'Structure of space-time'. In *Battelle Rencontres*, eds C. Dewitt and J. A. Wheeler. New York: Benjamin.

Penrose, R. (1969). 'Gravitational Collapse: the role of general relativity'. *Riv. Nuovo Cimento Soc. Ital. Fis.*, **1**, 252.

Penrose, R. (1972). *Techniques of Differential Topology in Relativity.* Philadelphia: AMS Colloquium Publications, SIAM.

Penrose, R. (1974a). 'Gravitational collapse'. In *Gravitational Radiation and Gravitational Collapse* (IAU Symposium No. 64), ed. C. DeWitt-Morettee. Dordrecht: Reidel.

Penrose, R. (1974b). 'Singularities in cosmology'. In *Confrontation of Cosmological Theories with Observational Data*, ed. M. S. Longair. Dordrecht: Reidel.

Penrose, R. (1979). 'Singularities and time asymmetry'. In *General Relativity: An Einstein Centenary Survey*, eds S. W. Hawking and W. Israel. Cambridge: Cambridge University Press.

Penrose, R. (1998). 'The question of cosmic censorship'. In *Black Holes and Relativistic Stars*, ed. R. M. Wald. Chicago: University of Chicago Press, p. 103.

Perlmutter, S., Aldering, G., Della Valle, M., *et al.* (1998). 'Discovery of a supernova explosion at half the age of the Universe and its cosmological implications'. *Nature*, **391**, 51.

Perlmutter, S., Aldering, G., Goldhaber, G., *et al.* (1999). 'Measurements of omega and lambda from 42 high redshift supernovae'. *Astrophys. J.*, **517**, 565.

Price, R. H. (1972). 'Non-spherical perturbations of relativistic gravitational collapse. I. Scalar and gravitational perturbations'. *Phys. Rev. D.*, **5**, 2419.

Randall, L. and Sundrum, R. (1999). 'An alternative to compactification'. *Phys. Rev. Lett.*, **83**, 4690.

Raychaudhuri, A. K. (1955). 'Relativistic cosmology'. *Phys. Rev.*, **98**, 1123.

Rein, G., Rendall, A. and Schaeffer, J. (1995). 'A regularity theorem for solutions of the spherically symmetric Vlasov–Einstein system'. *Commun. Math. Phys.*, **168**, 467.

Rendall, A. (1992). 'Cosmic censorship and the Vlasov equation'. *Class. Quant. Grav.*, **9**, L99.

Rendall, A. (2005). 'The nature of spacetime singularities', gr-qc/0503112.

Roberts, M. D. (1989). 'Scalar field counterexamples to the cosmic censorship hypothesis'. *Gen. Relat. Grav.*, **21**, 907.

Rocha, J. F. V. and Wang, A. (2000). 'Collapsing perfect fluid in higher-dimensional spherical spacetimes'. *Class. Quant. Grav.*, **17**, 2589.

Sachs, R. (1962). 'Gravitational waves in general relativity, VIII. Waves in asymptotically flat space-time'. *Proc. Roy. Soc. Lond. A.*, **270**, 103.

Saraykar, R. V. and Ghate, S. H. (1999). 'C1-stability of naked singularities arising in an inhomogeneous dust collapse'. *Class. Quant. Grav.*, **16**, 281.

Schoen, R. and Yau, S. -T. (1983). 'The existence of a black hole due to condensation of matter'. *Commun. Math. Phys.*, **90**, 575.

Seifert, H. J. (1967). 'Global connectivity by timelike geodesics'. *Zs. F. Naturfor.*, **22a**, 1356.

Senovilla, J. M. M. (2006). 'A singularity theorem based on spatial averages', gr-qc/0610127.

Shapiro, S. L. and Teukolsky, S. A. (1991). 'Formation of naked singularities: the violation of cosmic censorship'. *Phys. Rev. Lett.*, **66**, 994.

Shapiro, S. L. and Teukolsky, S. A. (1992). 'Gravitational collapse of rotating spheroids and the formation of naked singularities'. *Phys. Rev. D.*, **45**, 2006.

Sil, A. and Chaterjee, S. (1994). 'Singularity structure of a self-similar Tolman type model in a higher dimensional spacetime'. *Gen. Relat. Grav.*, **26**, 999.

Simmons, G. F. (1963). *Introduction to Topology and Modern Analysis.* Tokyo: McGraw-Hill Kogakusha Ltd.

Simpson, M. and Penrose, R. (1973). 'Internal instability in a Reissner–Nordstrom black hole'. *Int. J. Theor. Phys.*, **7**, 183.

Spivak, M. (1965). *Calculus on Manifolds.* New York: W. H. Freeman.

Stephani, H., Kramer, D., MacCallum, M., Hoenselaers, C. and Herlt, E. (2003). *Exact Solutions of Einstein's Field Equations.* Cambridge: Cambridge University Press.

Szekeres, P. (1960). 'On the singularities of a Riemannian manifold'. *Publ. Math. Debrecen*, **7**, 285.

Szekeres, P. (1975). 'Quasispherical gravitational collapse'. *Phys. Rev. D.*, **12**, 2941.

Szekeres, P. and Iyer, V. (1993). 'Spherically symmetric singularities and strong cosmic censorship'. *Phys. Rev. D.*, **47**, 4362.

Thorne, K. S. (1972). 'Non-spherical gravitational collapse: a short review'. In *Magic without Magic: John Archibald Wheeler*, ed. J. Clauder. New York: W. H. Freeman.

Tipler, F. (1977). 'Singularities in conformally flat spacetimes'. *Phys. Lett. A.*, **64**, 8.

Tipler, F., Clarke, C. J. S. and Ellis, G. F. R. (1980). 'Singularities and horizons'. In *General Relativity and Gravitation* Vol 2, ed. A. Held. New York: Plenum.

Tolman, R. C. (1934). 'Effect of inhomogeneity on cosmological models'. *Proc. Natl. Acad. Sci. USA*, **20**, 169.

Traschen, J. (1994). 'Discrete self-similarity and critical point behavior in fluctuations about extremal black holes'. *Phys. Rev. D.*, **50**, 7144.

Tsujikawa, S., Singh, P. and Maartens, R. (2004). 'Loop quantum gravity effects on inflation and the CMB'. *Class. Quant. Grav.*, **21**, 5767.

Vaidya, P. C. (1943). 'The external field of a radiating star in general relativity'. *Curr. Sci.*, **12**, 183.

Vaidya, P. C. (1951). 'The gravitational field of a radiating star'. *Proc. Indian Acad. Sci. A.*, **33**, 264.

Vaidya, P. C. (1953). 'Newtonian time in general relativity'. *Nature*, **171**, 260.

Vaz, C. and Witten, L. (1994). 'Evaporation of a naked singularity in 2D gravity'. *Phys. Lett. B.*, **325**, 27.

Vaz, C. and Witten, L. (1995). 'Soliton-induced singularities in 2D gravity and their evaporation'. *Class. Quant. Grav.*, **12**, 2607.

Wald, R. M. (1984). *General Relativity.* Chicago: University of Chicago Press.

Wald, R. M. (1997). 'Gravitational collapse and cosmic censorship', gr-qc/9710068.

Wald, R. M. and Iyer, V. (1991). 'Trapped surfaces in the Schwarzschild geometry and cosmic censorship'. *Phys. Rev. D.*, **44**, R3719.

Wang, A. and Wu, Y. (1999). 'Generalized Vaidya solutions'. *Gen. Relat. Grav.*, **31**, 1, 107.

Waugh, B. and Lake, K. (1988). 'Strengths of shell-focusing singularities in marginally bound collapsing self-similar Tolman space-times'. *Phys. Rev. D.*, **38**, 1315.

Waugh, B. and Lake, K. (1989). 'Shell-focusing singularities in spherically symmetric self-similar space-times'. *Phys. Rev. D.*, **40**, 2137.

Weinberg, S. (1972). *Gravitation and Cosmology.* New York: John Wiley.

Wheeler, J. A. (1962). 'Geometrodynamics'. New York: Academic Press.

Wheeler, J. A. (1964). 'Geometrodynamics and the issue of final state'. In *Relativity, Groups and Topology*, eds C. Dewitt and B. Dewitt. New York: Gordon and Breach.

Willard, S. (1970). *General Topology.* Reading, Mass.: Addison-Wesley.

Yodzis, P., Seifert, H. -J. and Muller zum Hagen, H. (1973). 'On the occurrence of naked singularities in general relativity'. *Commun. Math. Phys.*, **34**, 135.

Yodzis, P., Seifert, H. -J. and Muller zum Hagen, H. (1974). 'On the occurrence of naked singularities in general relativity II'. *Commun. Math. Phys.*, **37**, 29.

Zehavi, I. and Dekel, A. (1999). 'Constraints on the cosmological constant from flows and supernovae'. *Nature*, **401**, 252.

Index

achronal boundary, 139
achronal set, 139
advanced null coordinates, 47
affine parameter, 29
Alexandrov topology, 142
anisotropic de Sitter metric, 90
apparent horizon, 71, 79, 241
  in dust collapses, 98
  in higher-dimensions, 169, 173
arc length, 24
asymptotically flat spacetime, 162
atlas, 12

Bianchi identities, 35
Birkhoff's theorem, 44
blackhole, 3, 4, 54
blackhole region, 164
blue-shift instability, 178, 194
bouncing models, 67
brane-world models, 169

C^k-curve, 13
C^k-singularity, 161
C^r-function, 13
C^r-manifold, 12
Cauchy horizon, 146
  instability, 178, 195
  Reissner–Nordström case, 195
  in TBL metrics, 195
Cauchy surface, 45, 145
causal boundary, 147, 205
causal functions, 145
causal future (past), 45, 138
causal relations, 23
  and spacetime topology, 142
causal structure
  and naked singularity, 205
  near a singularity, 204
causality condition, 142
caustics, 156
chart, 12
Christoffel symbols, 34
chronological future (past), 45, 137
chronology condition, 142
commutator, 15
comoving coordinates, 64
conformal compactification, 47, 163
conformal flatness, 36
conformal transformation, 140
congruence
  hypersurface orthogonal, 32
  of non-spacelike curves, 154
  of null geodesics, 157
  of timelike geodesics, 154
conical singularity, 152
conjugate points, 157
connection, 25
  flat, 35
  integrable, 35
  torsion-free, 28, 34
contravariant vector, 13
convex normal neighborhood, 30
coordinate singularity, 149
cosmic censorship, 164
  in blackhole physics, 179
  and causal structure, 135
  and cosmological term, 129
  and global hyperbolicity, 144
  in higher dimensions, 169, 195

cosmic censorship (*Cont.*)
  from quantum gravity, 253
  in quantum gravity, 190
  vacuum version of, 179
cosmic censorship hypothesis, 5
cosmological constant, 43, 120
  and bouncing models, 121
  and singularity avoidance, 121
covariant derivative, 25
curvature scalar, 36

dark energy, 119
diffeomorphism, 13
differentiable manifold, 11
dispersal, 67
distinguishing condition, 142
domain of dependence, 145
dominant energy condition, 65, 156
dual basis, 15
dust collapse, 7
  bound, unbound, and marginally bound, 94
  and cosmic censorship, 91
  with cosmological constant, 121
  in five dimensions, 198
  generic analysis of, 91
  in higher dimensions, 131, 197
  self-similar, 106

Eddington–Finkelstein coordinates, 54
edge of a set, 144
Einstein cluster, 77
Einstein equations, 41
Einstein static universe, 48
Einstein tensor, 41
Einstein–Vlasov description, 185
embedded submanifold, 31
energy–momentum tensor, 38
  dust, 39
  perfect fluid, 39
equation of state, 70
  dust, 92
event horizon, 4, 55, 165
event horizon conjecture, 227
expansion, 154
exponential map, 30
extendible curves, 161
extrinsic curvature, 89, 157

first fundamental form, 88
free gravitational field, 222
Friedmann models, 53
future asymptotic predictability, 164
future end point, 140
future imprisoned curves, 143
future set, 139
future (past) timelike infinity, 46

generic condition, 159
geodesic deviation equation, 37, 156
geodesic equations, 30
geodesic incompleteness, 150
geodesics, 29
  affinely parametrized, 29
  complete, 30
  incomplete, 30
  maximal, 146
  timelike, null, spacelike, 29
global hyperbolicity, 144
global time function, 143
Gödel universe, 141
gravitational collapse, 1
  and energy conditions, 180
  and equation of state, 184
  of infinite cylinders, 226
  initial conditions, 68
  Misner–Sharp equations, 63
  and negative pressures, 238
  physical constraints, 179
  with purely tangential pressure, 69, 132
  and quantum effects, 3
  role of inhomogeneities, 225
  self-similar, 58
  and spacetime singularity, 60
gravitational focusing, 150, 154
gravitational red-shift, 52

homeomorphism, 13
homogeneous collapse, 52
hoop conjecture, 226
hypersurface, 31
  timelike, null, spacelike, 31

ideal points, 46, 147
immersed submanifold, 31
indecomposable past set, 147
inextendible causal curves, 46
inextendible curve, 140, 161
initial data for collapse, 182
Israel–Darmois conditions, 88

Jacobi equation, 37
Jacobi fields, 156

Killing equation, 40
Killing vector, 40
Klein–Gordon equation, 241
Kretschmann scalar, 79

limiting focusing condition, 234
line element field, 24
local causality, 40
local causality principle, 141
local conservation of energy and momentum, 39
local coordinate neighborhood, 11
loop quantum gravity, 236, 248
  discrete regime, 250
  Hamiltonian, 251
Lorentz group, 45
Lorentz transformations, 45
Lorentzian metric, 10

mass function, 66
massive stars
  continual collapse of, 215
  final fate of, 210
  life cycle of, 213
  and quantum gravity, 212
matching conditions, 244
metric tensor, 21
  non-degenerate, 22
  positive definite, 23
  signature of, 23
Minkowski spacetime, 23, 45

naked singularity, 3, 5, 177
  central, 81
  dissolution of, 212
  global visibility, 112
  gravitationally strong, 97
  gravitationally weak, 97
  and inhomogeneities, 216
  non-spacelike curves from, 208
  physical causes for, 225
  and quantum gravity, 186
  and spacetime shear, 216
  in the Vaidya metric, 58
negative pressure, 119
  and absence of trapped surfaces, 249
  and loop quantum gravity, 249
negative pressure in collapse, 244
neutron star, 213
non-spacelike curve, 24
non-spherical collapse, 225
normal coordinates, 30
null curve, 24
null infinity, 47
null vectors, 23

one-forms, 14
Oppenheimer–Snyder–Datt collapse
  special case of TBL metric, 199
Oppenheimer–Snyder–Datt model, 236

parallel transport, 24
partial Cauchy surface, 144
past set, 147
Penrose diagram, 48
perfect fluid collapse, 133
PIP and PIF, 148
Planck length, 2
Poincaré group, 45
points at infinity, 46
principle of equivalence, 38
principle of general covariance, 38
projection operator, 154

quantum effects
  semi-classical modifications, 251
quantum effects in collapse, 248
quantum gravity, 2, 236
  observational effects, 239

Raychaudhuri equation, 155, 224
regularity conditions, 69
retarded null coordinates, 47
Ricci tensor, 35
Riemann curvature tensor, 32
rotation of a congruence, 154

scalar field collapse, 202
  with a potential, 239
  quantum effects in, 248
Schwarzschild spacetime, 44, 49
  analytic extension of, 54
second fundamental form, 89, 158
self-similar collapse, 108, 181
shear, 154
  and delay of trapped surfaces, 216
  and distortion of apparent horizon, 220
shell-crossing
  avoidance of, 126
  extension of, 96
shell-crossing singularity, 66
shell-focusing singularity, 66
singularity
  curvature strength of, 112
  globally visible, 84
  locally visible, 84
  removable, 150
  scalar polynomial, 113
  stability and genericity, 191
singularity curve, 171
singularity theorems, 2, 160
smooth density profile, 182
space of causal curves, 146
spacelike curve, 24
spacetime
  asymptotically predictable, 164
  boundary, 17, 46
  causal structure of, 136
  compact, 142
  deterministic, 146
  extendible, 50
  finite measure on, 145
  geodesically incomplete, 30
  inextendible, 151
  and maximal length property, 147
  regular points, 147
  weakly asymptotically simple and empty, 163
spacetime manifold, 10, 23
  flat, 35
  orientable, 18
spacetime metric
  spatial part of, 154
spacetime singularity, 2, 149
  definition of, 150
  essential, 151
  removable, 50, 183
spacetime topology, 16, 137
  connected, 17
  and global hyperbolicity, 145
  Hausdorff, 17
  non-compact, 17
  paracompact, 17
spherical symmetry, 63
stable causality, 143
static limit, 51
static metric, 49
string theory, 236
strong causality, 142
strong cosmic censorship, 176, 178
  violations of, 178
strong curvature singularity, 229
  condition for, 183
  in the sense of Królak and Clarke, 114
  in the sense of Tipler, 114
strong energy condition, 156
strong limiting focusing condition, 234
supernova explosion, 214
surface gravity, 167
synchronous coordinates, 31
Szekeres collapse models, 230

tangent space, 14
Taub–NUT solutions, 153
TBL collapse models, 93
TBL metric, 93
tensor, 18
  addition, 20
  anti-symmetric, 21

contraction, 20
outer product, 20
symmetric, 21
symmetric part of, 21
tensor fields, 20
tensor product, 19
timelike and spacelike vectors, 23
timelike curve, 24
TIP and TIF, 160
topology
on the space of causal curves, 146
on the space of metrics, 193
torsion, 26
trapped surfaces, 55, 236, 241
absence of, 243
type I matter fields, 64
vacuum energy density, 120
Vaidya metric, 44, 56
charged, 90
generalized, 88, 244, 246
generalized mass function, 90
vector field, 15
volume expansion, 154

weak cosmic censorship, 176
weak energy condition, 65, 156
Weyl tensor, 36
electric, 222
magnetic, 222
white dwarf, 213
wormhole, 17